Aeolian Geomorphology

Aeolian Geomorphology

A New Introduction

Edited by

Ian Livingstone
University of Northampton, UK

Andrew Warren
University College London, UK

This edition first published 2019
© 2019 John Wiley & Sons Ltd

The right of Ian Livingstone and Andrew Warren to be identified as the authors of the editorial material in this work has been asserted in accordance with law.

Registered Office(s)
John Wiley & Sons, Inc., 111 River Street, Hoboken, NJ 07030, USA
John Wiley & Sons Ltd, The Atrium, Southern Gate, Chichester, West Sussex, PO19 8SQ, UK

Editorial Office
9600 Garsington Road, Oxford, OX4 2DQ, UK

For details of our global editorial offices, customer services, and more information about Wiley products visit us at www.wiley.com.

Wiley also publishes its books in a variety of electronic formats and by print-on-demand. Some content that appears in standard print versions of this book may not be available in other formats.

Library of Congress Cataloging-in-Publication Data

Names: Livingstone, Ian, 1957– editor. | Warren, Andrew, 1937– editor.
Title: Aeolian geomorphology : a new introduction / edited by Ian Livingstone
    (University of Northampton, Northampton, UK), Andrew Warren
    (University College London, London, UK).
Description: Hoboken, NJ : Wiley-Blackwell, 2018. | Includes bibliographical
    references and index. |
Identifiers: LCCN 2018026802 (print) | LCCN 2018033964 (ebook) | ISBN
    9781118945643 (Adobe PDF) | ISBN 9781118945636 (ePub) | ISBN 9781118945667 (hardcover)
Subjects: LCSH: Eolian processes. | Geomorphology. | Physical geography. | Erosion.
Classification: LCC QE597 (ebook) | LCC QE597 .A3425 2018 (print) | DDC
    551.3/7–dc23
LC record available at https://lccn.loc.gov/2018026802

Cover Design: Wiley
Cover image: © Giles Wiggs

Set in 10/12pt Warnock by SPi Global, Pondicherry, India

Printed in Singapore by C.O.S. Printers Pte Ltd

10 9 8 7 6 5 4 3 2 1

Contents

List of Contributors

Andreas C. W. Baas
King's College London
London, UK

Matthew Baddock
Loughborough University
Loughborough, UK

Matthew Balme
The Open University
Milton Keynes, UK

Mary C. Bourke
Trinity College
Dublin, Ireland

Charles Bristow
Birkbeck, University of London
London, UK

Joanna Bullard
Loughborough University
Loughborough, UK

Patrick A. Hesp
Flinders University
Adelaide, Australia

Paul Hesse
Macquarie University
Sydney, Australia

Jasper Knight
University of the Witwatersrand
Johannesburg, South Africa

Stephen Lewis
The Open University
Milton Keynes, UK

Ian Livingstone
University of Northampton
Northampton, UK

Ralph D. Lorenz
Johns Hopkins University
Laurel, Maryland, USA

Eric Parteli
University of Cologne
Cologne, Germany

Helen M. Roberts
Aberystwyth University
Aberystwyth, UK

Thomas A.G. Smyth
Liverpool Hope University
Liverpool, UK

David S.G. Thomas
University of Oxford
Oxford, UK

Andrew Warren
University College London
London, UK

Giles Wiggs
University of Oxford
Oxford, UK

Preface

Two decades after our first edition, we find aeolian geomorphology to be very much more vigorous. Our mandate is to organise and summarise the advances this vigour has brought.

While we alone wrote our earlier volume, we have now enlisted 16 new contributors, from four continents. The expansion in authors allows us to tap a much wider range of expertise and the insights of specialists. In some cases, as with 'dust', this allows us to separate contemporary lifting and transport of dust from its deposition 'loess', a process that took place over many thousands of years. In the case of 'dunes', we can now separate 'free' dunes (generally unvegetated) from 'anchored' dunes formed where there is a limited covering of plants. In the case of 'planetary landforms' we have introduced a topic that barely existed at the time of our first edition. All the chapters demonstrate huge advances in field observation, measurement, and mathematical modelling. Many, particularly Chapter 8 on sand seas, show the impact of greatly enhanced and accessible remote sensing; others (as in Chapter 6 on active dunes) bear the impact of improvements in field techniques; yet others (especially Chapter 5 on loess) show the power of greatly improved laboratory techniques.

Few aeolian processes operate in isolation from other geomorphological process domains, as on slopes and coasts, in rivers, glaciers, and periglacial environments, let alone the related geological processes that operate over much greater time-scales. The full understanding of most aeolian processes requires knowledge of the processes that pre-sort sediment to sizes that can be moved by the wind. In their turn, aeolian processes deliver sediment to fluvial, marine, or glacial processes. These interactions take place over a range of temporal or geographical scales. At a small scale, the deposition of aeolian dust stimulates other geomorphological processes such as the development of rock varnishes or stone pavements. At a very much larger scale, crustal uplift may stimulate erosion that may find its way to places from which it is removed by the wind. Our authors have highlighted many of these interactions and overlaps between geomorphological domains. Our hope is that a fuller appreciation of aeolian processes will broaden our understanding of the evolution of whole physical landscapes, the goal of all geomorphologists.

We believe that the 12 chapters provide a comprehensive new introduction to aeolian geomorphology. We are indebted to all our authors for contributing their expertise.

Ian Livingstone, 'Whitstable'
Andrew Warren, Farringdon

1

Global Frameworks for Aeolian Geomorphology

Andrew Warren

University College London, London, UK

1.1 Introduction

This chapter locates aeolian geomorphology in four global frameworks, arranged in increasing scale. The first is the framework of contemporary winds, from local to annual. The second is the changing erosivity of the wind and the erodibility of the surface during recent geological history. The third, digging deeper into the past, is the framework of the preparation and supply of sand and dust to the wind. The fourth, covered very briefly, is the yet wider and older framework of plate-tectonics. Box 1.1 is a short early history of aeolian geomorphology.

1.2 Wind

Chapter 2 of this book, 'Grains in Motion', covers the shortest, smallest and most fundamental of aeolian rhythms, leaving larger-scale processes and their longer rhythms to be introduced in this chapter.

1.2.1 Wind Systems with Daily Rhythm and Local Scale

1.2.1.1 Dust Devils

The scientific term, dust devil, is a close translation of *Dhūla kā śaitāna* in the North Indian languages from which it was borrowed. Most dust devils blow in daytime, reach diameters of a few metres, last for a few minutes and travel a few hundreds of metres; a few are larger, longer-lived and further-travelled. Some dust devils leave shallow tracks, most do not (see also Chapters 4 and 11).

1.2.1.2 Haboobs

A haboob (هَبوب, 'blustery wind' in Sudanese Arabic) is, in the contemporary climatological literature, a dusty, mobile thunderstorm that develops when and where warm, moist air is taken to otherwise hot, dry, sparsely vegetated, dust-yielding environments, as in central Sudan and Arizona. A 'cold pool' develops at ground level within the thunderstorm and this pulls a strong downdraft. When the downdraft hits the ground, it bursts out horizontally at velocities of up to $20\,ms^{-1}$. These gusts raise a large amount of dust from the surface, if it is available. The outer edge of the dust cloud is often a sharply-defined, mobile 'wall' of dust. Most haboobs develop late in the day and last for less than a day. A haboob in Arizona on 5 July 2011 maintained a width of ~10 km, but elongated to ~80 km, as it moved slowly east-south-eastward (Raman et al. 2014). Haboobs carry and redeposit large amounts of dust, some held down and redistributed by the accompanying rain (see also Chapter 4).

Other 'cold pool' phenomena operate and lift dust in three other situations. First, density currents flow down mountain slopes,

Box 1.1 A Short Early History of Aeolian Geomorphology

It was not until the late nineteenth century, even the early twentieth century, that a significant number of geo-scientists acknowledged that the wind was a major geomorphological agent. Rivers had been recognised in that role early in the century, particularly by Charles Lyell in his influential *Principles of Geology* (1875, first edition 1830). Lyell maintained that loess had been deposited by a flood, even the biblical Flood, not as dust, and ignored more obviously aeolian landforms. Despite pioneers like Udden (1894), Berg could still maintain that loess was a deep soil as late as 1916.

The principal catalyst to the aeolian enlightenment was the opening-up of the world in the late nineteenth century, particularly its deserts. But experience, alone, was insufficient to come to a defensible interpretation. Keyes (1909), well acquainted as he was with the deserts of the western USA, could still assert that the wind could level mountains. More astute interpretation of direct observation, now of the Chinese loess, led both by von Richthofen (e.g. 1882), and the more influential 'aeolianist', Obruchev (e.g. 1895, Box Figure 1.1), to conclude that it was a deposit of aeolian dust. Obruchev's observations on the deserts of central Asia, which he traversed on horseback, included wind-eroded terrain, and the dust-filled skies he experienced, strengthened his belief in the power of the wind. He went

Box Figure 1.1 Vladimir Afanasyevich Obruchev atop a mega-yardang in his 'Aeolian City' near the Chinese/Russian border at somewhat NE of 46°N, 85°30″E (Obruchev's coordinates; Obruchev 1911); and a portrait of the man himself.

further to propose that loess-sized particles could have been created either by glacial grinding, which is non-aeolian (much debated; Chapter 4); or by the grinding of wind-driven sand (a fully aeolian process), an idea which, although contested for decades, is enjoying rebirth (Crouvi et al. 2010). The location of some 'desert loess' (in southern Tunisia) is shown on Figure 1.12. The early study of loess is further discussed in Chapter 5.

First-hand experience and critical interpretation also had a crucial role in shaping opinions about desert dunes. The pioneers here were a succession of French observers of the sand deserts of northern Africa, beginning with Rolland (e.g. 1890), who had been sent to reconnoitre the route of a trans-Saharan railway through southern Algeria. Rolland argued that the dunes he saw, some large, were wholly the work of the wind, and had been aligned by the wind. He even speculated about the sources of the dune sand, and the distribution of sand seas, questions still asked. Rolland's ideas were developed by later French geomorphologists, particularly Aufrère (1930), whose work reached an international audience. There were still some blind alleys in the search for explanations of dunes, such as Vaughn-Cornish's wave theory (Goudie 2008), but in

the 1930s two major revolutions in aeolian geomorphology were about to break.

In the early twentieth century, although it was generally accepted that the wind could erode, move, and deposit sediment, few asked how it did so. The gap in knowledge was dramatically narrowed by Bagnold (e.g. 1941), who, like Obruchev, had traversed many wind-formed landscapes, in his case in south-west Asia and north-eastern Africa, and now travelling by motor car (Bagnold 1990). As an engineer, Bagnold focused mathematical modelling and experiments onto aeolian processes (Bagnold 1941; more in Chapter 2). He was probably the first to use a wind tunnel to study the movement of sand, and more certainly the first to study the mechanism of sand movement in the field (Box Figure 1.2). His wind tunnel was replicated by the Wind Erosion Research Unit in Kansas (see Chapter 12). In the Soviet Union, Znamenski (1958), who also replicated Bagnold's design of wind tunnel, quoting him as "Bagnolg". Bagnold's book is still the most commonly cited source in aeolian geomorphology. Development of many his ideas had to wait for decades (Chapter 2). One of his few excursions into application is described in Box 12.1 in Chapter 12. Bagnold is quoted in most of the chapters in this book.

Box Figure 1.2 Ralf Adger Bagnold measuring the wind-speed profile and the flux of sand, near 'Uweinat in southwestern Egypt in February 1938. Source: Photograph: R.F. Peel.

Wind measurements being made with my multiple manometer in a sand storm near Gilf Kebir, 1938. The sand collector is in the middle ground. Photograph by Ronald Peel.

The early twentieth century also saw a slower, more diffuse revolution, driven by the discovery of large areas of inactive aeolian terrain, as a by-product of the growing realisation of the extent of Pleistocene climatic change in other fields of geomorphology. Pioneers in the study of these terrains and deposits included Cailleux (e.g. 1936) in respect of mainly northern Europe (but also some more exotic places); others worked on evidence from West Africa (summarised in Grove and Warren 1968), and yet others in North America, for example, Leighton (1931) on loess. At first, the dating of these 'fossil' landscapes had to rely on stratigraphy and palaeontology, a problem that was overcome, first, by radiocarbon dating (Libby 1952), and three decades later by luminescence dating, which is better suited to generally inorganic aeolian material, and has a much greater historical reach (Chapters 5 and 10). Other methods, like the use of cosmogenic isotopes, came still later and have even greater historical reach (Vermeesch et al. 2010).

Many more techniques have now made significant contributions to aeolian geomorphology: remote sensing, which has opened new vistas on dune form, and the local generation and the global movement of dust (Chapter 4), mathematical modelling of the movement of sediment and of the form of dunes (Chapter 6) and much more, as explained in the chapters that follow.

and some of these raise dust on the plains below. On the southern slopes of the Atlas Mountains in Morocco, these events occur about eleven times in a year, between April and December; they have the same spatial scale as haboobs (Emmel et al. 2010). Second, similar density currents may bring momentum and feed dust to the landward phases of sea breezes (discussed later). Third, cold pool atmospheric outflows are the main mechanism for raising dust in the central Sahara (Allen et al. 2013). None of these events leaves an obvious pattern on the ground.

1.2.1.3 Low-Level Jets

Low-level jets are strong winds limited to a small area. The two examples described here are from the drylands. Both are constant in direction and strongest at night.

On the Pampa la Hoja in southern Peru (just south-west of Arequipa, marked on Figure 1.5), a low-level jet is sometimes active over a plain with an area of ~50 km downwind by ~4 km (Lettau and Lettau 1978). The Pampa is a plateau with very little relief that slopes gently up towards the east and which is virtually free of vegetation. The jet drives small, widely dispersed, very symmetrical, fast-moving barchan dunes eastward over the pampa (16°41'S; 71°51'W;[1] eye altitude 2.5 km²). The movement of the barchans is so closely related to the wind-speed, that the Lettaus could use their movement as anemometers (there is more on the movement of barchans dunes in Chapter 6).

The Bodélé low-level jet in northern Chad blows in another very dry part of the world (Figure 1.1, located on Figure 1.12). It boasts two aeolian superlatives: the dustiest place on Earth (Warren et al. 2007) and the fastest dunes on Earth (Vermeesch and Drake 2008). There is more on this in Chapter 4.

1.2.1.4 Sea Breezes

Sea breezes (Figure 1.2) blow onto any coast that has a large enough contrast between the thermal capacities of the land and of the water. They are thus more active in summer, at low latitudes, on warm coasts washed by cool currents, and on landmasses above some a minimum size, depending on local contrasts in temperature. They blow onto the coasts of oceans, seas, and lakes (selected locations on Figure 1.5). Sea breezes on the shores of smaller bodies of water, like Lake Erie, reach only a few kilometres inland (Sills et al. 2011). On the south-western coast of Western Australia, where, in summer, there is a strong contrast in temperature between

Figure 1.1 The Bodélé low-level jet in operation, raising a cloud of dust from the now-dry surface of a Holocene Lake. (MODIS true colour composite (channels 3, 4, and 1) image for 12:45UTC, 4 March 2005). See insert for colour representation of this figure.

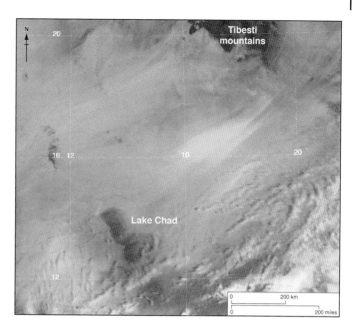

the ocean and the land, the sea breeze reaches over 600 km, east to Kalgoorlie.

Sea breezes blow on-shore in the morning, their velocity peaking at a sharply defined 'sea-breeze front'. At night, the wind is seaward and gentler (Furberg et al. 2002). Tsoar (1983) followed a linear seif dune (Chapter 6) in Sinai as it grew under the influence of a sea-breeze in summer and the Westerlies in winter. His site was some 60 km inland (in the direction of the breeze). In summer the sea-breeze front arrived dependably, suddenly and to a welcome, at lunch time.

Sea breezes help to build dunes behind many other coasts, as in Oregon (44° 04′15 N; 124° 07′33 W; Hunter and Richmond 1988); and Lençóis Maranhenses in north-eastern Brazil (02°30′S; 42°52′; eye altitude 20 km; Parteli et al. 2006). On the Makran coast in Pakistan, and probably on other steep coastal hinterlands, the seaward phase of a sea breeze may be strengthened by a cool density current (see Section 1.2.1.2), which carries small plumes of dust out to sea.

1.2.1.5 Hurricanes, Cyclones, Typhoons

Hurricanes (Carib: *Hurican*) (Figure 1.3) in the Atlantic, tropical cyclones in the Indian Ocean and South Pacific, or typhoons in north-western Pacific (台風 in Japanese) are everyday names for the same phenomenon. Diameters are ~300 km, varying considerably between hurricanes and their state of development. Wind speeds also vary widely, between and within hurricanes; in which, at their worst winds can reach at least $80 \, \mathrm{m \, s^{-1}}$.

Hurricanes form over the ocean where it is warm enough, which is, roughly, within 5° of the Equator; and where there is a source of vorticity, such as a thunderstorm. Some hurricanes sweep along shorelines, bringing different wind speeds from place to place and time to time, as did the very destructive Hurricane Sandy along the coasts of New Jersey and New York State in 2012 (Figure 1.3). Others travel at a more acute angle to the shore, their velocity dropping rapidly inland, as did Hurricane Katrina on the Louisiana coast in August 2005. If still at sea, hurricanes travel towards the appropriate pole and decelerate.

The high winds, huge waves, torrential rainfall, and storm surges of hurricanes can move great quantities of sand from coastal dunes, taking it inshore, along-shore or offshore. This damage is usually repaired before

Mean hourly hodographs averaged over all sea breeze days for each station

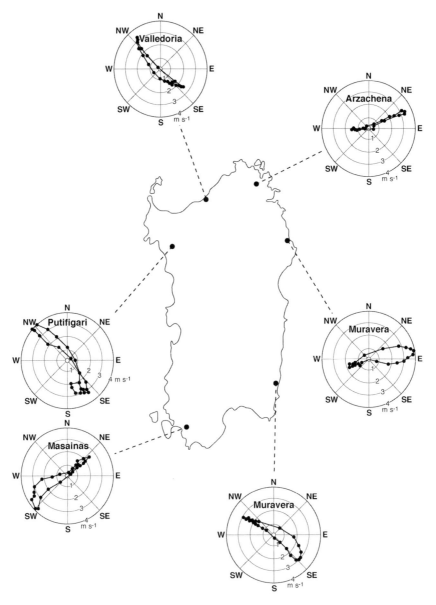

Figure 1.2 Sea breezes on the Sardinian Coast. Source: A selection from Furberg et al. (2002).

the next hurricane attacks by onshore sea breezes (see Section 1.2.1.4), as on the coast of Texas, after the attack by Hurricane Alicia in 1983 (Houser et al. 2015). But, as the Earth warms, attacks may have already become more frequent, and thus allow less time for recovery. The global distribution of hurricanes has already moved towards the poles (Stephens 2011). Wind-driven waves, despite their aeolian origin, are generally classified as 'coastal' rather than as aeolian geomorphological processes.

Figure 1.3 Hurricane Sandy off the Atlantic coast of the USA in October 2012. Glimpses of the east coast of the USA that it attacked can be seen.

1.2.1.6 Mountain Winds

Mountain winds may sometimes reach the same velocity as severe hurricanes but few do. They occur in all global wind systems. There are four types of mountain wind.

First, on high mountains, high winds may blow for days and many are strong. They lift and carry large quantities of snow, which they build into cornices in the lee of ridges. They also form wind-erosional features or *sastrugi* on ice-free and snow and ice surfaces (see Chapter 3). On gently sloping ground they build snow dunes.

Second, are the winds that are accelerated through mountain gaps. North of Rawlins, Wyoming, strong winds blow through a ~5 km-wide defile through a range of hills, through which they drive very fast-moving parabolic dunes (see Chapter 8) (42°12″N; 107°04″W eye altitude 12 km; Gaylord and Dawson 1987). Other defiles in Wyoming hold more fast-moving dunes.

Third, katabatic or 'downslope' winds, *sensu stricto*, are driven by differences in air density, most related to differences in temperature (as between high and low ground). Examples are the Santa Ana in southern California, and the Berg winds in South Africa (both named on Figure 1.5; the

association of high winds downwind of mountain ranges, worldwide, is shown on Figure 1.8). Chapter 6 briefly describes the effects some of these winds have on dune form. Velocity is greater in winds that have steeper descents, and in constricted valleys and passes. These winds are, in general, strongest in winter (Muhs et al. 1996 p. 129). East of the Rockies, katabatic winds reach ~700 km over the plains beyond. Chapter 6 briefly describes the effects on dune form of katabatic winds blowing over the steeply scarped edge of the African Plateau. The fiercest katabatic winds blow down the steep slopes of the Antarctic and Greenland ice plateaux (van den Broeke et al. 1994), driven by the great masses of very cold, very dense air on the plateaux (Figure 1.4).

In the San Gorgonio Pass in California, the Santa Anna hurled sand, and thus eroded blocks of Lucite in an 11-year experiment (Sharp 1980). Katabatic winds in Antarctica reach much greater velocities, and carve out huge fields of wind-erosional features in ice (*sastrugi*, see Chapter 3) (Bromwich et al. 1990) and drive a unique aeolian landform on the coast: dunes formed of pebbles with diameters of about 1.9 cm, which, it is claimed, could only have been moved by

Wind roses for Halley and automatic weather stations C2 and C4

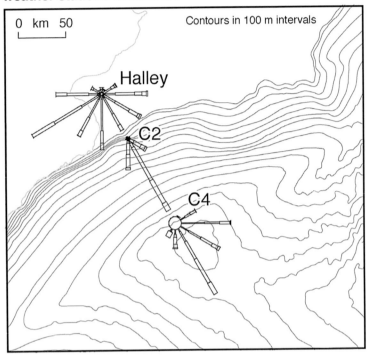

Figure 1.4 A wind rose for a station experiencing katabatic winds descending from the Antarctic Plateau (Renfrew 2004). Wind speeds in $5\,\mathrm{m\,s^{-1}}$ bins from 0.1 to $30\,\mathrm{m\,s^{-1}}$.

winds of $>54\,\mathrm{m\,s^{-1}}$ (Bendixen and Isbel 2007). In the Pleistocene, katabatic winds formed ventifacts on the margins of the expanded ice sheets (see Chapter 3).

Fourth, föhns are *sensu lato* katabatic, in that they blow downhill, but they develop only where and when warm, wet air, as in a cyclonic system, releases its moisture as it climbs up the upwind, generally western or southern, slope of a mountain massif, to descend in the lee as a warm, dry wind, which may carry dust (as in the Alps, where the name belongs). Föhns on the western Canadian prairies (locally: '*Chinooks*') have eroded distinctive gullies on the bluffs of the Old Man River near Lethbridge in Alberta (49° 40′N; 112°51′W, eye altitude 14 km). The gullies on either side of the river are aligned with the Chinook, rather than with usual the dendritic pattern of fluvial tributaries. Beaty (1975) believed that the

gullies were relics of stronger Chinooks of the past. Many Alpine föhns carry and deposit dust taken from North Africa by cyclonic systems.

1.2.2 Wind Systems with Annual Rhythms and Semi-Global Scale

1.2.2.1 Westerlies

The Westerlies are active between approximately 30° and 60°, north and south (Figure 1.5). They carry cyclonic (frontal, low-pressure) systems, in which the winds are strongest in winter. In these systems, winds are very variable in direction (Figure 1.6). Most Westerlies carry enough moisture to ensure that the land surfaces over which they blow are protected by vegetation, which reduces erosion to a very low level (Figure 1.7).

There are places and situations, nonetheless, where and when the westerlies move

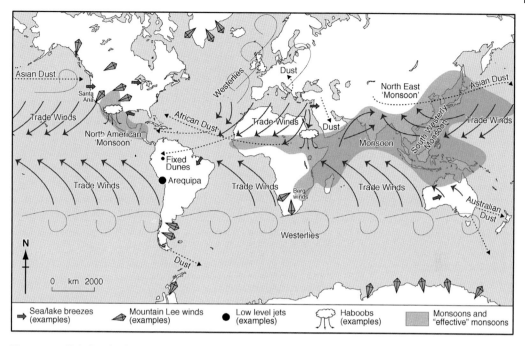

Figure 1.5 Global and selected local wind systems, dust pathways, etc.

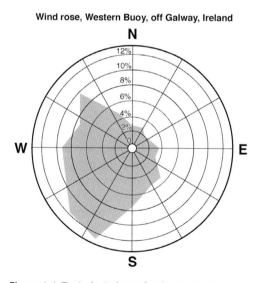

Figure 1.6 Typical wind rose for the Westerlies: at the 'Western Buoy' off Galway on the west coast of Ireland, showing a dominantly south-westerly flow, but also the great spread of wind directions, associated with the passage of low-pressure systems. Percentage figures on this, and subsequent wind roses, are for the percentage of annual observations of winds from that sector.

sand and raise dust. Large amounts are moved, for example, on west-facing macro-tidal shores, as on the coasts of Oregon, Chile, Ireland, and France (Figure 1.8). Many of these coasts also experience macro-tidal seas, which regularly expose large areas of bare sand to the wind. Much of this sand is blown ashore to coastal dunes. In south-western France, Late Pleistocene and Holocene Westerlies drove coastal sand 125 km inland (Bertran et al. 2011). Inland, smaller amounts of sand are blown to dunes in the dry parts of Iceland and Patagonia (Mountney and Russell 2004; Del Valle et al. 2008). The westerlies also move sand and dust when they encounter sandy soil on fields cleared for cultivation (see Chapter 12).

On their southern flanks, the Westerlies also carry large quantities of dust, although much less than the Harmattan. The first of these areas is in the Mediterranean, where the tracks of the frontal systems shift south, most often in the winter or spring. There, they meet the dry, sparsely vegetated,

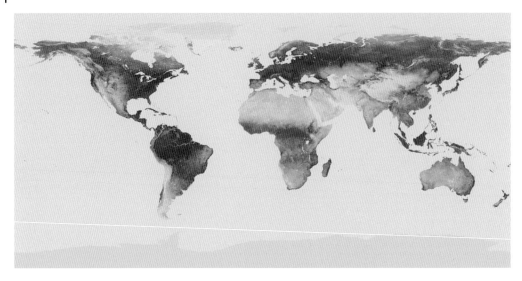

Figure 1.7 Global land cover, the principal global control of wind erosion. Darker areas have more land cover. Unless disturbed, all land except those with the sparse vegetation in arid lands, is protected from wind erosion and transport. Source: Based on: http://www.esa-landcover-cci.org/?q=node/158.

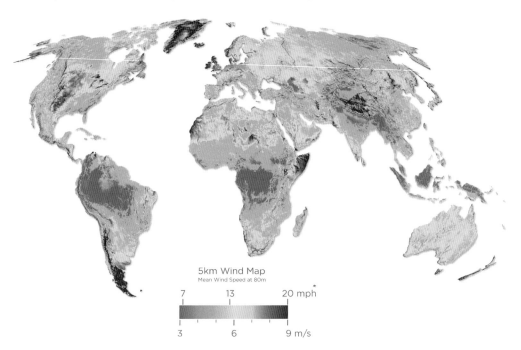

Figure 1.8 Global distribution of wind speed. The map shows average annual wind speed at 80 m above-ground at a 5 km spatial resolution. Wind-speed declines towards the ground, at a rate determined by, among other things, surface roughness (a much fuller explanation is given in Chapter 2). Thus, the map shows only the global pattern of relative differences in wind speed. The map is available at: http://www.3tier.com/static/ttcms/us/images/support/maps/3tier_5km_global_wind_speed.pdf. The average wind speeds are based on over 10 years of hourly data developed with computer model simulations that create realistic wind fields. The wind speed dataset compares well with observations from over 4000 NCEP-ADP network stations. Source: The second section of this caption is a shortened version of text provided by Vaisala Inc., 2001. See insert for colour representation of this figure.

dust-yielding parts of North Africa (Figure 1.5.). When they reach Europe, the best-known term for these dusty winds is the *Sirocco* (of which more later in Section 1.4.4); or in Libya they are *Ghibli* (جبلــــي); and in Egypt and in the Levant, *Khamsin* (خمســـين) or the *Sharav* (שרב). Dust is also raised by cyclonic systems on their southern flanks in North America and Australia (see Chapter 4). The main beneficiaries of this dust are soils on the northern flanks of the tracks of these cyclones (Yaalon 1997). Some of the dust joins föhn winds (see Section 1.2.1.6), and some escapes further north to southern England, as it did on 12 December 2015, and sometimes beyond.

Winter frontal dust storms also plague Iraq, Iran, and northern Arabia, but the better-known and dustier wind, also blowing from the west, is the summer *Shamal* (شمال), a wind that is driven by the difference between high pressure in the west and monsoonal low-pressure in the east. Winter and summer in these areas, westerly winds are accelerated as they come up against the mountains of southern Turkey and Iran before they meet the very dry and dusty alluvial plain of southern Iraq (Hamidi et al. 2013), as the intruders in the Second Gulf War discovered to their cost.

Yet more dust is raised on the pole-ward flanks of the Westerlies, as in eastern Iceland (Mountney and Russell 2004), and Patagonia. In the Patagonian winter, some of this dust joins cyclonic systems in the Westerlies to become the dusty *Pampero,* which is strongest in early summer, and may reach southern Brazil.

1.2.2.2 The Trade Winds

The Trades blow between approximately 30° and 10°, north and south (Figures 1.5 and 1.9). ('Trade' is apt in an aeolian-geomorphological context as it meant 'movement' in old English.) In the sub-tropics, their pathways curve towards the equator in both hemispheres, clockwise in the north, anticlockwise in the south. In the eastern Sahara, the curve is made visible by the borders of sand seas,

barchan-dune 'trains', and wind-eroded features, which are evident even on small-scale scenes on Google Earth, creating the most unmistakable aeolian imprint of the wind on the Earth's land surface.

The Trades probably move much more sand than any other wind system, local or global (their nearest competitor is probably the north-east monsoon in western China, see Section 1.4.4). They are fastest and steadiest on the Atlantic coasts of Morocco and Western Sahara (Figure 1.8), where they drive trains of fast-moving barchans (Elbelrhiti 2012; Figure 1.5 and Chapter 6); and in parts of the Western Desert of Egypt (Stokes et al. 1999), where they blow sand into the oases of El Kharga (Chapter 12, Box 12.1), and blew for the first field measurement of blowing sand by Brigadier Bagnold in 1938 (Box Figure 1.2).

The Trades are strongest in winter, when their trajectories move bodily towards the equator. They weaken in summer, as they are nudged towards the poles by the monsoons (Section 1.2.2.4). Figure 1.9 shows seasonal variation in direction and strength of the Trades, which may explain the extensive

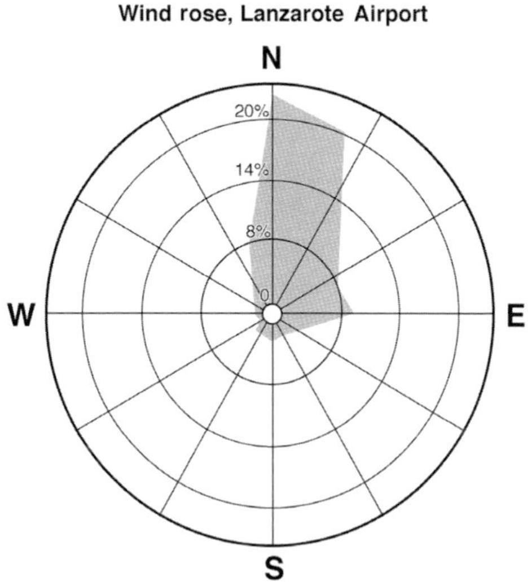

Figure 1.9 Wind rose for the airport in Lanzarote, showing the extraordinarily directional steadiness of the Trade Winds. The two close peaks represent a small, seasonal shift in wind direction.

areas of linear seif dunes in much of the central Sahara (Warren 1972) (seif dunes are explained in Chapter 6). Climate change appears to be accelerating the Trades (Zheng et al. 2016).

1.2.2.3 The Harmattan

The word 'Harmattan' comes from Akan, the language of parts of Ghana and Côte d'Ivoire. In the southern Sahara/northern Sahel and in the early part of the year, the Trades curve round to become easterlies, and become the Harmattan (Figure 1.5). This wind lifts dust from many surfaces, but most comes from dry lakes, the biggest being the Bodélé (see Section 1.2.1.3). This dust: (i) darkens the sky over much of West Africa (find Accra on Google Earth; if the date when the image was taken was in March or April, it will probably be hazy); (ii) fertilises otherwise poor Sahelian soils, adding to them, by one measurement, $\sim 2000\,kg\,ha^{-1}\,yr^{-1}$ (Drees et al. 1993; Herrmann et al. 1996; Chapter 12); (iii) crosses the Atlantic; (iv) to fertilise the Amazon rainforest; and (v) occasionally reaches Ecuador, Florida, and even the American Mid-West (Bristow et al. 2010). Finally, (vi) dust accumulated on the floor of the Atlantic shows that the Harmattan has been blowing since the Cretaceous (Lever and McCave 1983).

1.2.2.4 Monsoons

Monsoons (Figure 1.5) (موسم, 'rainy season' in the original Arabic) are now in most climatological terminologies: low-latitude weather systems in which winds reverse in direction during the annual cycle, bringing rain summers and dry winters. The summer wind is the 'wet monsoon'; the winter wind, the 'dry monsoon'. In effect, the monsoons are sea breezes (above) writ large.

In the summer, haboobs (Section 1.2.1.2) are common on the dry margins of wet monsoons. For example, the North American monsoon (Figure 1.5) breaks over the north of Central America and parts of the dry US south-west. On the equatorial sides of the monsoons, heavy rains feed dense canopies of vegetation (Figure 1.7), which, if undisturbed, allow little loss of dust to the wind. Quite the reverse: many wet monsoons wash dust out of the atmosphere.

There are two dry monsoons: one warm, one cold. The warm one blows in summer, over north-eastern Somalia and southern Oman (Figure 1.5) where it drives small fields of barchan dunes (Wiggs 1993). It is an extension of the East African monsoon or a precursor of the Indian monsoon.

In north-eastern Asia, the North-West Monsoon comes from the east, from the dry heart of Asia (Figure 1.5). It carries dust raised by cyclonic systems (Jugder 2005). In the Gobi Desert of Mongolia, long-term analysis of weather data shows that dust storms occur, on 30–37 days in the year (Natsagdorj et al. 2003). A study using the Nd–Sr ratio of dust has shown that many of the main sources are sand seas (Jiedong et al. 2009).

When the north-west monsoon reaches north-eastern China in winter, it is dry, bitingly cold, and very dusty. Figure 1.10 shows the annual alternation of wind directions at Beijing Airport (the dry north-west monsoon in winter and the wet 'south-western'

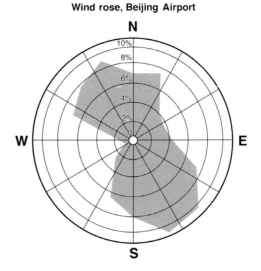

Figure 1.10 Wind rose for Beijing Airport, showing the alternation between the dry and dusty northeast monsoon in winter and the wet South-western monsoon (here south-easterly) in summer.

monsoon in summer, twisted in at its northern limit here to a south-easterly). The phenomenally thick loesses in parts of China were built by this wind in the Holocene, Pleistocene, and before (see Chapter 5). On its contemporary traverse over north-eastern China, the wind picks up industrial pollution. The toxic mix is then swept over Korea and Japan, where, as, Kosa (黃砂) or 'yellow sand', it has a long written-history (Uematsu et al. 2002). It then crosses the North Pacific to western North America, much depleted by its contribution to deep sea sediments, but with some pollution in tow (Figure 1.5).

1.2.3 The Calmer Globe

On Figure 1.8, the calmness of most of the equatorial belt is striking, but even if there were strong winds, they could move little sediment beneath dense, tall vegetation (Figure 1.7), watered by abundant rainfall. Yet, as recently as the Holocene, sand was moved in places now covered by these dense forests (discussed shortly).

1.3 Rhythms of Erosivity and Erodibility from the Semi-Decadal to Hundreds of Thousands of Years

1.3.1 Multiannual Rhythms

Multiannual rhythms, most of which are driven by ocean-surface processes, create variations of wind speed, dust transport, and rainfall at any one site. The strongest and best known of these rhythms is the El Niño Southern Oscillation (ENSO), in which sea-surface temperature oscillates between the western and eastern Pacific. The periodicity of severe ENSO events is very variable, but usually in the range of two to seven years. Other multiannual ocean-based oscillations include the North Atlantic Oscillation (NAO), the Arctic and the Antarctic Oscillations.

Good correlations have been found between severe ENSO events and increased

deposition of dust on the floor of the Pacific itself (Marx et al. 2009), and with dust-raising in the south-western United States (Okin and Reheis 2002); Chapter 12 discusses the relationship between the ENSO and the Dust Bowl in the Great Plains. Its effects may be felt in Europe and beyond. The North Atlantic Oscillation (NAO), like the ENSO, has also been found to have increased the activity of coastal dunes in Europe (Clarke and Rendell 2006); as has the Southern Oscillation in northern Brazil (Maia et al. 2005). The Antarctic Oscillation, surprisingly, pulses in the same rhythm as dustiness in parts of China (Ke and Wang 2004).

1.3.2 Century-Scale Rhythms

Century-scale rhythms or events with return periods of centuries, like the 'Little Ice Age' (*c*.1350–*c*.1850), have a spatial range that is uncertain, but may be at least hemispherical (Szkornik et al. 2008; Jinhua and Wang 2014). Accelerated winds in these systems built coastal and inland dunes and raised dust, some of which reached Antarctica (Mosley-Thompson et al. 1990).

1.3.3 Orbitally-Forced Rhythms

Very much longer climatic rhythms are driven by the combined effect of several of Earth's movements in space, known sometimes as the 'Milankovich' cycles. Although challenged by others, Milankovich's theory of temperature variations and his calculations, by and large, have been upheld (Fagan 2009). Individual cycles are regularly cyclic, but their combination creates complex temporal patterns of solar radiation, and hence temperature, on the Earth's surface. The Pleistocene was a period when the rhythms collaborated to create a sequence of cooler periods, each lasting ~100 000 years (the Glacials), in which ice sheets covered large parts of North America and Europe. They alternated with shorter, warmer periods in some interglacials. There were many smaller variations in temperature within each of these periods, especially those at 40 000-year and 20 000-year cycles, not all of which were directly related to the solar input. The climatic impacts of shorter cool events, such as the 'Heinrich Events', which lasted ~1000 years, include dune formation in the Negev of Israel (Roskin et al. 2011).

Aeolian activity was radically changed in these fluctuations. First, winds that were accelerated by the topographic obtrusion of the enlarged ice sheets in both Europe and North America, lifted, carried, and deposited more loose sediments than did the gentler winds of the interglacials (more in Chapters 5 and 10). They also excavated huge wind-parallel ridges or 'mega-yardangs' in the American Mid-West (Zakrewska 1963), and Hungary (Sebé et al. 2011; Figure 1.11). Second, the directions of some of the winds near the ice caps were reversed (Chapter 3).

Third, and more significantly, massive alternations of wet and dry conditions drove massive shifts in global vegetation zones, which then drove major alternations in aeolian activity. In some of the wet periods, many areas that now are desert were covered in savannah or grassland, allowing early man to move out of Africa; in other places, deserts shrank or disappeared. In the drier periods, even in the Holocene, places now covered by closed-canopy rainforest were dry and free enough of vegetation to allow dune formation, as in the upper Amazon Basin (Carneiro Filho et al. 2002, located on Figure 1.5).

The chronology of these dry/humid sequences in southern Africa, and the uncertainties encountered when trying to date them, are described in Chapter 10. The following account is much less detailed (and less analytical) and refers only to northern Africa and western Europe.

In northern Africa, the least ambiguous evidence for these fluctuations is in the aeolian sands and dune patterns that are now in areas where rainfall is greater than ~250 mm (annually). If undisturbed, these areas are capable of sustaining wooded savannah (Figure 1.12) and even agriculture (zoom into and away from 12°53′N; 30°41′E, to see mega-dune patterns in an agricultural landscape, well south of the 250 mm isohyet) (Warren 1970).

Close to the present edge of the desert, in Mauritania, there is much more accurate evidence of three Late Pleistocene and Holocene periods of dune formation (Figure 1.13). Dune-building periods were interspersed with periods of stability and soil formation. The winds in each dune-forming phase came from a slightly different direction (Lancaster

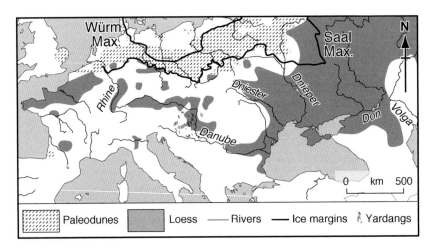

Figure 1.11 Europe: Distribution of sandy and loessic soils, associated rivers, selected limits of glaciations in part of Europe and some yardangs of Pleistocene Age (various sources).

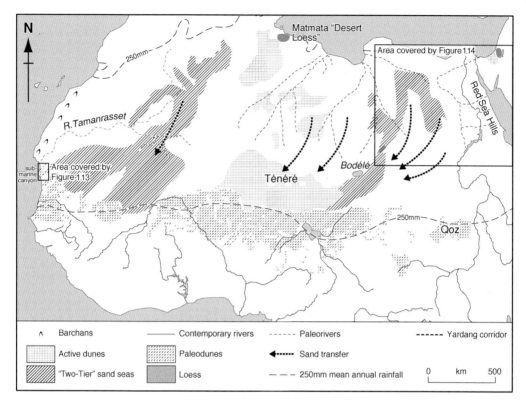

Figure 1.12 Some effects of Pleistocene climatic changes in northern Africa, showing, among other things: sand seas, including 'two-tier' sand seas (only those for which there is undoubted evidence of two generations of dune). Many, if not most, other sand seas contain evidence of repeated periods of dune formation, but more ambiguously. The figure also shows the course of the Tamanrasset River. Source: Skonieczny et al. 2015.

et al. 2002). In the drier Sahara, superimposed generations of dunes are almost universal, but very few are as well dated or as evident. The most obvious are shown on Figure 1.13 as 'two-tier sand seas'. Even more convincing evidence now for the wetter phases (as a major river that traversed the western Sahara) is described in Section 1.4.2.1.

In Europe, these climatic changes had profound effects on aeolian activity. On the north European plain, large quantities of aeolian sand were taken from the alluvium of massive rivers, fed by meltwaters from the continental and mountain ice (Figure 1.11). The rivers flowed westward from the Ukraine to The Netherlands. Their strong seasonality and the coarseness of their load determined that they flowed in shallow, braided channels, in which islands and banks were sparsely

covered with vegetation, given the cool, periglacial environment of the time. Exposed and desiccated, sand was blown to many small and a few large bodies of aeolian dunes. These sands now form the 'European Sand Belt' (Figure 1.11; Zeeberg 1998). One consequence is the distribution of the very erodible agricultural soils of the belt (see Figure 12.3).

The surface textures of individual sand grains in the European Sand Belt have distinct signatures of either fluvial or aeolian transport (Manikowska 2000). The 'aeolisation' (the degree of aeolian abrasion) is greater in the sands of successive Pleistocene ice ages, suggesting repeated reworking by the wind; and the presence of 'aeolised' grains in the sands of successive fluvio-glacial deposits is evidence of repeated interchanges between rivers and the wind.

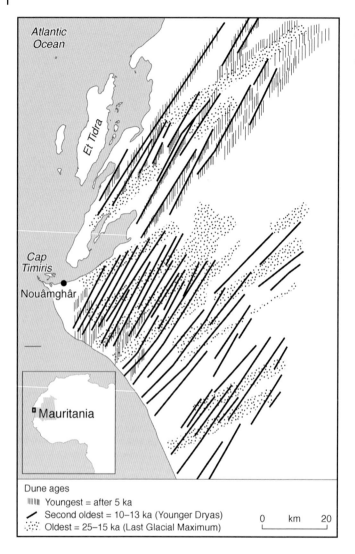

Figure 1.13 A well-dated example of multiple periods of dune growth in Mauritania. Source: Lancaster et al. 2002.

Atlantic
Ocean

Et Tidra

Cap
Timiris

Nouâmghâr

Mauritania

Dune ages
||||||| Youngest = after 5 ka
/ Second oldest = 10–13 ka (Younger Dryas)
∴ Oldest = 25–15 ka (Last Glacial Maximum)

0   km   20

1.4 Frameworks of Sedimentary Supply

1.4.1 Hard Rock

Hard rock is a very minor contributor of aeolian sand or dust. The wind can and does abrade cohesive rock (Chapter 3), but the amount of sand that is released is minute compared to the other sources of aeolian sand discussed shortly. A possible exception might be the sand that must have been released by the erosion of mega-yardangs in ancient sandstones in the Tibesti area of the Sahara, but even that erosional episode (or episodes) is unlikely to have released more than a small proportion of the sand in the southern Sahara, specifically to the dune-fields in the Ténéré, downwind of the Tibesti (Figure 1.12). Other large fields of mega-yardangs, as in the Kalut in Iran, have no major deposit of sand downwind. Soft rock, however, was eroded and yielded dust.

1.4.2 Sand (63–2000 μm)

This section describes the mineralogies of aeolian sand in the order of the area that each covers. The section ignores exotic sands, which cover smaller areas than the rough threshold adopted here, such as those of volcanic debris (as in Hawaii; Tirsch et al. 2012).

Gypsum dissolves in wet climates and fragments in aeolian transport, and is thus found, as a sand, only in dry climates and closely downwind of the ephemeral lakes in which it is deposited. White Sands in New Mexico (in a National Park) is the best-known location. There are others in Tunisia (Coque 1962), Australia, Saudi Arabia, northern Egypt, and Mars (Szynkiewicz et al. 2010; and Chapter 11).

Calcium carbonate sands undoubtedly cover more area than gypsum sands, even if it is difficult to estimate their global coverage. Most are coastal, the outcome of a process that begins with the growth of shells in shallow seas; proceeds to their breakdown by waves; and their transport shoreward to beaches; and ends when they are then blown from beaches to dunes. Most carbonate dune sands are features of tropical coasts, where shell production is most prolific (Brooke 2001). Higher-latitude coastal sands are much smaller in volume, and only 20–80% carbonate, as on the Machair of the western coasts on Scotland and Ireland, also derived from shell-sand beaches (Hansom and Angus 2005) (56°19′49″N;06°59′29 W: Machair on Iona, western Scotland and its origin in a shell-sand beach). Carbonate sands are soluble, but being much less soluble than gypsum, they are lithified in humid climates and some may be dissolved to form karst. Many are of venerable age, and some preserve the bones of very early hominids. They make up whole islands such as Bermuda, the Bahamas and several in the Pacific, and dominate many other tropical mainland coasts (such as the Lençóis Maranhenses, see Section 1.2.1.4).

Given the total area of the Greenland and Antarctic ice sheets at ~5 million km^2 (other ice sheets are very much smaller), ice dunes cannot compete with the total area covered by quartzose sands even if the ice sheets were covered in dunes. Nonetheless, a field of transverse 'mega-dunes' at ~3000 m above sea level on the Antarctic Plateau has been reported to cover 500 000 km^2 (Anschütz et al. 2006), which puts it in second place by area among global sand seas after the Rub' al Khali (the largest sand sea by most measures at ~560 000 km^2, although the stabilised sand sea in the Kalahari is probably bigger). These ice mega-dunes are much smaller than those of most mega-dunes built of siliceous sand. Whether this ice 'sand sea' is unique is unknown. Much smaller areas are covered by ice and snow dunes in northern Alaska, Canada, and Siberia, but there are two uncertainties about the area covered by these 'sands': the area they cover (probably small); and their seasonality.

Given a conservative estimate of the area covered by dunes dominantly of quartz sand (active and stabilised) at 8 million km^2 (after Wilson 1973), their coverage vastly exceeds that of dunes of all the other mineralogies. By volume, quartz sands are even more dominant, given the greater heights and widths of their dunes.

Sands of other mineralogies in mixtures with dominantly aeolian quartzitic sands rarely make up more that 20%, and are generally very much less common. Nonetheless, concentrations of heavy minerals in coastal dunes in New South Wales, and South Africa, at about 8%, are enough to attract commercial mining. Heavy-mineral analysis of aeolian sands (and loess) at much smaller concentration is frequently used to identify their origin (see Chapter 8). Yet smaller concentrations of iron contribute to the red colour of many desert sands, as in the Namib (White et al. 2007).

The reasons for the dominance of quartz in aeolian sands are discussed in Chapter 9. The proportion of quartz in aeolian sands increases as they 'mature' in alternating humid periods, as other minerals are dissolved in moist climates, or in dry and windy environments, where they are winnowed away, over many thousands of years (Muhs 2004).

1.4.2.1 The Fluvial Origins of Most Quartz Sand

By far the greatest quantity of aeolian sand has been delivered to dunes by four, largely fluvial processes.

First, non-fluvial processes deliver material to rivers from the sides of their valleys, where weathering has increased the content of quartz, as less durable minerals are removed from the regolith. Many quartz crystals in igneous rocks such as granite are already of sand size. Where sandstones outcrop in river basins, they have delivered ready-made sands to rivers, as in the case of the extensive outcrop of the Nubian Sandstone in the eastern Sahara.

Second, the confinement of runoff to a channel concentrates its energy and sorts its load. Thus, the earliest bodies of quartzose aeolian sand, in the Pre-Cambrian, may not have appeared until bank-side vegetation had confined runoff to defined courses, in which sand could be sorted, concentrated, and made wind-ready (Eriksson and Simpson 1998).

Third, rivers create sand by grinding down the boulders, cobbles, and gravel that has been delivered to them from valley-sides. In their longitudinal profiles, most large rivers reach an abrupt point at which the mean size of the bed-load declines from gravel to sand. On the bed of the Fraser River in British Columbia, which is 1375 km long, there is a very sharply defined transition from gravel to sand, 210 km downstream of its source (Church 2010). Some transitions are more diffuse (Frings (2001) discusses transitions, their various characters, their position in the course of a river, and their causes). Downstream of the cross-over point, low river-level flow exposes sand on bars and banks, ready to be blown to dunes. Exposure is most frequent in rivers with marked variation in flow, and where vegetation is sparse, these being rivers in very cold climates, such as Banks Island in the Canadian Arctic (Good and Bryant 1985; 72°28′N; 123°45′ W, 6 km eye altitude), or in semi-arid or arid conditions, such as those of the Oued Saoura

in Algeria (30°15′52″N; 02°16′30″W, 5 km eye altitude).

More evidence for the importance of rivers as suppliers of aeolian sand is the association of the mouths of many large rivers and large deposits of aeolian sand: as in the Gran Desierto in Mexico (near the mouth of the Colorado River; Lancaster 1995); the Sinai-Negev Erg (the mouth of the Nile; Muhs et al. 2013); and the Namib sand sea (the mouth of the Orange River; Garzanti et al. 2012). There are also many examples of the relationship in the past, such as the Landes and the mouth of the Garonne in south-western France.

Only in large deserts, such as the central Sahara, central Australia, and perhaps central Asia, do aeolian sands break away from their fluvial network, although even here fluvial origins can be detected from their mineralogy (Pell et al. 2000), or inferred in other ways. More details in Chapter 7.

1.4.3 Coarse Dust (10–63 μm)

Coarse dust (most of which ends up in loess, Chapter 5) is another product of sorting and grinding by rivers. The link between large rivers and loess, composed largely of coarse dust, was first made by Obruchev (Box 1.1). The link has been most explicitly stated, with more up-to-date and world-wide examples, by Smalley et al. (2009). In Europe, Figure 1.11 shows that most thick deposits of loess are indeed east (downwind) of the large rivers, such as the Rhine, Danube, Dniester, Dnieper, Don, and Volga. Part of the explanation of this distribution is that coarse dust, the bulk of loess, travels close to the ground, so that it cannot be taken far from river banks before it is trapped and thus accumulates (Tsoar and Pye 1987).

1.4.4 Fine Dust (<10 μm)

Fine mineral dust is another fluvial product, much of which has been taken further downstream in rivers than sand, to accumulate in seasonal or more ephemeral lakes, in which the dust is exposed seasonally or less frequently. Such dry lakes in the Chihuahuan

Desert in the south-west of the USA cover only 4% of the surface, but have been found to be responsible for 48% of dust plumes (Baddock et al. 2011) (see Figure 12.3). A high proportion of the fine dust in the Sirocco (Section 1.2.2.1) has been shown to come from the Chott ech Chergui in north-central Algeria, another seasonally dry lake (Dulac et al. 1992). Artificially desiccated lakes are major contributors to contemporary fine atmospheric dust: as in the Aral Sea in central Asia (Micklin 2007) and Owens Lake in California (Reheis 1997). However, it seems much of the dust in the North-West Monsoon in China comes from sand seas (Section 1.2.2.2). In the wetter climates of the late Pleistocene and Holocene there were many more of these lakes in areas now desert, and these yielded large quantities of dust when they dried up, leaving behind small yardangs (Chapter 3).

In short, the network of large rivers is the framework for the location of most aeolian material. Coasts and ice sheets provide networks for large, but volumetrically much smaller quantities of these sediments.

1.5 Plate-Tectonic Frameworks: A Glimpse

All the processes so far examined have operated through geological time. At longer, geological scales they are joined by the last of our frameworks: plate tectonics.

In Egypt, about 60 m years ago in the Late Palaeocene, the Arabian Continental Plate began to tear away from the African Plate (Figure 1.14). By between 40 and 20 million years ago (mostly in the Oligocene), the Red Sea Rift began to widen, and its western shoulders, the Red Sea Hills, were uplifted more rapidly than anywhere else in Africa in that era, as determined by the fission-track method. Uplift gave the many right-bank tributaries of a Proto-Nile (in one of its many courses) the energy to cut down rapidly and thus to deliver large amounts of sediment to the main river, which then followed in a course well to the west of its present course, especially in the north (MacGregor 2012). The large aeolian landform that was created by this protracted and complex process is the Great Sand Sea in

Figure 1.14 The Cainozoic geography of tectonic uplift, rapid fluvial erosion, and an ancient course of the Nile: a possible pre-history of the Great Western Sand Sea in Egypt (the position of the proto-Nile of its time). Source: MacGregor 2012.

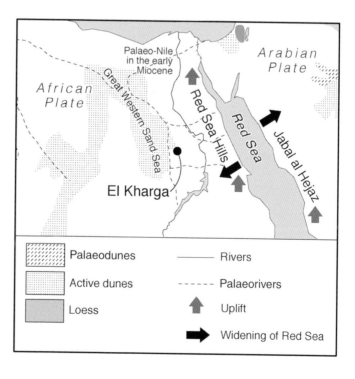

Egypt. There is more on the connection of tectonics, rivers, and dune sand in Chapter 8.

1.6 Conclusion

Some active aeolian landforms have been shaped by the seasonal rhythms in wind systems of semi-global scale. Many more are products either of two or more of these global systems, as they alternate seasonally, or of seasonal alternation of a global with wind systems at smaller scales. The distribution of most aeolian sand and dust is controlled by the location of their immediate source, which, away from the Polar regions, is dominantly fluvial. Most, if not all, large bodies of aeolian sediment can ultimately be connected to inputs of energy from tectonic movements, many millions of years ago. Thus, depositional aeolian geomorphology is like the icing on the terrestrial pudding: dusting (some thick, as in loess), fringes (coastal dunes), and filigree (inland dunes, especially star dunes, Figure 6.3). Erosional aeolian geomorphology scratches a small part of the rest of the pudding (Figure 3.7).

Acknowledgements

Thanks to Paul Hesse for the Japanese for 'Typhoon'; to Martin Todd, University of Sussex, for Figure 1.1; and to Vaisala Inc. for the use of Figure 1.6.

Note

1 References like this are the coordinates of sites on Google Earth.

Further Reading

Good introductions to global climatic frameworks are provided by Rohli (2008) and Warner (2004). The book by Fagan provides a good introduction to how climate has changed in the recent geological past (Fagan 2009). Chapters 3, 'How the Ice Ages began', and 4, 'The Climate Roller-Coaster', both by Maslin, are especially useful.

References

Allen, C.J.T., Washington, R., and Engelstaedter, S. (2013). Dust emission and transport mechanisms in the central Sahara; fennec ground-based observations from Bordj Badji Mokhtar, June 2011. *Journal of Geophysical Research, Atmospheres* 118 (D12): 6212–6232. doi: 10.1002/jgrd.50534.

Anschütz, H., Eisen, O., Rack, W., and Scheinert, M. (2006). Periodic surface features in coastal East Antarctica. *Geophysical Research Letters* 33: L22501. doi: 10.1029/2006GL027871.

Aufrère, L. (1930). L'orientation des dunes continentales. In: 12th International Geographical Congress, Cambridge, 1928, Report of Proceedings, pp. 220–231.

Baddock, M.C., Gill, T.E., Bullard, J.E. et al. (2011). Geomorphology of the Chihuahuan Desert based on potential dust emissions. *Journal of Maps* 7: 249–259. doi: 10.4113/jom.2011.1178.

Bagnold, R.A. (1941). *The Physics of Blown Sand and Desert Dunes*. London: Methuen (reprinted 1954; 1960; 2005, by Dover, Mineola, NY).

Bagnold, R.A. (1990). *Sand, Wind, and War: Memoirs of a Desert Explorer*. Tucson: University of Arizona Press.

Beaty, C.B. (1975). Coulee alignment and the wind in southern Alberta, Canada. *Geological Society of America Bulletin* 86 (1): 119–128. doi: 10.1130/0016-7606(1975) 86<119:CAATWI>2.0.CO;2 (Discussion: Rahn, P.H. 1976. Geological Society of America Bulletin 87(1): 157.).

Bendixen, Q.D. and Isbell, J.L. (2007). Gravel dunes formed by aeolian processes in a cold, dry environment, Allan Hills, Antarctica: a possible Mars analogue. *Geological Society of America, Abstracts with Programs* 39 (6): 205.

Bertran, P., Bateman, M.D., Hernandez, M. et al. (2011). Inland aeolian deposits of South-West France: facies, stratigraphy and chronology. *Journal of Quaternary Science* 26 (4): 374–388. doi: 10.1002/jqs.1461.

Bristow, C.S., Hudson-Edwards, K.A., and Chappell, A. (2010). Fertilizing the Amazon and equatorial Atlantic with west African dust. *Geophysical Research Letters* 37: L14807. doi: 10.1029/2010GL043486.

Bromwich, D.H., Parish, T.R., and Zorman, C.A. (1990). The confluence zone of the intense katabatic winds at Terra-Nova Bay, Antarctica, as derived from airborne sastrugi surveys and mesoscale numerical modeling. *Journal of Geophysical Research: Atmospheres* 95 (D5): 5495–5509. doi: 10.1029/JD095iD05p05495.

Brooke, B.P. (2001). The distribution of carbonate eolianite. *Earth-Science Reviews* 55 (1–2): 135–164. doi: 10.1016/ S0012-8252(01)00054-X.

Cailleux, A. (1936). Les actions éoliennes périglaciaires en Europe. *Bulletin de la Société géologique de France* 5 (5): 102–104.

Carneiro Filho, A., Schwartz, D., Tatumi, S.H., and Rosique, T. (2002). Amazonian paleodunes provide evidence for drier climate phases during the late Pleistocene-Holocene. *Quaternary Research* 58 (2): 205–209. doi: 10.1006/ qres.2002.2345.

Church, M. (2010). Gravel-bed rivers. In: *Sediment Cascades* (ed. T.P. Burt and R.J. Allison), 241–269. Chichester: Wiley.

Clarke, M.L. and Rendell, H.M. (2006). Effects of storminess, sand supply and the North Atlantic Oscillation on sand invasion and coastal dune accretion in western Portugal. *The Holocene* 16 (3): 341–355. doi: 10.1191/0959683606hl932rp.

Coque, R. (1962). *La Tunisie présaharienne: étude géomorphologique*. Paris: Colin.

Crouvi, O., Amit, R., Enzel, Y., and Gillespie, A.R. (2010). Active sand seas and the formation of desert loess. *Quaternary Science Reviews* 29 (17–18): 2087–2098. doi: 10.1016/j.quascirev.2010.04.026.

Del Valle, H.F., Rostagno, C.M., Coronato, F.R. et al. (2008). Sand dune activity in north-eastern Patagonia. *Journal of Arid Environments* 72 (4): 411–422. doi: 10.1016/j.jaridenv.2007.07.011.

Drees, L.R., Manu, A., and Wilding, L.P. (1993). Characteristics of aeolian dusts in Niger, West Africa. *Geoderma* 59 (1–4): 213–233. doi: 10.1016/0016-7061(93) 90070-2.

Dulac, F., Tanré, D., Bergametti, G. et al. (1992). Assessment of the African airborne dust mass over the western Mediterranean Sea using Meteosat data. *Journal of Geophysical Research: Atmospheres* 97 (D2): 2489–2506. doi: 10.1029/91JD02427.

Elbelrhiti, H. (2012). Initiation and early development of barchan dunes: a case study of the Moroccan Atlantic Sahara Desert. *Geomorphology* 138 (1): 181–188. doi: 10.1016/j.geomorph.2011.08.033.

Emmel, C., Knippertz, P., and Schulz, O. (2010). Climatology of convective density currents in the southern foothills of the Atlas Mountains. *Journal of Geophysical Research: Atmospheres* 120 (12): D11115. doi: 10.1029/2009JD012863.

Eriksson, K.A. and Simpson, E.L. (1998). Controls on spatial and temporal distribution of Precambrian eolianites.

Sedimentary Geology 120 (1–4): 275–294. doi: 10.1016/S0037-0738(98)00036-0.

Fagan, B.M.F. (ed.) (2009). *The Complete Ice Age: How Climate Change Shaped the World*. London: Thames & Hudson.

Frings, R.M. (2001). Sedimentary characteristics of the gravel–sand transition in the river Rhine. *Journal of Sedimentary Research* 81 (1): 52–63. doi: 10.2110/jsr.2011.

Furberg, M., Steyn, D.G., and Baldi, M. (2002). The climatology of sea breezes on Sardinia. *International Journal of Climatology* 22 (8): 917–932. doi: 10.1002/joc.780.

Garzanti, E., Andò, S., Vezzoli, G. et al. (2012). Petrology of the Namib Sand Sea: long-distance transport and compositional variability in the wind-displaced orange delta. *Earth-Science Reviews* 112 (3–4): 173–189. doi: 10.1016/j.earscirev.2012.02.008.

Gaylord, D.R. and Dawson, P.J. (1987). Airflow-terrain interactions through a mountain gap, with an example of eolian activity beneath and atmospheric hydraulic jump. *Geology* 15 (9): 789–792. doi: 10.1130/0091-7613(1987)152.0.CO;2.

Good, T.R. and Bryant, I.D. (1985). Fluvio-aeolian sedimentation – an example from Banks Island, N.W.T., Canada. *Geografiska Annaler A* 67 (1–2): 33–46. doi: 10.2307/520464.

Goudie, A.S. (2008). Aeolian Processes and Forms. In: *The History of the Study of Landforms or the Development of Geomorphology*, vol. 4 (ed. T.P. Burt, R.J. Chorley, D. Brunsden, et al.), 767–804. London: Geological Society of London.

Grove, A.T. and Warren, A. (1968). Quaternary landforms and climate on the south side of the Sahara. *The Geographical Journal* 134 (2): 194–208. doi: 10.2307/1792436.

Hamidi, M., Kavianpour, M.R., and Yaping, S. (2013). Synoptic analysis of dust storms in the Middle East. *Asia-Pacific Journal of Atmospheric Science* 49 (3): 279–286. doi: 10.1007/s13143-013-0027-9.

Hansom, J.D. and Angus, S. (2005). Machair of the Western Isles: Scottish landform example no. 36. *Scottish Geographical Journal* 121 (4): 401–412. doi: 1080/00369220518737247.

Herrmann, L., Jahn, R., and Stahr, K. (1996). Identification and quantification of dust additions in peri-Saharan soils. In: *The Impact of Desert Dust across the Mediterranean* (ed. S. Guerzoni and R. Chester), 173–182. Dordrecht: Kluwer.

Houser, C., Wernette, P., Rentschlar, E. et al. (2015). Post-storm beach and dune recovery: implications for barrier island resilience. *Geomorphology* 234: 54–63. doi: 10.1016/j.geomorph.2014.12.044.

Hunter, R.E. and Richmond, B.M. (1988). Daily cycles in coastal dunes. *Sedimentary Geology* 55 (1): 43–67. doi: 10.1016/0037-0738(88)90089-9.

Jiedong, Y., Li, G., Wenbo, R., and Ji, J. (2009). Isotopic evidences for provenance of East Asian dust. *Atmospheric Environment* 43 (29): 4481–4490. doi: 10.1016/j.atmosenv.2009.06.035.

Jinhua, D. and Wang, X. (2014). Optically stimulated luminescence dating of sand-dune formed within the little ice age. *Journal of Asian Earth Sciences* 91: 154–162. doi: 10.1016/j.jseaes.2014.05.012.

Jugder, D. (2005). Discriminate analysis for dust storm prediction in the Gobi and steppe regions in Mongolia. *Water, Air, and Soil Pollution: Focus* 5 (3–6): 37–49. doi: 10.1007/s11267-005-0725-0.

Ke, F. and Wang, H. (2004). Antarctic oscillation and the dust weather frequency in North China. *Geophysical Research Letters* 31: L10201. doi: 10.1029/2004GL019465.

Keyes, C.R. (1909). Base-level of eolian erosion. *The Journal of Geology* 17 (7): 659–663. doi: 10.1086/621666.

Lancaster, N. (1995). Origin of the Gran Desierto sand sea, Sonora, Mexico: evidence from dune morphology and sedimentology. In: *Desert Aeolian Processes* (ed. V.P. Tchakerian), 11–26. London: Chapman and Hall.

Lancaster, N., Kocurek, G., Singhvi, A. et al. (2002). Late Pleistocene and Holocene dune activity and wind regimes in the western Sahara Desert of Mauritania. *Geology* 30 (11): 991–994. doi: 10.1130/0091-7613 (2002)030<0991:LPAHDA>2.0.CO;2 Comment: Swezey, C.S. Geology 31(1): 18, doi:10.1130/0091-7613-31.1.e18.

Leighton, M.M. (1931). The Peorian loess and the classification of the glacial drift sheets of the Mississippi Valley. *The Journal of Geology* 39 (1): 45–53. doi: 10.1086/623787 (Comment: Leighton, M.M. Science 77(1980): 168.).

Lettau, H. and Lettau, K. (eds) (1978). Exploring the world's driest climate. University of Wisconsin at Madison, Institute for Environmental Studies, IES Report 101.

Lever, A. and McCave, I.N. (1983). Eolian components in cretaceous and tertiary North Atlantic sediments. *Journal of Sedimentary Petrology* 53 (3): 811–832. doi: 0.1306/212F82C9-2B24-11D7-8648000102C1865D.

Libby, W.F. (1952). *Radiocarbon Dating*, 1e. Chicago: University of Chicago Press.

Lyell, C. (1875). *Principles of Geology*, 12e, vol. 2. London: John Murray.

MacGregor, D. (2012). Rift shoulder source to pro-delta sink: the Cenozoic development of the Nile Drainage System. *Petroleum GeoScience* 18 (4): 417–431. doi: 10.1144/petgeo2011-07418.

Maia, L.P., Freire, G.S.S., and Lacerda, L.D. (2005). Accelerated dune migration and aeolian transport during El Niño events along the NE Brazilian coast. *Journal of Coastal Research* 21 (6): 1121–1126. doi: 10.2112/03-702A.1.

Manikowska, B. (2000). Stages of the activity of aeolian processes during the Vistulian period in Central Poland. In: *Aeolian Processes in Different Landscape Zones* (ed. R. Dulias and G.J. Pełka), 70–79. Sosnowiec: University of Silesia.

Marx, S.K., McGowan, H.A., and Kamber, B.S. (2009). Long-range dust transport from eastern Australia; a proxy for Holocene aridity and ENSO-type climate variability. *Earth and Planetary Science Letters* 282

(1–4): 167–177. doi: 10.1016/j.epsl.2009.03.013.

Micklin, P. (2007). The Aral Sea disaster. *Annual Review of Earth and Planetary Sciencels* 35: 47–72. doi: 10.1146/annurev.earth.35.031306.140120, online.

Mosley-Thompson, E., Thompson, L.G., Grootes, P.M., and Gundestrup, N. (1990). Little ice age (neoglacial) paleoenvironmental conditions at Siple Station, Antarctica. *Annals of Glaciology* 14: 199–204. doi: 10.3198/1990AoG14-1-199-204.

Mountney, N.P. and Russell, A.J. (2004). Sedimentology of cold-climate aeolian sandsheet deposits in the Askja region of Northeast Iceland. *Sedimentary Geology* 166 (3–4): 223–244. doi: 10.1016/j.sedgeo.2003.12.007.

Muhs, D.R. (2004). Mineralogical maturity in dunefields of North America, Africa and Australia. *Geomorphology* 59 (1–4): 247–269. doi: 10.1016/j.geomorph.2003.07.020.

Muhs, D.R., Roskin, J., Tsoar, H. et al. (2013). Origin of the Sinai-Negev erg, Egypt and Israel: mineralogical and geochemical evidence for the importance of the Nile and sea level history. *Quaternary Science Reviews* 69: 28–48. doi: 10.1016/j.quascirev.2013.02.022.

Muhs, D.R., Stafford, T.W., Cowherd, S.D. et al. (1996). Origin of the late quaternary dune fields of northeastern Colorado. *Geomorphology* 17 (1–3): 129–149. doi: 10.1016/0169-555X(95)00100-J.

Natsagdorj, L., Jugder, D., and Chung, Y.-S. (2003). Analysis of dust storms observed in Mongolia during 1937–1999. *Atmospheric Environment* 37 (9–10): 1401–1411. doi: 10.1016/S1352-2310(02)01023-3.

Obruchev, V.A. (1895). O protsessakh vyvetriviya i rasvevaniya v tsentral'noy Azii (On the processes of the decomposition and deflation in Central Asia). *Vlervye napechtago v. Zapiski Mineralogicheskago Obshchestva 2 Seriya* 33 (1): 229–272. (in Russian) (Reproduced in Izbrannye raboty po geografii Azii [*Selected Transactions on*

Geography of Asia] 3 pp 131–160; (*Abstract in Neues Jahrbuch für Mineralogie* (1897), pp. 466–471.).

Obruchev, V.A. (1911). Eolovyi Gorad (An aeolian city). Zemlevedenie (3): 1–22. (Summary: 'Eine aeolische Stadt' in Pettermanns geographische Mitteilungen 1912 (58): 289; and de Hutorowicz, H. (1912). *Bulletin of the American Geographical Society* 44(12): 916–917.)

Okin, G.S. and Reheis, M.C. (2002). An ENSO predictor of dust emission in the southwestern United States. *Geophysical Research Letters* 29 (9): 1332. doi: 10.1029/2001GL014494.

Parteli, E.J.R., Schwämmle, V., Herrmann, H.J. et al. (2006). Profile measurement and simulation of a transverse dune field in the Lençóis Maranhenses. *Geomorphology* 81 (1–2): 29–42. doi: 10.1016/j. geomorph.2006.02.015.

Pell, S.D., Chivas, A.R., and Williams, I.S. (2000). The Simpson, Strzelecki and Tirari deserts: development and sand provenance. *Sedimentary Geology* 130 (1–2): 107–130. doi: 10.1016/S0037-0738(99)00108-6.

Raman, A., Arellano, A.F. Jr., and Brost, J.J. (2014). Revisiting haboobs in the southwestern United States: an observational case study of the 5 July 2011 Phoenix dust storm. *Atmospheric Environment* 89: 179–188. doi: 10.1016/ j.atmosenv.2014.02.026.

Reheis, M.C. (1997). Dust deposition downwind of Owens (dry) Lake, 1991–1994: preliminary findings. *Journal of Geophysical Research: Atmospheres* 102 (D22): 25999–26008.

Renfrew, I.A. (2004). The dynamics of idealised katabatic flow over a moderate slope and ice shelf. *Quarterly Journal of the Royal Meteorological Society, part A* 130: 1023–1045. doi: 10.1256/qj.05.148.

Robson, D.F. and Sampath, N. (1977). Geophysical response of heavy-mineral sand deposits at Jerusalem Creek, New South Wales. *BMR Journal of Australian Geology and Geophysics* 2 (2): 149–154.

Rohli, R.V. (2008). *Climatology*. London: Jones and Bartlett.

Rolland, G. (1890). Sur les grandes dunes de sable du Sahara. *Comptes rendus des séances de la Société de Géographie, Paris* 110 (12): 659–662.

Roskin, J., Porat, N., Tsoar, H. et al. (2011). Age, origin and climatic controls on vegetated linear dunes in the northwestern Negev Desert (Israel). *Quaternary Science Reviews* 30 (13–14): 1649–1674. doi: 10.1016/j.quascirev.2011.03.010.

Sebé, K., Csillag, G., Ruszkiczay-Rüdiger, Z. et al. (2011). Wind erosion under cold climate: a Pleistocene periglacial mega-yardang system in Central Europe (western Pannonian Basin, Hungary). *Geomorphology* 134 (3–4): 470–482. doi: 10.1016/j. geomorph.2011.08.003.

Sharp, R.P. (1980). Wind-driven sand in Coachella valley, California – further data. *Geological Society of America Bulletin* 91 (12): 724–730. doi: 10.1130/0016-7606(1980) 91<724:WSICVC>2.0.CO;2.

Sills, D.M.L., Brook, J.R., Levy, I. et al. (2011). Lake breezes in the southern Great Lakes region and their influence during BAQS-Met 200. *Atmospheric Chemistry and Physics* 11: 7955–7973. doi: 10.5194/ acp-11-7955-2011.

Skonieczny, C., Paillou, P., Bory, A. et al. (2015). African humid periods triggered the reactivation of a large river system in Western Sahara. *Nature Communications* 6: 8751. doi: 10.1038/ncomms975.

Smalley, I.J., O'Hara-Dhand, K., Wint, J. et al. (2009). Rivers and loess: the significance of long river transportation in the complex event-sequence approach to loess deposit formation. *Quaternary International* 198 (1–2): 7–18. doi: 10.1016/j.quaint. 2008.06.009.

Stephens, G. (2011). Storminess in a warming world. *Nature Climate Change* 1: 252–253. doi: 10.1038/nclimate1176.

Stokes, S., Goudie, A.S., Ballard, J. et al. (1999). Accurate dune displacement and morphometric data using kinematic GPS.

Zeitschrift für Geomorphologie, Supplementbände 116: 195–214.

Szkornik, K., Gehrels, W.R., and Murray, A.S. (2008). Aeolian sand movement and relative sea-level rise in Ho Bugt, western Denmark, during the 'Little Ice Age'. *Holocene* 18 (6): 951–965. doi: 10.1177/0959683608091800.

Szynkiewicz, S., Ewing, R.C., Moore, C.H. et al. (2010). Origin of terrestrial gypsum dunes—implications for Martian gypsum-rich dunes of Olympia Undae. *Geomorphology* 121 (1–2): 69–83. doi: 10.1016/j.geomorph.2009.02.017.

Tirsch, D., Craddock, R.A., Platz, T. et al. (2012). Spectral and petrologic analyses of basaltic sands in Ka'u Desert (Hawaii): implications for the dark dunes on Mars. *Earth Surface Processes and Landforms* 37 (4): 434–448. doi: 10.1002/esp.2266.

Tsoar, H. (1983). Dynamic processes acting on a longitudinal (seif) sand dune. *Sedimentology* 30 (4): 567–578. doi: 10.1111/j.1365-3091.1983.tb00694.x.

Tsoar, H. and Pye, K. (1987). Dust transport and the question of desert loess formation. *Sedimentology* 34 (1): 139–154. doi: 10.1111/j.1365-3091.1987.tb00566.x.

Udden, J.A. (1894). Erosion, transport and sedimentation performed by the atmosphere. *The Journal of Geology* 2 (2): 318–331. doi: 10.1086/606957.

Uematsu, M., Yoshikawa, A., Muraki, H. et al. (2002). Transport of mineral and anthropogenic aerosols during a Kosa event over East Asia. *Geophysical Research Letters* 29 (7): 4059. doi: 10.1029/2001JD000333.

van den Broeke, M.R., Duynkerke, P.G., and Oerlemans, J. (1994). The observed katabatic flow at the edge of the Greenland ice sheet during GIMEX-91. *Global and Planetary Change* 9 (1–2): 3–15. doi: 10.1016/0921-8181(94)90003-5.

Vermeesch, P. and Drake, N. (2008). Remotely sensed dune celerity and sand flux measurements of the world's fastest barchans (Bodélé, Chad). *Geophysical Research Letters* 35: L24404. doi: 10.1029/2008GL035921.

Vermeesch, P., Fenton, C.R., Kober, F. et al. (2010). Sand residence times of one million years in the Namib Sand Sea from cosmogenic nuclides. *Nature Geoscience* 3 (12): 862–865. doi: 10.1038/ngeo985.

von Richthofen, F. (1882). On the mode of origin of the loess. *The Geological Magazine, Decade II* 9 (7): 293–305. doi: 10.1017/S001675680017164X.

Warner, T.T. (2004). *Desert Meteorology*. Cambridge: Cambridge University Press.

Warren, A. (1970). Dune trends and their implications in the Central Sudan. *Zeitschrift für Geomorphologie Supplementbände* 10: 154–179.

Warren, A. (1972). Observations on dunes and bimodal sands in the Ténéré Desert. *Sedimentology* 19 (1): 37–44. doi: 10.1111/j.1365-3091.1972.tb00234.x.

Warren, A., Chappell, A., Todd, M.C. et al. (2007). Dust-raising in the dustiest place on Earth. *Geomorphology* 92 (1–2): 25–37. doi: 10.1016/j.geomorph.2007.02.007.

White, K., Walden, J., and Gurney, S.D. (2007). Spectral properties, iron oxide content and provenance of Namib dune sands. *Geomorphology* 86 (3–4): 219–229. doi: 10.1016/j.geomorph.2006.08.014.

Wiggs, G.F.S. (1993). Desert dune dynamics and the evaluation of shear velocity: an integrated approach. In: *The Dynamics and Environmental Context of Aeolian Sedimentary Systems*, vol. 72 (ed. K. Pye), 37–46. London: Geological Society of London Special Publication, doi:10.1144/GSL.SP.1993.072.01.05.

Wilson, I.G. (1973). Ergs. *Sedimentary Geology* 10 (1): 77–108. doi: 10.1016/0037-0738(73)90001-8.

Yaalon, D.H. (1997). Soils in the Mediterranean region: what makes them different? *Catena* 28 (3–4): 157–169. doi: 10.1016/S0341-8162(96)00035-5.

Zakrewska, B. (1963). An analysis of landforms in a part of the central Great Plains. *Annals of the Association of American Geographers*

53 (4): 536–568. doi: 10.1111/j.1467-8306. 1963.tb00465.x.

Zeeberg, J.J. (1998). The European sand belt in eastern Europe and comparison of Late Glacial dune orientation with GCM simulation results. *Boreas* 27 (2): 127–139. doi: 10.1111/j.1502-38845.1998. tb00873.x.

Zheng, H., Pan, J., and Chongyin, L. (2016). Global oceanic wind speed trends. *Ocean and Coastal Management* 129: 15–24. doi: 10.1016/j.ocecoaman.2011.04.002.

Znamenski, A.I. (1958). Eksperimentalnye issledovaniya protsessov vetrovoy erozii peskov (Experimental study of the wind erosion process) i voprosy zashchity ot peschanykh zanosov. *Materialy Issledovanii v pomoshch' proektirovaniiu stroitl'stvu Karakumskogo Kanala, Ashkhabad* 1: 7–14.

2

Grains in Motion

Andreas C.W. Baas

King's College London, London, UK

2.1 Introduction

Atmospheric circulation moves currents of air around the globe, creating wind systems over the Earth's surface (see Chapter 1). Turbulence in these winds trickles down in an eddy cascade from storms of semi-continental size to wafts that ruffle a blade of grass. Where winds of sufficient speed touch a loose surface, they transport sediments and change the shape of the landscape, producing ripples, dunes, and the variety of other aeolian landforms described in other chapters of this book.

This chapter is about the detailed mechanics of how wind moves sediment: how grains are put in motion by the airflow, and how they hop and skip over the surface as they are accelerated forward. We start with an overview of the driving force, the turbulent wind, and then consider the initiation of movement and our understanding of the forces and mechanics of grains in motion in an idealised steady-state framework. Real-world conditions in natural environments are, however, much more complex and introduce a number of complications and constraints that are discussed subsequently.

2.2 The Nature of Wind

2.2.1 Turbulent Boundary Layers

As air flows over the ground, the roughness of the surface imposes a friction that slows the horizontal wind speed down close to the ground, right down to a very thin layer of air that is not moving at all, immediately above the surface (the 'no-slip condition'). The resultant vertical gradient in wind speeds at the bottom of the airflow reflects a drain of momentum from the larger-scale 'free-stream' flow overhead to the solid ground below, and defines a 'boundary layer' where the airflow is adjusting to the roughness of the underlying surface. In almost all surface winds that can initiate grain motion and sediment transport, this vertical transfer of momentum through the flow is achieved via turbulence and eddies of air parcels – scaling from metres to tens of metres high up in the airflow to centimetres and millimetres very close to the ground. When close to the ground, this exchange of turbulent kinetic energy is lost to frictional heat or geomorphological work, such as sediment transport at the surface.

Aeolian Geomorphology: A New Introduction, First Edition. Edited by Ian Livingstone and Andrew Warren.
© 2019 John Wiley & Sons Ltd. Published 2019 by John Wiley & Sons Ltd.

As air travels over a stretch of ground – a sand patch in the desert or a beach surface, for example – a turbulent boundary layer of adjusted flow develops, starting from the leading edge and growing with distance downwind (known as 'fetch') upward into the free-stream overhead. The structure of the turbulent boundary layer shows two distinct zones: an 'inner region' in the bottom 10% of the layer, where the wind speed profile is controlled by the roughness of the surface; and an 'outer region' where the profile is also influenced by the free-stream overhead (Figure 2.1).

There are many equations for estimating the height of the turbulent boundary layer, δ, as a function of fetch length, x, and also of the roughness of the surface that the flow is adjusting to, represented by the roughness length, z_0 (defined further below). A frequently-used formula is that of Elliot (1958) (Eq. 2.1)

$$\delta = 0.75\, z_0 \left(x/z_0\right)^{0.8} \qquad (2.1)$$

which, for example, for a fetch of 100 m and a roughness length of 1 mm (a characteristic order of magnitude for sandy surfaces) predicts a δ of roughly 8 m. This is known to be a high estimate compared with other formulas, however (Walmsley 1989). Note that the above conceptualisation of boundary layer development represented in Figure 2.1 and

Eq. (2.1) is highly simplified: as air flows over the Earth's surface in the over-arching atmospheric boundary layer as a whole (the bottom 1000 m or so of the troposphere), every change in surface roughness triggers a new flow adjustment and the development of its own 'internal' or nested boundary layer growing from the bottom upward into the remnant flow behaviour of the preceding boundary layer overhead. Because most land surfaces have highly variable roughness patterns, a simple boundary layer profile cannot usually develop. Examples of surfaces where roughness remains similar for long distances so that boundary layer profiles approaching the theoretical one shown in Figure 2.1 can develop include oceans, snow fields, and wide sandy beaches.

Simplifications notwithstanding, the inner region of the local boundary layer (the bottom ~10%) is also known as the constant-stress layer: it is here that the vertical momentum transfer towards the ground exhibits itself as a 'shearing' or shear stress, τ, with units of $(N\,m^{-2})$ – a force parallel with the horizontal streamlines – within the airflow that is constant at all heights and that is also therefore acting on the ground surface where it may initiate grain motion. The wind speed profile (horizontal wind speed U as a function of height z) associated with this constant shear stress layer is described by the Prandtl-Von Kármán logarithmic velocity

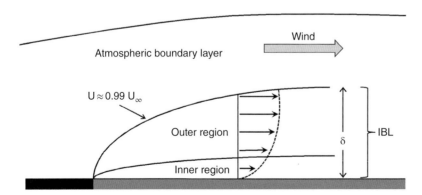

Figure 2.1 The structure of the turbulent boundary layer: an internal boundary layer (IBL), with height δ, growing inside the atmospheric boundary layer downwind of a surface transition. IBL shows logarithmic wind speed profile. The bottom 10% of the IBL is usually defined as the inner region, also known as the constant-stress-layer.

profile equation, developed in the 1930s and traditionally referred to as the Law-of-the-Wall (Eq. 2.2):

$$U(z) = \frac{U_*}{\kappa} \ln\left(\frac{z}{z_0}\right) \qquad (2.2)$$

where U_* is the so-called *shear velocity* (sometimes also referred to as 'friction velocity') with units of $m\,s^{-1}$, in Eq. (2.3) defined as:

$$U_* \overset{\text{def}}{=} \sqrt{\frac{\tau}{\rho}} \qquad (2.3)$$

relationship therefore represents the shear stress in the flow, though normalised by the air density. The shear velocity (U_*) is scaled by κ: the Von Kármán constant, usually assigned a value of 0.4. The roughness length, z_0, is nominally the height above the bed beneath which the wind speed is assumed to be zero, although this is effectively a theoretical construct or simplification, since the precise details of the micro-scale flow mechanics so close to the bed (within a few grain diameters) are complex. The roughness length may be thought of as the height above the surface where the flow transitions from momentum exchange via eddies – pockets of air rotating and moving up and down – to momentum exchange via molecular viscosity – individual molecules colliding with each other (in the so-called viscous sub-layer). In small-scale engineering applications z_0 has been traditionally estimated as 1/30th of the grain diameter, based on historical pipe flow studies by Nikuradse (1933). In applications to aeolian geomorphology, however, it is found that the roughness of sandy beds, particularly with active sand transport, is of the order of millimetres instead.

The Law-of-the-Wall can be used to determine the shear velocity (or shear stress) acting on the surface, as well as the roughness length, by measuring time-averaged wind speeds at various heights in the inner region of the boundary layer and subsequently fitting a least-squares linear regression through these data on a plot of U versus $\ln(z)$ (e.g.

Wilkinson 1984; Bauer et al. 1992). The flow parameters can then be determined from the slope, m, and the intercept, b, of the fitted regression line (Eq. 2.4):

$$U_* = \kappa\, m \qquad (2.4)$$

$$z_0 = e^{-b/m}$$

The determination of z_0 is, however, particularly sensitive to the regression fitting and prone to large error bars (Sherman and Bauer 1993).

The advent of high-frequency 3D sonic anemometry has facilitated another method for determining the shear stress in the flow. This method relies on Prandtl's mixing length theory which relates the vertical transfer of momentum (which we can measure as shear stress) to the vertical exchange, or mixing, of low-speed air parcels towards higher regions and high-speed air parcels towards lower regions of the flow. High-frequency time-series of streamwise, $u(t)$, spanwise, $v(t)$, and vertical, $w(t)$, wind speed components of the full 3D airflow velocity vector can be split into a time-average, e.g. $\bar{u}$, and a time-dependent fluctuating component, e.g. $u'(t)$, via what is known as Reynolds decomposition (Eq. 2.5):

$$u(t) = \bar{u} + u'(t) \qquad (2.5)$$

$$v(t) = \bar{v} + v'(t)$$

$$w(t) = \bar{w} + w'(t)$$

where the prime indicates the fluctuating component. The total resultant horizontal shear stress can then be estimated by the covariances between each of the horizontal components and the vertical component (i.e. $\overline{u'w'}$ and $\overline{v'w'}$) (Eq. 2.6), as:

$$\tau = -\rho \sqrt{\left(\overline{u'w'}\right)^2 + \left(\overline{v'w'}\right)^2} \qquad (2.6)$$

In an idealised boundary layer, the shear stress measured via Reynolds decomposition is theoretically identical to the shear stress determined from a Law-of-the-Wall analysis. However, several field studies have found significant discrepancies (King et al. 2008; Lee and Baas 2015), casting doubt on the correspondence.

This section has focused on those characteristics of turbulent boundary layer flow that are relevant to aeolian sediment transport. Much more detailed and authoritative discussions on all aspects of boundary layers can be found in Schlichting and Gersten (2000) from the perspective of aerodynamic engineering, and in Kaimal and Finnigan (1994) within a meteorology context. Furthermore, the above overview does not consider the complex changes in wind dynamics and the structure of the boundary layer for flow over relief, such as sand dunes (obviously very relevant to aeolian geomorphology). When air flows over the typical profile of a bare-sand dune, the near-surface wind experiences a sequence of decelerations and accelerations, as it first approaches the toe of the dune, accelerates over the stoss (windward) slope and over the crest, separates from the brink, creating a wake behind the dune with a recirculation zone, and re-attaches to the ground surface some distance beyond the dune. These flow dynamics have dramatic impacts on surface shear stress at different locations along a dune and related sediment transport potentials. Chapters 6 and 7 delve deeper into this issue.

2.2.2 Turbulence and Coherent Flow Structures

The boundary layer and the associated shear stress essentially represent time-averaged turbulence, but in the new millennium it has become increasingly clear that the passage of discrete, individual turbulent motions or coherent flow structures (CFSs) are likely to have some relation to sediment transport at the surface (Bauer et al. 2013). In aerodynamic engineering so-called burst-sweep events, a form of CFS, were identified early on (Kline et al. 1967; Robinson 1991). In this process pockets of slow-moving fluid at the surface eject or 'burst' upwards into the flow, flanked by streamwise vortices, to be replaced by an inrush or 'sweep' of high-speed fluid from the overlying regions of the flow towards the bottom. The potential link between burst-sweep events and the initiation of sediment transport was quickly noted, particularly in rivers (Jackson 1976; Best 1992), and their flow signatures have been reputedly identified in aeolian field studies (Bauer et al. 1998; Sterk et al. 1998), as well as related to the variability of sand transport (Leenders et al. 2005; Wiggs and Weaver 2012). Experimental work and numerical studies have also found burst-sweep sequences that generate packets or trains of multiple vortex structures rising up and advecting with the flow (Adrian et al. 2000).

However, the scaling of CFSs is problematic if their relevance to aeolian sand transport is to be demonstrated. Burst-sweep events and vortex trains are thought to manifest themselves well below $z^+ = 100$ (wall units), where z^+ is a non-dimensional height found by using the so-called 'inner-layer scaling': $z^+ = \dfrac{zU_*}{\nu}$, with ν kinematic viscosity. For moderate aeolian sand transport conditions (e.g. $U_* = 0.4$), a hundred wall units translate to a physical height of only a few millimetres, which might make CFSs relevant to grain-scale forces, but not easily linked to larger-scale transport dynamics. Recent studies, however, have found other types of CFSs at larger scales within the atmospheric boundary layer: 'Very Large Scale Motions' (VLSMs, see Figure 2.2) or 'Super-Structures' (Marusic and Hutchins 2008; Marusic et al. 2010), with streamwise lengths of 15–20 δ (so-called outer-layer scaling), which can therefore be several tens of metres in aeolian geomorphology context. The origins of these super-structures are not yet clear. While most studies assume that CFSs originate at the bed (inner-layer scaling) or that they at least grow upwards into the boundary layer (e.g. super-structures), an alternative approach is the top-down

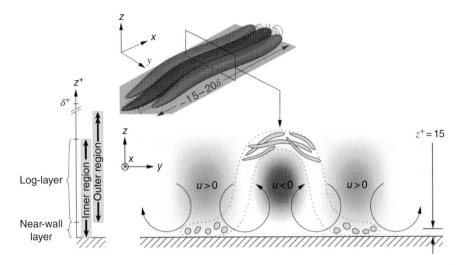

Figure 2.2 Very large-scale motions or 'super-structures' in the inner boundary layer. Grey patches indicate vortex filaments, both very close to the surface (cross-sections) and at the top of the inner region (sideways). *Source:* Modified from Marusic et al. (2010). See insert for colour representation of this figure.

turbulence model proposed by Hunt and his collaborators (Hunt and Morrison 2000; Hunt and Carlotti 2001). In this model, large eddies in the outer region of the flow travel downward and impinge onto the surface layer, where they are blocked and sheared and break down into sub-structures that may have similar characteristics as other CFSs near the bed.

The application of models and findings on boundary layer flows and coherent flow structures to the aeolian sediment transport environment remains challenging because we have to navigate between two fluid mechanics research fields that work on very different length scales: that of aerodynamic engineering, which has historically been focused on flow dynamics at a thickness of a few millimetres over airfoils in wind tunnel studies, and that of 'micro'-meteorology, which concerns airflow in the mixing layer, the bottom 1000 m or so of the planetary atmospheric boundary layer. At the boundary layer scales of metres to tens of metres, which wind-driven sand transport over beaches and desert surfaces occupy, theories and experimental findings often unravel.

2.3 Transport Modes

Four distinct types (or modes) of wind-driven grain movement are commonly recognised: (i) saltation; (ii) creep; (iii) suspension; and (iv) reptation (Figure 2.3).

Saltation is a process of repeated bouncing over the bed in which grains follow parabolic trajectories (the term derives from the Latin verb *saltare*, 'to leap'). Grains are ejected from the bed by the impact of preceding saltating grains or by direct entrainment into the flow (see Section 2.5). The grains are then accelerated by the wind and follow a ballistic trajectory to return to the bed, where they either impact, burrowing into the bed packing and splashing up other grains, or ricochet through a near-elastic collision into another saltation trajectory. Saltation is the dominant mode of grain motion over sandy beds and is also the main driver for the other three transport modes. Theoretical reasoning and analysis of saltation entrainment and trajectories underlie most bulk sand transport models (see Section 2.6).

Creep is a rolling and sliding motion of grains over the surface in near-continuous contact with the bed. Creep usually involves

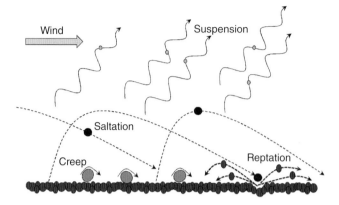

Figure 2.3 The four transport modes of grain transport: saltation, creep, suspension, and reptation.

coarse grains that are too heavy to go into saltation, being pushed along the bed by both the shear stress of the airflow, as well as the bombardment of saltating grains. Estimates of the contribution of creep to the total wind-driven mass flux vary considerably, from 25% (Bagnold 1941) to only 5% (Nickling and McKenna Neuman 2009). Recent wind tunnel studies using high-speed photography (Yang et al. 2009) as well as specialised sand traps (Cheng et al. 2015) indicate that the contribution of creep to total sand transport decreases with increasing wind speed, while the creep mass flux scales with cubic shear velocity.

Suspension is the complete lifting of grains away from the bed, carried along in the airflow and held aloft by turbulence for extended periods of time. Suspension can be maintained as long as vertical turbulent fluctuations in the airflow exceed the fall-velocity of the grains. As a consequence, suspension is the dominant mode for clay and silt particles ($< 62\,\mu m$) which have small fall-velocities. The initiation (and therefore mass flux) of suspension are primarily controlled by saltation, as the clay and silt particles need to be released from their inter-particle cohesion forces by the splashing impact of saltating grains in order to become entrained into the airflow. Clay and silt particles, collectively referred to as *dust*, are therefore the particles which can be carried the longest distances by the wind (see Chapter 4). Sand-size grains are rarely suspended, except at locations of

exceptional turbulence and strong airflow accelerations, such as over the crests of steep dunes with flow detachment, leading to trajectories in *modified saltation*, which has the characteristics of partial suspension, and subsequent grains fall over extended distances along the slip face of transverse dunes (Nickling et al. 2002) and behind coastal foredunes (Arens 1996b).

The fourth mode of transport was identified with the advent of high-speed photography and associated with understanding the development of ripples (Anderson 1987). *Reptation* is the low-energy, short-distance splashing of grains ejected from the bed by the ballistic impacts of high-energy saltating grains. Reptation occurs over distances of only tens of grain diameters, on the scale of millimetres, and is best represented as probability distributions or splash functions (Ungar and Haff 1987) of heights and lengths in all azimuth directions (including 'backwards' or upwind).

2.4 Ripples

The migration of *ripples* may be considered a contribution to the total movement of sand grains over a surface, and is directly linked to the population of reptating grains. Ripples form on nearly all unconsolidated sedimentary beds under active transport, not only in air but also under water in rivers and oceans. On sand surfaces exposed to wind, typical

Figure 2.4 Ripple formation driven by reptation.

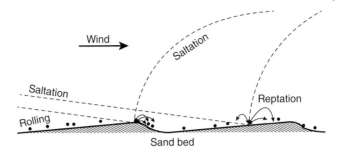

ripple wavelengths are on the order of tens of centimetres, with trough-to-crest heights of a few centimetres. Celerity (or speed) of ripple migration is on the order of a few centimetres per minute (Sharp 1963; Andreotti et al. 2006) and shows a linear relationship with the shear velocity of the sand-driving wind. Ripples form over the course of tens of minutes and while their mean wavelength is strongly related to the shear velocity, the key control on their spacing is the grain size of the sediment, with coarser sands forming larger, more widely spaced ripples. Numerical modelling by Pelletier (2009) suggests that the grain-size control on ripple size is mediated by the increased aerodynamic roughness length of the rippled surface. In extreme cases, very coarse sedimentary beds can develop into so-called *mega-ripples* with wavelengths of up to several metres (Sakamoto-Arnold 1981; Yizhaq 2004). In beds with a mix of sand sizes, the coarser grains are found concentrated on ripple crests and the finer grains in the troughs.

Despite earlier attempts to link ripple formation to saltation trajectories, the current model relates ripple formation, as well as the size segregation between crest and trough, to the reptation mode. Numerical modelling and theoretical analysis, first by Anderson (1987) and since confirmed by other studies (e.g. Durán et al. 2014), show that the low-energy splash of grains in the reptation mode, propelled by the impacts of high-energy saltating grains, drives an inherent instability of flat granular beds, so that initially small random perturbations will grow into ripples at a wavelength of roughly six times the mean reptation length. The instability arises from the fact that the reptation flux on the windward side of small undulations is greater than that on the lee side due to the slanting impact angle of the saltating grains that drive the reptation mode (see Figure 2.4), and this yields a combined lateral migration and vertical growth process at an optimal wavelength. The size-segregation between crest and trough in mixed beds is a consequence of the difference in reptation distances for fine grains and coarse grains (Anderson and Bunas 1993): reptation of the coarse fraction is very short and cannot escape the crest area of the developing ripple, while the finer grains can reptate from the windward slope to the next trough. This leads to an accumulation of coarse grains which are re-circulated on and within the crests of ripples, and a natural segregation process much akin to the self-organisation of pattern formation. Self-organisation has also been invoked as a fundamental process for the development and interactions of ripples and bedforms in general (Werner and Gillespie 1993; Kocurek et al. 2009), including the dynamics of Y-junctions and crest terminations in migrating ripple patterns (Werner and Kocurek 1999).

2.5 Initiation of Grain Motion

2.5.1 Thresholds

As saltation is the dominant transport mode, most analyses of the initiation of grain movement consider the context of sand grains dislodging from the bed and launching into saltation. The main approach to understanding the initial dislodgement is an analysis of forces acting on a grain perched atop or in between two neighbouring grains, as in Figure 2.5. The fluid forces acting on the grain include a vertical lift force due to the Bernoulli effect (sometimes referred to as the 'Saffman' (1965) lift force), and a horizontal aerodynamic drag force acting on the particle protruding into the flow. Theoretical and numerical studies suggest that the lift force may be substantially weaker than the aerodynamic drag. These fluid forces are resisted by the pull of gravitation keeping the grain in place, as well as potential inter-particle cohesive forces at contacts between the surfaces of adjacent grains, such as the surface tension of any moisture films that may be present and the electrostatic forces existing between clay particles. The grain can be dislodged when the fluid forces exceed the resisting forces, acting around a pivoting point at the contact with the neighbouring grain, launching the grain at a significant angle from the horizontal. For unconsolidated sediment, assuming no cohesive forces, the initiation of grain motion is usually expressed as a threshold shear velocity that needs to be exceeded, which is principally a function of grain size, D (Bagnold 1936) (Eq. 2.7):

$$U_{*t} = A \sqrt{\frac{\rho_s - \rho}{\rho} g D} \qquad (2.7)$$

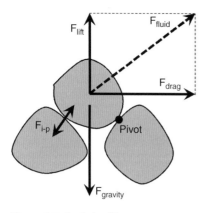

Figure 2.5 Analysis of forces on resting sand grain perched on neighbouring grains, for determining threshold for initiation of motion.

where ρ and ρ_s are densities of respectively air and the sand mineral (usually assumed to be quartz) and g is the gravitational acceleration. The coefficient A is equivalent to the square root of Shield's (1936) parameter and relates partly to grain packing and size sorting and also incorporates the Bernoulli effect. Its value was originally set to 0.1 by Bagnold (1936), but studies since have suggested quite different ranges of values (e.g. 0.08–0.22, or 0.11–0.12) with varying impact on the predicted threshold (Ellis and Sherman 2013). While the values for A reflect the initiation of movement by fluid forces alone (the 'fluid' or 'static' threshold), the entrainment of grains when there is already active saltation is facilitated by the impacting and splashing of preceding saltation grains and hence a lower value for A of 0.082 is applied (the 'impact' or 'dynamic' threshold).

While the force analysis, equation, and Shield's parameter suggest a relatively simple, singular, and universal threshold, experimental observations and numerical simulations indicate much more complex and fuzzy conditions around the precise initiation of grain motion. High-speed filming of grains in wind tunnel studies at around the threshold shear velocity shows grains vibrating in situ within the grain packing before violently ejecting from the bed, and McMenamin et al. (2002) showed a power-law avalanching behaviour of the initiation of saltation near the threshold shear velocity. This suggests that the initiation of grain motion needs to be treated as a stochastic (or random) process with probability distributions rather than a discreet boundary.

Instead of a theoretical prediction, a threshold may also be determined empirically, i.e. retrospectively, based on synchronous time-series measurements of concurring wind and sand transport, using the so-called time-fraction equivalence method, first introduced by Stout and Zobeck (1996, 1997) and refined further by Stout (2004). This method determines the wind speed threshold for which the time-fraction of speeds above it is equivalent to the recorded time-fraction of sand transport activity. The method is somewhat dependent on parameters of the time-series, such as measurement frequency and duration, but it has been applied in a large number of studies – with some variants – as reviewed by Barchyn and Hugenholtz (2011). The latter also review some other empirical threshold determination methods, such as the instantaneous matching of wind speeds for which sand transport begins and/or ends, and statistical regression methods.

2.5.2 Grain Size Control and Dust Emission

The principal controlling variable in the threshold equation is the grain size, according to equation 2.7 graphed in Figure 2.6. The model breaks down, however, when considering silt and clay particles ($<63\,\mu m$) because inter-particle cohesive forces are stronger and can no longer be assumed negligible. Van der Waals forces and electrostatic forces are also not insignificant between silt and clay particles and the associated threshold shear velocity therefore rapidly increases again with smaller particle size below $63\,\mu m$. This is only relevant to the fluid (or 'static') threshold, however: as mentioned under

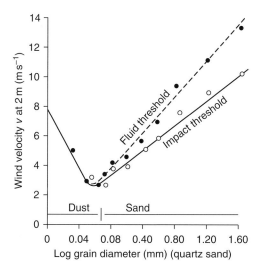

Figure 2.6 The shear velocity threshold for initiation of motion, U_{*t}, as a function of grain size, showing both the fluid threshold and the impact threshold.

transport modes (section 2.3), the entrainment of silt and clay particles from the bed is achieved by the bombardment of sand-size grains and thus related to sand fluxes over mixed sediment beds. Building on the foundational work by Gillette et al. (1972), Gillette and Stockton (1986) first established the assumption that (vertical) dust emission flux, F_d (traditionally in units of $g\,m^{-2}\,s^{-1}$) is proportional to the (horizontal) bulk sand transport rate, Q_s (in units of $g\,m^{-1}\,s^{-1}$ in this context), later formalised by Gillette et al. (2004) for emissions of PM_{10} dust ('Particulate Matter smaller than 10 microns'). Shao et al. (1993) used this assumption in their analysis of dust entrainment as a function of the abrasive force of the saltating sand grains versus the cohesive strength of the sedimentary bed to model the dust emission flux (Eq. 2.8) as:

$$F_d = \frac{m_d\,g\,Q_s}{\psi}\left(\frac{c_N}{2c}\right)\left(\frac{U_1+U_0}{U_*}\right) \qquad (2.8)$$

where m_d is the mass of a dust particle, ψ is the cohesive bonding energy between dust particles in the bed, c_N and c are proportionality constants relating to energy transfer and transport rates respectively, and the ratio $\left(\dfrac{U_1+U_0}{U_*}\right)$ is a dimensionless wind parameter including ejection (U_0) and impact (U_1) velocities of saltating grains, with a magnitude on the order of 10. This model may be simplified by substituting a bulk sand transport rate formula to:

$$F_d = \alpha\,U_*^3\left(1-\frac{U_{*t}^2}{U_*^2}\right) \qquad (2.9)$$

with α a dimensional parameter representing the efficiency of the sand abrasion.

Recent field observations and wind tunnel experiments have shown that dust emission takes place in the absence of saltation and at wind speeds thought to be far below the fluid threshold for the particle size involved (e.g. Kjelgaard et al. 2004; extended review in Gillies 2013), which may reflect the influence of instantaneous shear stress fluctuations at the bed combined with a more stochastic nature of the entrainment threshold. All aspects of dust emission and transport are discussed in detail in Chapter 4.

2.5.3 Other Sedimentary Controls

Apart from grain-size as a principal control on initiation of motion, there are other sedimentary controls. First and foremost is the effect of surface moisture on the retention of grains by increasing the cohesive forces (Namikas and Sherman 1995; Cornelis and Gabriels 2003). Small pockets of water occupying the interstices between grains exert strong surface tension along the air-water interface, binding particles together. Very small amounts of surface moisture have significant impact on the transport threshold. Belly's (1964) empirical work, for example, showed that only 0.6% moisture content by weight is sufficient to double the transport threshold. His model for estimating the 'wet' threshold shear velocity, U_{*tw}, as a function of moisture is still frequently used (Eq. 2.10):

$$U_{*tw} = U_{*t}\left(1.8 + 0.61 \log W\right) \qquad (2.10)$$

where W is the % moisture content by weight. Wind tunnel and theoretical studies since then have further suggested that a moisture content of 1% is sufficient to halt aerodynamic entrainment of sand grains from the bed (McKenna Neuman and Nickling 1989; Wiggs et al. 2004). However, some of this work is challenged by field studies that show sand transport where moisture contents are above 8% (Sherman et al. 1998), and many observations of active wind-blown sand over beaches even during periods of intense rainfall (Arens 1996a; Jackson and Nordstrom 1998). The latter can be explained by raindrop splash and subsequent aerial entrainment of grains into wind-driven saltation (Cornelis et al. 2004; Erpul et al. 2009).

It is the moisture in a thin surface layer, only a few grain diameters thick, that impacts on sand transport, with implications for empirical measurement, particularly in the field. This thin layer can quickly dry out, subject to aerodynamically enhanced evaporation and the relative humidity of the air (Ravi et al. 2006), leading to a high spatio-temporal variability. Traditional (and problematic) methods of surface sampling using a scraper followed by moisture determination in a laboratory have now been superseded by in-situ measurement using electronic probes (Edwards and Namikas 2009) and remote sensing techniques based on variations of surface brightness in photography (Darke et al. 2009) and laser-scanning (Nield et al. 2011).

Other sedimentary controls include such things as mixed grain sizes (naturally graded sand), biological, mechanical, or saline crusts, and the possible effects of compaction and grain packing at the surface. Most of these controls have not been fully measured or quantified.

2.6 Sand Transport: Steady-State

2.6.1 Saltation Trajectories

Once a sand grain has been launched from the bed, it is accelerated by the airflow and follows a parabolic or ballistic trajectory back to the surface, while it is subjected to several different forces. There are two main forces. The force of gravity acts against the upward launch speed, pulling the grain back down to the bed while the aerodynamic drag of the wind accelerates the grain forward throughout most of its trajectory. Other forces include: an aerodynamic lift force due to the grain tumbling about its axis, much as the spin of a dimpled golf ball carries it further, known as the *Magnus effect* (White and Schultz 1977): the Saffman lift force (described previously); and more recently recognised electrostatic forces (Kok and Rennó 2008). The Magnus effect and the Saffman lift force are thought to be minor relative to gravity and aerodynamic drag however (McEwan and Willetts 1993), while the origins of electrostatic forces are unclear and measurements are as yet limited. The forces acting on the grain in flight can be numerically solved to model exact characteristic trajectories (e.g. Shao and Li 1999). Much of this analysis depends, however, on the initial launch speed and angle of the grain when it is first ejected from the bed, and also on the precise modelling of the wind profile close to the bed. The final component of a saltation model relates to the impact of the saltating grain onto the bed, the resultant splash and entrainment of other grains, as well as the potential rebound of the original saltator into a further subsequent trajectory. These grain-bed collisions cannot be solved analytically and are instead treated in the context of populations of saltating grains, with probability distributions representing ejections (so-called *splash functions*) and rebound in terms of angles and speeds (see Kok et al. 2012, for a detailed review). Numerical simulations of such so-called self-regulating saltation systems have been compared with detailed field measurements using compartmentalised horizontal and vertical traps by Namikas (2003) to investigate the impact of the various parameters, showing, for example, that the launch speed of a grain may be treated as a constant. These studies also indicate that steady-state transport rate is primarily controlled by the splash and ejection of grains under ongoing

saltation bombardment, rather than direct entrainment by lift and drag forces of the airflow. The current state-of-the-art is the comprehensive numerical saltation model COMSALT by Kok and Rennó (2009). This model predicts typical saltation trajectories, for medium-sized sand grains, that traverse 20–30 cm in horizontal length and reach heights on the order of 2–3 cm at their apex.

Measurements of saltation trajectories and impact, splash, and rebound mechanics derive primarily from photogrammetry techniques, most notably particle tracking with high-speed film cameras in wind tunnels (Figure 2.7), or from numerical experiments. These studies indicate that saltating grains impact the bed at quite shallow angles of 10–15° from horizontal (Anderson and Hallet 1986), while rebound angles are considerably steeper at 30–45° and occasionally even near-vertically into the airflow (McEwan and Willetts 1991; Nalpanis et al. 1993; Dong et al. 2002). The flight speed of saltating grains depends on the point along their trajectory. Field measurements by Greeley et al. (1996) showed that most mid-trajectory grains at 2 cm above the surface have speeds of 1.5–2 m s^{-1}, with descending particles slightly faster than ascending ones. Recent wind tunnel measurements (Rasmussen and Sorensen 2008; Creyssels et al. 2009; Ho et al. 2013), however, indicate much higher horizontal flight speeds on the order of 2–4 m s^{-1}

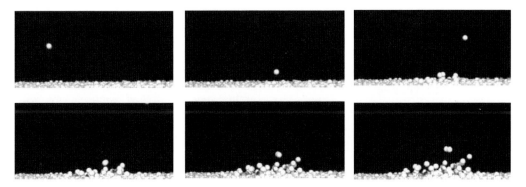

Figure 2.7 Particle tracking of a saltation impact with high-speed camera: successive snapshots, 4 milliseconds apart (from top-row left-to-right, then bottom row, left-to-right). Top row shows incoming saltating grain and rebound; bottom row shows splash and reptation. *Source:* Reproduced from Beladjine et al. (2007).

at 2 cm elevation. The speed at impact upon the bed for large grains is close to their speed at the top of their trajectory (Nalpanis et al. 1993), while smaller grains are somewhat decelerated during their final descent before impact by the relatively low wind speeds close to the bed. Most impact speeds are below $4\,\mathrm{m\,s^{-1}}$ (Anderson and Haff 1988). Upon impact, 80–90% of the saltators rebound, retaining about 50–60% of their prior momentum and with rebound speeds on the order of $1\,\mathrm{m\,s^{-1}}$ (Wang et al. 2008; Gordon and McKenna Neuman 2011), while a small number of saltators dig into and get buried within the particulate bed (Rice et al. 1996). About 15% of the momentum lost by the saltator is transferred to splashing of other grains (Rice et al. 1995). The number of such 'ejecta' scales linearly with the impact momentum of the saltator and in general their launch speeds are an order of magnitude less than that of the impacting grain, lifting off at 40–60° from horizontal (Anderson and Haff 1988). Most of these splashed grains thus form the reptation population (driving wind ripple formation, see Section 2.4), while a smaller fraction gains sufficient launch speed to enter into a proper saltation trajectory, replacing the 10–20% of saltators that are lost upon impact, to sustain a steady state saltation flux.

The saltation characteristics described above principally reflect medium-sized sand and moderate sand-driving wind conditions (where U_* is roughly twice U_{*t}). They vary considerably and in complicated ways depending on grain size and shear velocity.

2.6.2 The Vertical Profile

The horizontal flux of sand over a surface can be thought of as the product of the volume of grains in motion and their forward velocity, and so the height of the saltation layer and its vertical profiles of sand concentration and flight speeds can be used to model transport rate. Conversely, the saltation flux profile is a reflection of the

fundamental saltation mechanics – launch speeds and angles, trajectory geometries and velocities, impact, rebound, and splash dynamics – so that measurements of the flux profile have been used to infer these saltation parameters. Individual saltating grains can reach a wide range of heights above a surface. While some grains fly at heights of metres above the ground (as any common experience of sand stinging the eyes during strong winds on a beach can attest), the bulk of saltation occurs over a gradient of just a few centimetres above the bed. Saltation does not exhibit a sharp upper boundary, but the height of the saltation layer is commonly defined as the height below which 50% of the mass flux takes place. Early theoretical work by Bagnold (1941) and Owen (1964) predicted that the height of the saltation layer would scale with U_*^2, but modern measurements and numerical studies show that it is largely constant at approximately 3.5 cm (Greeley et al. 1996; Namikas 2003), or only very weakly proportional with U_* (Kok and Rennó 2009, Dong et al. 2012). Studies of the saltation layer have undergone an explosion in the last decade, particularly with several wind tunnel groups in China (see also Ellis and Sherman 2013, p. 101). These have measured vertical profiles of sand concentration, $c(z)$ [in $\mathrm{g\,cm^{-3}}$], principally as exponential decay functions of height: $c(z) = a\,e^{b\,z}$, with intercept a ranging from $5\cdot10^{-5}$ to $3\cdot10^{-4}$ and slope ranging from -0.18 to -0.20 with increasing wind speed, for medium-sized sand (Liu and Dong 2004). Relative volumetric particle concentrations are of the order of 10^{-4}, while relative mass concentrations amount to 10^{-1} (Dong et al. 2003b). Vertical profiles of the horizontal flight speed of grains generally follow a power law with greater speeds at higher elevations. Flight speeds in the upper part of the trajectory increase with higher U_*, but wind tunnel experiments and numerical modelling crucially show that the mean grain speed within a few millimetres from the bed surface is largely independent of wind forcing and roughly constant at $\sim1\,\mathrm{m\,s^{-1}}$ (Kok et al. 2012).

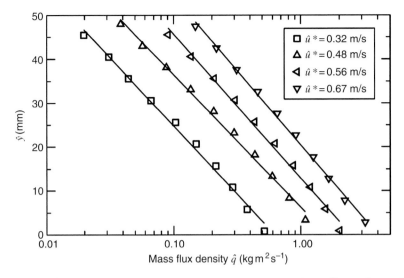

Figure 2.8 Vertical profiles of particle mass flux density under four different shear velocities, measured in a wind tunnel using particle-tracking velocimetry. *Source:* Reproduced from Creyssels et al. (2009).

The consequent vertical profile of particle mass flux density (the combination of particle concentrations and horizontal speeds, see Figure 2.8) has been variably represented as power, logarithmic and exponential decay functions, with a wide variation in profile coefficients.

Comparisons are hampered by methodological differences as well as scaling problems between wind tunnel studies and field measurements (see also Section 2.6.3). Recent field studies suggest an exponential decay profile, with Li et al. (2009) and Ellis et al. (2009) proposing a flux profile as:

$$Q_n(z) = a\,e^{bz} \qquad (2.11)$$

where Q_n is a normalised flux (in % of total flux, per 10 mm of height), z is in millimetres, the intercept a ranges from 13.5–24.8, and slope b ranges from −0.01 to −0.02.

The vertical profile characteristics cited above principally provide a representative sample of the multitude of wind tunnel studies on grain concentrations, flight speeds, and transport fluxes. Comparison of results is, however, difficult because of differences in laboratory methods and specifications and the small-scale nature of wind tunnel measurements (Rasmussen et al. 2015).

2.6.3 Modification of the Wind

It was soon recognised that extraction of momentum from the airflow by saltating sand grains alters the wind speed profile inside the saltation layer, deviating from the standard logarithmic form. Bagnold (1936) found in his wind tunnel studies that the wind profile inside the saltation layer showed a distinct 'kink', or inflection point, at some fixed height above the bed. Wind speed profiles under different values of U_* all intersect at this same point, where speed is independent of U_*, but a function only of the transport threshold (i.e. a constant related to sediment size), and hence the point has been called the *Bagnold focus* (McEwan 1993; Rasmussen et al. 1996). The height of the focus is slightly uncertain: for 250 μm-sized sand the wind tunnel study of Rasmussen et al. indicates a focus at 6.3 mm height, whereas numerical modelling by Kok et al. (2012) suggested a height of ~15 mm and a scaling proposed by Durán et al. (2011) of 15 times the grain diameter implies a height of 3.8 mm.

Owen (1964) argued that inside the saltation layer the total shear stress of the airflow overhead is partially transferred to the bed by the saltating grains (a grain-borne shear stress) and he hypothesised that a negative feedback mechanism between this so-called shear stress partitioning and particle entrainment meant that the shear stress at the bed must be constant and always at the impact threshold for the initiation of motion. Owen's hypothesis assumes that saltation is sustained principally by the direct entrainment of grains into the airflow, rather than by the splash process, but numerical modelling (e.g. Kok et al. 2012) and the abundant evidence of the importance of splash entrainment (see Section 2.3) indicate that the hypothesis is incorrect.

The shear-stress partitioning concept, however, has also been used in considering the saltation layer to present an increased roughness to the overall airflow in the boundary layer, comparable to roughness imposed by vegetation elements (Raupach 1991). Owen (1964) proposed the saltation-induced roughness length, z_0', as:

$$z_0' = C \frac{U_*^2}{2g} \qquad (2.12)$$

with C now commonly referred to as the Charnock constant, recognising that Charnock (1955) earlier proposed the same relationship for roughness induced by wind-generated ripples on a sea surface. The 'constant' varies considerably between wind tunnel and field studies; a comprehensive meta-analysis by Sherman and Farrell (2008) put its value from field experiments at 0.085, while wind tunnel studies exhibit values an order of magnitude smaller. The field value predicts a saltation roughness length of ~0.6 mm for shear velocities at twice the transport threshold for 250 μm size sand, two orders of magnitude greater than Nikuradse's grain roughness (se Section 2.2.1).

2.6.4 Bulk Transport Models

Many studies have been dedicated to establishing and testing predictive bulk sand transport relationships, resulting in several dozen different equations. Table 2.1 lists some of the most commonly used ones. These models are typically based on a force or momentum balance analysis of a control volume in the saltation layer, assuming varying kinds of relationships between shear stress, wind velocity, thresholds, splash and

Table 2.1 Four commonly used bulk sand transport rate equations.

Reference	Model	Notes
Bagnold (1936)	$q = C \sqrt{\dfrac{d}{D}} \dfrac{\rho}{g} U_*^3$	$C = 1.8$ for naturally graded sand
Kawamura (1951)	$q = C \dfrac{\rho}{g} \left(U_* - U_{*t} \right)\left(U_* + U_{*t} \right)^2$	$C = 2.78$
Hsu (1971)	$q = C \left(\dfrac{U_*}{\sqrt{g\,d}} \right)^3$	$C = \exp(-0.47 + 4.97\, d') \times 10^{-5}$, with d' in mm
Lettau and Lettau (1978)	$q = C \sqrt{\dfrac{d}{D}} \dfrac{\rho}{g} U_*^2 \left(U_* - U_{*t} \right)$	$C = 6.7$

Notes: Bulk transport rate, q (kg m^{-1} s^{-1}), d is grain diameter (m), D is reference grain diameter of 0.00025 m, ρ is air density (kg m^{-3}), g is gravitational acceleration (9.81 m s^{-2} on Earth), U_* is shear velocity (m s^{-1}), U_{*t} is threshold shear velocity (m s^{-1}).

entrainment, and the velocities and geometries of the mean saltation trajectory, and they include proportionality coefficients that have often been derived from wind tunnel studies. The bulk sand transport rate, Q (sometimes q), is usually expressed as the mass (or volume) of sand passing across a unit transverse width of surface per unit time, hence (kg m^{-1} s^{-1}). Bagnold's (1936) equation first established a basic proportionality with the cube of the shear velocity (U_*^3), which appears in various guises in many other equations. Bagnold's equation is strictly applicable only when shear velocity is well above the transport threshold, and hence most subsequent equations include the threshold shear velocity explicitly (e.g. those of Kawamura 1951 and Lettau and Lettau 1978). Kawamura's equation has been incorporated into many global dust emission models (e.g. Marticorena and Bergametti 1995; Zender et al. 2003), while Lettau and Lettau's equation is used in the popular Fryberger and Dean (1979)-type sand rose analysis (see Section 2.7.5). Dong et al.

(2003a) reviewed and categorised 19 different sand transport equations, including alternative formulations that relate simply to wind speed rather than shear velocity. In most models the cubic function of U_* rests on an assumption that the speed of saltating grains is proportional to U_*. As reviewed above, this appears to be only partially true and so some models relate sand transport rate to U_*^2 instead, e.g. the model of Ungar and Haff (1987) and those of Kok and Rennó (2008) and Durán et al. (2011) (see Table 2.1).

Accurate prediction of bulk sand transport rates is vital for practical applications to sediment erosion and deposition, dust emission, and dune dynamics. It is therefore unfortunate that a direct comparison of a number of common transport equations, using the best estimates of empirical coefficients and threshold values, shows a great divergence between predictions, ranging over nearly an order of magnitude for any given shear velocity (Sherman and Li 2012; see Figure 2.9). Even more disconcerting is the fact that the majority of field studies testing predictive

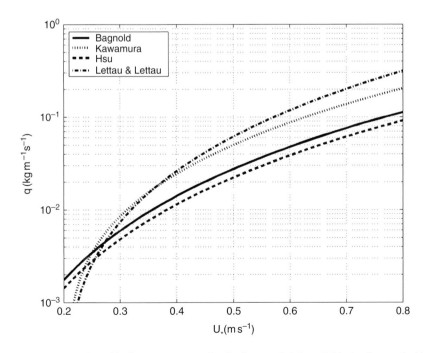

Figure 2.9 Predicted bulk transport rates for the four models from Table 2.1, for sand with a grain diameter of 0.25 mm (medium/fine sand). Shear velocity threshold, $U_{*t} = 0.2$ m s^{-1} (Eq. (2.7)).

equations against direct measurements indicate a very poor performance of all these equations (e.g. Bauer et al. 1990, Arens 1996a; Wiggs 2011), with measured transport rates ranging anywhere from 65 to 300% of predicted rates (Ellis and Sherman 2013). The comparisons are confounded by the fact that the empirical constants and coefficients that are incorporated in the transport models themselves are poorly constrained, and by the fact that accurate and synchronous empirical measurement of sand transport as well as wind-forcing in field experiments is very challenging (Baas 2008; see also Box 2.1), i.e. both sides of the comparison suffer from significant error bars. A far more fundamental root of the problem, however, is that aeolian sand transport in real-world

---

Box 2.1 Measuring Sand Transport by Wind

Attempts to measure wind-blown sand transport in the field (and indeed in wind tunnels) have a long history, yielding a great variety of instruments. These can be classified according to three working principles: (i) sand-capturing devices (traps); (ii) optical devices; and (iii) impact responders.

Sand transport has traditionally been measured with traps, either through manual operation of collecting and weighing the captured sand after a distinct trapping period, or through continuously recording electronic weighing systems inside an accumulating trap. Capture designs include horizontal traps, such as the 'Bagnold sand box', the funnel trap, or the water trap, and vertical traps, such as the Leatherman trap, the BSNE sampler, the wedge-shaped or Guelph trap, and the Fryberger trap (Box Figure 2.1a). Some of the vertical traps can be compartmentalised to determine the distribution of sand flux with height. Traps have some serious limitations though: nearly all traps are manufactured by individual researchers to varying designs and are often deployed on a per-experiment basis. Calibration is difficult and rare and so the trapping efficiency, and hence the accuracy of the measurements, are usually unknown. Furthermore, many traps are rendered inoperable by wet field conditions because trap openings get clogged with moist sand, and traps require physical handling to empty accumulated sand, thus limiting temporal resolution or measurement period.

Optical electronic sensors obviate the need for capturing sand and can operate at high measuring frequencies. Industrial laser systems have been used in wind tunnels to project a sheet of light perpendicular to the sand flux so that the degree of interference or obstruction of the laser sheet caused by the passage of saltating grains can be measured. A more recent improvement has been the testing and deployment of robust and

Box Figure 2.1a Vertical slot (Fryberger) trap on a wind vane turntable in the Hobq Desert, China; sand moves from lower-left into trap.

Box Figure 2.1b Transverse measurement array with two Wenglor particle counters at either end ('forks' pointing upwind), a compartmentalised hose trap, a vertical stack of Miniphones (black probes with silver tips pointing upwind), and two sonic anemometers, at Jericoacoara, Brazil; sand moves from left to right.

commercially available laser particle sensors ('Wenglors', Box Figure 2.1b) to count passing grains in field experiments. These sensors can measure over just a few centimetres and close to the bed and can be combined in vertical arrays. They do require careful calibration, however, in a wind tunnel or a sand fall flume, and the optics can get dusty and dirty over prolonged measuring periods.

Impact responders are devices with small sensitive elements that generate electric signals when they are hit by saltating sand grains. Like optical sensors they can measure over very small regions (on the order of a centimetre) and at high frequencies. Some impact responders, such as the Saltiphone and the 'miniphone', use an acoustic microphone element and complex filtering and processing electronics or software to essentially record and evaluate the sound of the saltation flux.

Others are based on piezo-electric technology, small crystals that release a measurable electronic pulse when the crystal lattice is disturbed by the impact of a saltating grain. Designs include the Sensit, the Safire (Box Figure 2.1c), and the Buzzer Disk. The key limitation of impact responders is that detection and measurement are based on the kinetic energy of the saltating grains, the product of mass and velocity, and hence a mass flux is difficult to extract without very detailed site-specific calibrations.

Apart from the ubiquitous calibration issue, a common challenge that affects all devices (apart perhaps from the unwieldy Bagnold sand box) is that they only provide point-source measurements of a spatio-temporally varying transport system. Most field experiments therefore deploy multiple traps and sensors in order to try and capture a bulk representation.

Box Figure 2.1c A transverse array of Safires, black tubes with the 2 cm-high sensitive ring slightly protruding near the base, at Windy Point, California; a number of cup-anemometers also visible; sand moves from right to left

environments rarely satisfies the three central tenets that underlie all transport equations. First, they relate transport to one simple parameter of the airflow, usually shear velocity (some use wind speed at a reference height). Second, they assume a steady and uniform flow field for this wind forcing. Third, they assume a flat and homogeneous sand surface, with respect to the transport threshold, for example. These models thus treat aeolian sand transport as a uniform blanket of saltating grains that is in equilibrium with a time-averaged shear velocity of the wind. The reality is that boundary-layer airflow is highly unsteady and variable in space and time, with eddies, gusts, and coherent flow structures playing out over multiple scales (Bauer et al. 2013), while the sediment surface is rarely flat and homogeneous, but instead exhibits micro-relief and spatial variability in surface erodibility controlled by grain size and packing differences,

moisture content, and crusting. These complexities pose primary controls on the local transport regime (see Section 2.7) that are not included in predictive models. Despite a long history of attempts, accurate and reliable prediction of sand transport in real-world environments is thus still one of the biggest challenges in aeolian geomorphology.

2.7 Sand Transport: Natural Environments

Most theoretical development and understanding of saltation mechanics and sand transport processes have been based on measurement and calibration in laboratory wind tunnels, where conditions and parameters can be carefully controlled and where complex and often fragile high-precision instrumentation can be deployed. There are, however, significant discrepancies between wind tunnels and the real-world natural envi-ronments of wind-blown sand that we are trying to understand (White 1996). Most notably, wind tunnels are restricted in size, with 'working sections' that are usually a few metres in length and only up to a maximum of 1 m or so in width (many wind tunnels have widths < 30 cm). This precludes much of the horizontal spatial variability and com-plexity in sand transport observed in the real world. In particular, the turbulent behaviour of airflow in wind tunnels is very restricted and lacks the larger-scale gusts and eddies that play an important role in natural winds (Spies and McEwan 2000). While fluid dynamics can be scaled down in wind tun-nels to study airflow around miniature models of dunes or wind breaks, for exam-ple, saltation by sand grains cannot be min-iaturised, and so grains follow the same typical trajectories reaching heights of sev-eral centimetres and jump distances of 20–30 cm as under natural conditions. In wind-tunnel studies the height of the inter-nal boundary layer at the working section is only on the order of 10 cm, leading to con-ditions where the 2–3 cm high saltation layer intrudes into a significant portion of the simulated boundary layer, much more strongly affecting the local flow dynamics than is possible in real-world environments (Farrell and Sherman 2006).

Natural environments and real-world con-ditions introduce a number of complications and adjustments to the standard idealised steady-state sand transport model described in Section 2.6. Most importantly, natural sand surfaces differ significantly from the *transport-limited* conditions that are typical of models and wind tunnel measurements by the fact that during various periods of time and over various areas of the surface, sand transport is in fact subject to *supply-limited* conditions, where the wind's maximum sand transport capacity is never reached (De Vries et al. 2014). Surface moisture is a good exam-ple of a supply-limiting control in natural environments that can be highly variable in both space and time.

2.7.1 Fetch and Saturation

The idealised steady-state sand transport models describe a saturated saltation layer, driven by a constant wind over a limitless uniform surface, whereas natural sand sur-faces are patchy and bounded. From the upwind boundary of a new sand surface, or after a transition in roughness, it takes some-times a considerable distance before saltation reaches its maximum, saturated, state. The impact of field length on agricultural wind erosion was already recognised by Chepil and Milne (1939), and Svasek and Terwindt (1974) found a similar distance control on beaches. The so-called *fetch effect* (Gillette et al. 1996) includes three spatio-temporal adjustment mechanisms: (i) the growth of saltation from zero at the upwind (leading) edge of a sand surface requires a downwind saltation cascade, starting with aerodynamic entrainment and progressively developing to predominantly splash and impact entrain-ment; (ii) the associated saltation-induced roughness increases the shear stress inside the internal boundary layer, hence leading to a positive feedback mechanism until a momentum balance is reached; and (iii) the downwind increase in sand-blasting poten-tial can progressively release more surface grains in supply-limited situations where, for example, moisture control or crusting limit initial entrainment. The fetch effect has been measured in field experiments (Stout 1990; Fryrear and Saleh 1996) and in wind tunnels (Shao and Raupach 1992), as well as in numerical simulations (Spies and McEwan 2000); a comprehensive review is provided by Delgado-Fernandez (2010). The critical fetch, the distance required to reach equilib-rium saltation, can be several tens of metres on beaches or even 100 s of metres in agricultural situations, with higher wind speeds generally requiring longer distances (Davidson-Arnott and Law 1990). In many key instances the critical fetch may never be reached, fundamentally restricting sand transport rates. Bauer and Davidson-Arnott (2003) developed a framework for predicting transport rates in fetch-limited environments, such as beaches, depending on wind direction and the shape of the fetch control function, while Delgado-Fernandez and Davidson-Arnott (2011) included the fetch effect in a meso-scale magnitude-frequency model for long-term prediction of wind-blown sand inputs to coastal dunes.

Independent of the fetch effect literature, the term *saturation length* was coined in physics literature (e.g. Sauermann et al. 2001; Andreotti et al. 2002), to describe a relaxation length-scale as the saltation layer reached its steady state, after a transition via the saltation cascade (the first mechanism described by Gillette et al. 1996). Different physical parameters have been linked to the saturation length, including: the grain hop length; the distance needed to accelerate new grains (a 'drag length'); an entrainment cascade length; or a distance associated with the saltation-induced roughness feedback. The saturation length appears to be on the order of 1 m and Andreotti et al. (2010) argued that it is fundamentally different from the fetch effect. Some confusion remains, however, about the length scale (Kok et al. 2012), as different papers present differing estimates (e.g. Andreotti et al. 2002 present a drag length of 9 m) and the parameter is at the core of a debate about a potential minimum dune size.

2.7.2 Spatio-Temporal Variability

Wind in natural internal boundary layers is strongly turbulent and highly variable over a range of time-scales, and this causes both spatio-temporal variability in transport rate, as well as an intermittent occurrence of saltation altogether. The influence of gusts on sand transport was first investigated by Lee (1987), and the response time of sand transport to rapid fluctuations in wind speed or shear velocity has subsequently been estimated at 1–2 seconds, from field experiments by Butterfield (1991), Bauer et al. (1998) and Baas and Sherman (2005), from wind tunnel studies (Butterfield 1993, 1998; Spies et al.

2000), and from numerical modelling (McEwan and Willetts 1991). This response time-scale is, of course, also intimately related to the saturation length discussed above, and a response time of 1–2 seconds implies a response distance of 1–10 m. Spatial variability in transport rate is usually quantified as a Coefficient of Variation ('CoV', being the standard deviation of a sample divided by its mean), and it has been measured with suites of sand traps over length scales of metres to tens of metres, mostly on beaches. Gares et al. (1996) showed CoVs of 25% on a scale of 1 m and 50% over distances up to 50 m, for 15-minute measurements. Jackson et al. (2006) compared transport rates measured in five sand traps spaced at 1 m intervals and found variability commonly exceeded 150% within the array. Baas and Sherman (2006) found transport variability ranging from 30% over a spanwise distance of 0.1 m and periods of 120 seconds to 266% over a spanwise distance of 4.0 m and periods of 1 second, while Ellis et al. (2012) reported variability ranging from 17–68% over 4 m under varying conditions. In the temporal domain the intermittent nature of sand transport at a fixed point has been measured and analysed by Stout and Zobeck (1997) to relate intermittency levels to a relative wind strength.

The most recognisable and near-universal manifestation of spatio-temporal sand transport variability is the presence of aeolian streamers, longitudinal concentrations of saltating sand, weaving and meandering sideways as they move downwind (Figure 2.10). Elaborate field measurements by Baas and Sherman (2005) have shown that streamer formation and behaviour are linked to turbulent eddies in the boundary layer which scrape over the sand bed, initiating and propelling saltation along their tracks. A surface homogenisation experiment meanwhile demonstrated that potential small-scale surface controls such as differentiations in grain-size, moisture content, or micro-relief, are not necessary for streamer formation. Baas and Sherman identified the surface-scraping eddies with the top-down

Figure 2.10 Streamers moving and meandering over a beach (image with enhanced colour contrasts). See insert for colour representation of this figure.

turbulence model of Hunt and his colleagues (Hunt and Morrison 2000; Hunt and Carlotti 2001), but boundary layer 'super-structures' (Marusic et al. 2010) or other types of coherent flow structures may be equally responsible. Baas and Sherman (2005) distinguished three types of streamer patterns: isolated streamers, nested streamers, and saltation clouds with embedded streamers, appearing at increasing shear velocities respectively. Streamers are typically 0.1–0.2 m wide, propagating downwind at speeds of around 3.6 m s^{-1}, and exhibit an at-a-point lifetime on the order of 0.5–1 seconds, yielding streamwise lengths on the order of 1.8–3.6 m (Baas 2003). A discrete-particle numerical saltation model coupled with a Large Eddy Simulation (LES) of boundary layer airflow by Dupont et al. (2013) has reproduced and confirmed the development of different streamer patterns, and their fundamental length and time scales.

2.7.3 Slope

Most sand transport occurs over sloping surfaces of some kind, both up-sloping as well as down-sloping. Beaches, for example, exhibit varying slopes from 2° to 10° at different parts of a cross-shore profile (e.g. shore-face, or back-beach), while windward (stoss) slopes of desert dunes are typically inclined at around 11° (Hesp and Hastings 1998), with steeper slopes on the flanks. Slopes have an impact on the transport threshold as well as the transport magnitude: an up-sloping surface yields a steeper pivot angle in the force-balance relationships for grain entrainment (see Figure 2.5), making it harder to dislodge a grain, and the sloping surface changes the momentum transfer through splash and reptating and saltating impacts, all of which affect the dynamic threshold. Saltation trajectories meanwhile are curtailed by a rising surface (upslope) or lengthened under a falling slope.

The slope effect on the transport threshold has been quantified by Iversen and Rasmussen (1994) as the ratio of the sloped threshold, U_{*ts}, relative to that over a horizontal surface (Eq. 2.11):

$$\frac{U_{*ts}}{U_{*t}} = \cos\theta + \frac{\sin\theta}{\tan\varphi} \tag{2.13}$$

where: θ is the surface slope and φ is the angle of internal friction, a grain-interaction parameter typically assumed equal to the angle of repose, which is around 33° for dry sand. This relationship (Figure 2.11) shows that the threshold on the stoss slope of a typical dune is roughly 28% greater than that over a horizontal surface. The direct impact of slope on transport rate is usually expressed in the form of a simple correction factor. Bagnold (1973) proposed it as Eq. (2.12):

$$B_s = \frac{\tan\varphi}{\cos\theta\left(\tan\varphi + \tan\theta\right)} \quad (2.14)$$

which predicts that sand transport rates over a typical 5° back-beach (uphill) are reduced by about 12% compared to a horizontal surface. The impact of slope on sand transport is therefore potentially significant, but in application and modelling much of this is essentially offset by the fact that a wind blowing over an uphill slope, such as back beaches and dunes, generally speeds up, thus generating an increased sand-driving force that

counteracts and/or overwhelms the slope impacts (see Chapter 7).

The downhill end of the slope-impact spectrum is also shown in Figure 2.11, for downward slopes that approach the angle of repose, as on the slip-face of a dune, which indicates that the transport threshold then drops to zero. This has a regime of slope failure and avalanching. Avalanching is not strictly an aeolian (wind-driven) process and as such is not further discussed here, but it is a universal feature of landforms in granular materials, and has a huge literature of its own (see Sutton et al. 2013 for an entry from an aeolian perspective).

2.7.4 Vegetation

Vegetation is usually a first-order control on sand transport by wind, but much depends on the structure, porosity, geometry, size, and growth form of the plants (grasses, herbs, shrubs, trees) as well as their spatial distribution, lateral density, and potential patterning relative to wind directions.

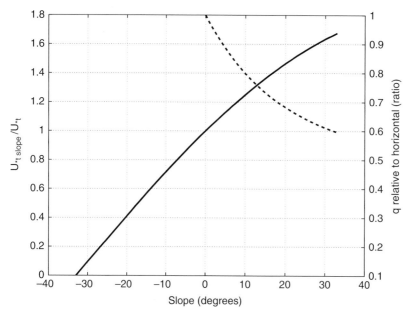

Figure 2.11 Effect of surface slope on the shear velocity threshold (Eq. 2.11), solid line and left-hand axis, and the transport rate (Eq. 2.12), dashed line and right-hand axis. Negative angles are sloping downward with the wind (e.g. lee slopes), positive angles are sloping upward with the wind (e.g. stoss slopes).

Vegetation affects sand transport both by reducing the amount of bare surface available for transport (the 'footprints' occupied by vegetation elements), as well as by slowing down the airflow close to the ground, thus extracting momentum from the wind, which is then no longer available to move sand (Wolfe and Nickling 1993). For uniform ground cover, such as grasses, vegetation introduces an upward displacement of the typical logarithmic velocity profile in the boundary layer, such that the wind speed appears to be reduced to zero at some height within the canopy, at an apparent plane of momentum absorption, leading to a modified Law-of-the-Wall in the airflow above the vegetation (Eq. 2.13):

$$U(z) = \frac{U_*}{\kappa} \ln\left(\frac{z-d}{z_0} \right) \qquad (2.15)$$

where d is the zero-plane displacement length (other symbols defined earlier in Eq. 2.2). Within the canopy the momentum absorption is distributed between the plant (roughness) elements and the bare ground surface, in a shear stress (or drag) partitioning (Gillette and Stockton 1989). The impact of vegetation on sand transport can be quantified in two different ways. The older approach is through a 'soil flux ratio', the ratio of sediment transport in the presence of roughness elements versus that over bare soil (Chepil 1944; Fryrear 1985), fairly well represented by an exponential decay function of soil cover fraction. The more contemporary approach is less direct: through quantifying the threshold shear velocity required for initiating sand transport over the vegetated surface, relative to that over a bare surface, based on the drag partitioning principle. The model of Raupach et al. (1993), the one most widely adopted, predicts a threshold shear velocity ratio as a function of roughness density λ (Eq. 2.14) as:

$$\frac{U_{*t}}{U_{*tR}} = \frac{1}{\sqrt{(1 - m\sigma\lambda)(1 + m\beta\lambda)}} \qquad (2.16)$$

where U_{*tR} is the threshold shear velocity for the surface with roughness elements, β is the ratio of the drag coefficient of an individual roughness element to the drag coefficient of the bare surface (of order 100), σ is the basal-to-frontal area ratio of a roughness element (equal to $(\pi d/4 h)$ for cylindrical elements of height h and diameter d), and m is a coefficient (< 1) representing spatial variability in surface shear stress. The controlling variable, roughness density λ, also referred to as the frontal area index (FAI) or roughness element lateral cover, is defined as the cumulative frontal or silhouette area of the plants per unit surface area, e.g. (Eq. 2.15):

$$\lambda = \frac{n A_s}{S} \qquad (2.17)$$

where ground surface area S is occupied by n roughness elements with individual silhouette area A_s. Though commonly applied (e.g. Arens et al. 2001) the model is problematic because the m coefficient is poorly constrained, and has a variety of values in literature from 0.18–0.58. A more sophisticated and computational model has been developed by Okin (2008), which includes estimating spatial variability of shear stress on the ground surface between vegetation elements.

For the relatively uniform ground cover of low-growing vegetation (typically grasses), field measurements by Lancaster and Baas (1998) showed that the effective threshold shear velocity is twice as large as that for a bare surface when lateral cover reaches ~20%, and that most long-term sand transport is effectively suppressed when lateral cover is only ~15% (Figure 2.12). Even at 20% cover there are still significant episodes of sand transport during storm conditions. The situation is more complex for more distinct, clump-like vegetation (such as shrubs). At low densities the turbulence and eddy shedding behind individual shrubs can in fact enhance sediment transport over the intervening ground surface (Hesp 1981), and the orientation and patterning of elements can

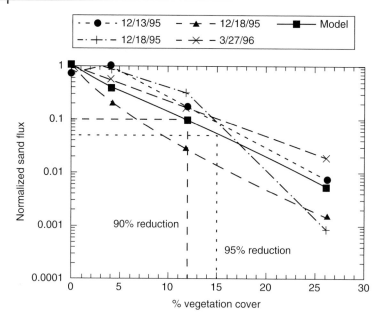

Figure 2.12 Reduction in sand transport rates as a consequence of vegetation cover, for four distinct wind events measured at Owens Lake, CA, and for the Raupach et al. (1993) model (Eq. 2.14). *Source:* Reproduced from Lancaster and Baas (1998).

create 'streets' in certain alignments that can offer a relatively long bare surface fetch for wind to accelerate and sand transport to surge under certain wind directions (Okin and Gillette 2001; King et al. 2006).

2.7.5 Sand Roses

Natural wind climates can have great directional variability on a seasonal or sometimes even daily time-scale (see Chapter 1), usually represented in the form of a wind rose. For any given site or region, this has direct bearing on understanding and predicting spatial patterns of sand transport, erosion and deposition, and related dune form and migration. The directional variability of sand transport can be captured by the 'Fryberger and Dean (1979) method' for calculating *sand roses*. The method uses an adaptation of the Lettau and Lettau (1978; Chapter 1) transport equation (see Table 2.1) to determine a *sand drift potential* (DP) for each compass direction from a long-term, usually annual, time-series

record of wind speeds and directions (Eq. 2.16), so that:

$$DP_{compass} = \sum_i \frac{\bar{U}_i^2 (\bar{U}_i - U_t)}{100} t_i \qquad (2.18)$$

where the summation is over i categories of wind speed above the transport threshold U_t, with $\bar{U}_i$ being the mean wind speed of the category, and t_i being the percentage of time the wind blew in that category. The vector sum of drift potentials from all compass directions yields a *resultant drift potential* (RDP) vector with both a magnitude and a direction, representing the *net* potential sand movement over the period covered. The ratio of the RDP magnitude to the (scalar) total DP magnitude across all compass directions (the RDP/DP ratio) can be used to quantify a directional variability in sand transport – with a value of one indicating a perfectly uni-modal regime and values close to zero indicating very complex regimes with very little net sand movement – which has been

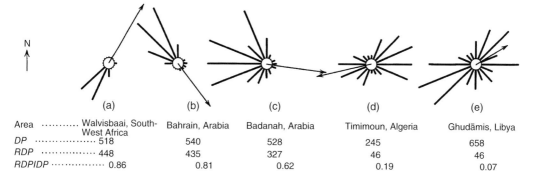

Area	..........Walvisbaai, South-West Africa	Bahrain, Arabia	Badanah, Arabia	Timimoun, Algeria	Ghudāmis, Libya
DP	518	540	528	245	658
RDP	448	435	327	46	46
RDP/DP	...............0.86	0.81	0.62	0.19	0.07

Figure 2.13 Sand roses showing five patterns of directional variability in sand transport: (a) narrow unimodal, (b) wide unimodal, (c) acute bimodal, (d) obtuse bimodal, (e) complex. *Source:* After: Fryberger and Dean (1979). Resultant drift vector indicated with thin arrow. DP, RDP, and ratios indicated.

linked to the occurrence of specific desert dune types, such as barchans (uni-modal), linear dunes (bi-modal), and star dunes (complex), by Wasson and Hyde (1983). The resultant drift potential vector has a bearing on the crest orientation of bed forms in general and the concept of maximum gross bed-form-normal transport (Rubin and Hunter 1987; Rubin and Ikeda 1990), with the application of deriving wind regimes from observing dune field patterns on other planets (Fenton et al. 2014).

Fryberger presented DP and RDP results in 'vector units' (implying dimensionless values), but his worked examples and his definitions of low, intermediate, and high wind energy environments were based on wind speeds measured in knots, which are not equivalent to DP values derived from metric data (Bullard 1997). The drift potential Eq. (2.16) shows that the units of DP are in fact equivalent to the wind speed units cubed, e.g. $(m/s)^3$ (Figure 2.13).

2.8 Conclusion

This chapter has shown the great extent and extraordinary details of recent research on wind-blown sand transport, from new thoughts on the role of turbulence in the airflow, to grain-scale observations in wind tunnels, comprehensive numerical models of self-regulatory saltation systems, and exploration of the many complications and constraints of natural environments.

However, this review also implies that research in recent decades has been dominated by wind tunnel work and numerical modelling. Field experiments in natural environments are more challenging and difficult to accomplish, but they often generate the most innovative insights and open up new research questions. They also show that there are still great improvements to be made in our ability to understand and predict aeolian sand transport in real-world applications.

Further Reading

A good introduction to atmospheric boundary layers and the treatment and analysis of wind, shear stress, and turbulence – from a meteorological perspective – is found in Stull (1988). More rigorous and advanced discussion of boundary layer airflow is found in Kaimal and Finnigan (1994), while boundary

layers within a sand transport context are reviewed by Bauer (2013). For more in-depth coverage of all aspects of sand transport mechanics, Kok et al. (2012) provide an excellent and very thoroughly researched review. Rasmussen et al. (2015) discuss and compare a range of wind tunnel studies on vertical profiles in the saltation layer. An overview, history, and re-calibration of bulk sand transport equations are provided by Sherman and Li (2012). The impact of vegetation on aeolian sand transport is extensively reviewed by Okin (2013).

References

Adrian, R.J., Meinhart, C.D., and Tomkins, C.D. (2000). Vortex organization in the outer region of the turbulent boundary layer. *Journal of Fluid Mechanics* 422: 1–54. doi: 10.1017/S0022112000001580.

Anderson, R.S. (1987). A theoretical model for aeolian impact ripples. *Sedimentology* 34 (5): 943–956. doi: 10.1111/j.1365-3091.1987.tb00814.x.

Anderson, R.S. and Bunas, K.L. (1993). Grain size segregation and stratigraphy in aeolian ripples modelled with a cellular automaton. *Nature* 365 (6448): 740–743. doi: 10.1038/365740a0.

Anderson, R.S. and Haff, P.K. (1988). Simulation of eolian saltation. *Science* 241 (4867): 820–823. doi: 10.1126/science.241.4867.820.

Anderson, R.S. and Hallet, B. (1986). Sediment transport by wind: toward a general model. *Geological Society of America Bulletin* 97 (5): 523–535. doi: 10.1130/0016-7606(1986)97<523:STBWTA>2.0.CO;2.

Andreotti, B., Claudin, P., and Douady, S. (2002). Selection of dune shapes and velocities. Part 1: dynamics of sand, wind and barchans. *European Physical Journal B* 28 (3): 321–339. doi: 10.1140/epjb/e2002-00236-4#.

Andreotti, B., Claudin, P., and Pouliquen, O. (2006). Aeolian sand ripples: experimental study of fully developed states. *Physical Review Letters* 96 (2): 028001. doi: 10.1103/PhysRevLett.96.028001.

Andreotti, B., Claudin, P., and Pouliquen, O. (2010). Measurements of the aeolian sand transport saturation length. *Geomorphology* 123 (3–4): 343–348.

Arens, S.M. (1996a). Rates of aeolian transport on a beach in a temperate humid climate. *Geomorphology* 17: 3–18.

Arens, S.M. (1996b). Patterns of sand transport on vegetated foredunes. *Geomorphology* 17 (4): 339–350. doi: 10.1016/0169-555X(96)00016-5.

Arens, S.M., Baas, A.C.W., van Boxel, J.H., and Kalkman, C. (2001). Influence of reed stem density on foredune development. *Earth Surface Processes and Landforms* 26 (11): 1161–1176. doi: 10.1002/esp.257.

Baas, A.C.W. (2003). The formation and behavior of aeolian streamers. Doctoral thesis, University of Southern California, Los Angeles.

Baas, A.C.W. (2008). Challenges in aeolian geomorphology: investigating aeolian streamers. *Geomorphology* 93 (1–2): 3–16. doi: 10.1016/j.geomorph.2006.12.015.

Baas, A.C.W. and Sherman, D.J. (2005). Formation and behavior of aeolian streamers. *Journal of Geophysical Research. Earth Surface* 110: F03011. doi: 10.1029/2004JF000270.

Baas, A.C.W. and Sherman, D.J. (2006). Spatiotemporal variability of aeolian sand transport in a coastal dune environment. *Journal of Coastal Research* 22 (5): 1198–1205. doi: 10.1098/rspa.1954.0186of aeolian sand.

Bagnold, R.A. (1936). The movement of desert sand. *Proceedings of the Royal Society of London A* 157 (892): 594–620. doi: 10.1098/rspa.1936.0218.

Bagnold, R.A. (1941). *The Physics of Blown Sand and Desert Dunes*. London: Methuen (reprinted 1954; 1960; 2005, by Dover, Mineola, NY).

Bagnold, R.A. (1973). The nature of saltation and of 'bed-load' transport in water. *Proceedings of the Royal Society of London A: Mathematical, Physical and Engineering Sciences* 332 (591): 473–504. doi: 10.1098/rspa.1973.0038.

Barchyn, T.E. and Hugenholtz, C.H. (2011). Comparison of four methods to calculate aeolian sediment transport threshold from field data: implications for transport prediction and discussion of method evolution. *Geomorphology* 129 (3–4): 190–203. doi: 10.1016/j.geomorph. 2011.01.022.

Bauer, B.O. (2013). Fundamentals of aeolian sediment transport: boundary-layer processes (chapter 11.2). In: *Treatise on Geomorphology, Volume 11. Aeolian Geomorphology* (ed. J.F. Schroder, N. Lancaster, A.C.W. Baas and D.J. Sherman), 7–22. Rotterdam: Elsevier.

Bauer, B.O. and Davidson-Arnott, R.G.D. (2003). A general framework for modeling sediment supply to coastal dunes including wind angle, beach geometry and fetch effects. *Geomorphology* 49 (1–2): 89–108. doi: 10.1016/S0169-555X(02)00165-4.

Bauer, B.O., Sherman, D.J., Nordstrom, K.F., and Gares, P.A. (1990). Aeolian transport measurement and prediction across a beach and dune at Castroville, California. In: *Coastal Dunes: Form and Process* (ed. K. Nordstrom, N. Psuty and R.W.G. Carter), 39–56. New York: Wiley.

Bauer, B.O., Sherman, D.J., and Wolcott, J.F. (1992). Sources of uncertainty in shear stress and roughness length estimates derived from velocity profiles. *Professional Geographer* 44 (4): 453–464. doi: 0.1111/j.0033-0124.1992.00453.x.

Bauer, B.O., Walker, I.J., Baas, A.C.W. et al. (2013). Critical reflections on the coherent flow structures paradigm in aeolian geomorphology. In: *Coherent Flow Structures at the Earth's Surface* (ed. J.G. Venditti, J.L. Best, M. Church and R.J. Hardy), 111–134. Chichester: Wiley-Blackwell.

Bauer, B.O., Yi, J.C., Namikas, S.L., and Sherman, D.J. (1998). Event detection and conditional averaging in unsteady aeolian systems. *Journal of Arid Environments* 39 (3): 345–375. doi: 10.1006/jare.1998.0380.

Beladjine, D., Ammi, M., Oger, L., and Valance, A. (2007). Collision process between an incident bead and a three-dimensional granular packing. *Physical Review*

E – Statistical, Nonlinear, and Soft Matter Physics 75: 061305. doi: 10.1103/PhysRev E.75.061305.

Belly, P.-Y. (1964). Sand movement by wind. United States Army, Corps of Engineers, Coastal Engineering Research Center, Washington DC, Technical Memorandum TM-1.

Best, J. (1992). On the entrainment of sediment and initiation of bed defects – insights from recent developments within turbulent boundary-layer research. *Sedimentology* 39 (5): 797–811. doi: 10.1111/j.1365-3091.1992.tb02154.x.

Bullard, J.E. (1997). A note on the use of the "Fryberger method" for evaluating potential sand transport by wind. *Journal of Sedimentary Research A* 67 (3): 499–501. doi: 10.1306/D42685A9-2B26-11D7-8648000102C1865D.

Butterfield, G.R. (1991). Grain transport rates in steady and unsteady turbulent airflows. In: *Aeolian Grain Transport 1: Mechanics* (ed. O.E. Barndorff-Nielsen and B.B. Willets), 97–122. Vienna: Springer-Verlag.

Butterfield, G.R. (1993). Sand transport response to fluctuating wind velocity. In: *Turbulence: Perspectives on Sediment Transport* (ed. N.J. Clifford, J.R. French and J. Hardisty), 305–335. Chichester: Wiley.

Butterfield, G.R. (1998). Transitional behaviour of saltation: wind tunnel observations of unsteady winds. *Journal of Arid Environments* 39 (3): 377–394. doi: 10.1006/jare.1997.0367.

Charnock, H. (1955). Wind stress on a water surface. *Quarterly Journal of the Royal Meteorological Society* 81 (350): 639–640. doi: 0.1002/qj.49708135027.

Cheng, H., He, J., Zou, X. et al. (2015). Characteristics of particle size for creeping and saltating sand grains in aeolian transport. *Sedimentology* 62 (5): 1497–1511. doi: 10.1111/sed.12191.

Chepil, W.S. (1944). Utilization of crop residues for wind erosion control. *Scientific Agriculture* 24 (7): 307–319.

Chepil, W.S. and Milne, R.A. (1939). Comparative study of soil drifting in the field and in a wind tunnel. *Scientific Agriculture* 19 (5): 249–257.

Cornelis, W.M. and Gabriels, D. (2003). The effect of surface moisture on the entrainment of dune sand by wind: an evaluation of selected models. *Sedimentology* 50 (4): 771–790. doi: 10.1046/j.1365-3091.2003.00577.x.

Cornelis, W.M., Oltenfreiter, G., Gabriels, D., and Hartmann, R. (2004). Splash-saltation of sand due to wind-driven rain. *Soil Science Society of America Journal* 68 (1): 41–46. doi: 10.2136/sssaj2004.4100.

Creyssels, M., Dupont, P., El Moctar, A.O. et al. (2009). Saltating particles in a turbulent boundary layer: experiment and theory. *Journal of Fluid Mechanics* 625: 47–74. doi: 10.1017/S0022112008005491.

Davidson-Arnott, R.G.D. and Law, M.N. (1990). Seasonal patterns and controls on sediment supply to coastal foredunes, long point, Lake Erie. In: *Coastal Dunes: Form and Process* (ed. K. Nordstrom, N. Psuty and R.W.G. Carter), 177–200. Chichester: Wiley.

Darke, I.R., Davidson-Arnott, R., and Ollerhead, J. (2009). Measurement of beach surface moisture using surface brightness. *Journal of Coastal Research* 25 (1): 248–256. doi: 10.2112/07-0905.1.

Delgado-Fernandez, I. (2010). A review of the application of the fetch effect to modelling sand supply to coastal foredunes. *Aeolian Research* 2 (2–3): 61–70. doi: 10.1016/j.aeolia.2010.04.001.

Delgado-Fernandez, I. and Davidson-Arnott, R. (2011). Meso-scale aeolian sediment input to coastal dunes: the nature of aeolian transport events. *Geomorphology* 126 (1–2): 217–232. doi: 10.1016/j.geomorph.2011.04.001.

De Vries, S., van Thiel de Vries, J.S.M., van Rijn, L.C. et al. (2014). Aeolian sediment transport in supply limited situations. *Aeolian Research* 14: 75–85. doi: 10.1016/j.aeolia.2013.11.005.

Dong, Z., Liu, X., Li, F. et al. (2002). Impact-entrainment relationship in a saltating cloud. *Earth Surface Processes and Landforms* 27 (6): 641–658. doi: 10.1002/esp.341.

Dong, Z., Liu, X., Wang, H., and Wang, X. (2003a). Aeolian sand transport: a wind tunnel model. *Sedimentary Geology* 161 (1–2): 71–83. doi: 10.1016/S0037-0738(02)00396-2.

Dong, Z., Lv, P., Zhang, Z. et al. (2012). Aeolian transport in the field: a comparison of the effects of different surface treatments. *Journal of Geophysical Research: Atmospheres* 117: D09210. doi: 10.1029/2012JD017538.

Dong, Z., Wang, H., Zhang, X., and Ayrault, M. (2003b). Height profile of particle concentration in an aeolian saltating cloud: a wind tunnel investigation by PIV MSD. *Geophysical Research Letters* 30 (19): doi: 10.1029/2003GL017915.

Dupont, S., Bergametti, G., Marticorena, B., and Simoëns, S. (2013). Modeling saltation intermittency. *Journal of Geophysical Research: Atmospheres* 118 (13): 7109–7128. doi: 10.1002/jgrd.50528.

Durán, O., Claudin, P., and Andreotti, B. (2011). On aeolian transport: grain-scale interactions, dynamical mechanisms and scaling laws. *Aeolian Research* 3 (3): 243–270. doi: 10.1016/j.aeolia.2011.07.006.

Durán, O., Claudin, P., and Andreotti, B. (2014). Direct numerical simulations of aeolian sand ripples. *Proceedings of the National Academy of Sciences of the United States of America* 111 (44): 15665–15668. doi: 10.1073/pnas.1413058111.

Edwards, B.L. and Namikas, S.L. (2009). Small-scale variability in surface moisture on a fine-grained beach; implications for modeling aeolian transport. *Earth Surface Processes and Landforms* 34 (10): 1333–1338. doi: 10.1002/esp.1817.

Elliot, W.P. (1958). The growth of the atmospheric internal boundary layer. *Transactions of the American Geophysical Union* 39: 1048–1054. doi: 10.1029/TR039i006p01048.

Ellis, J.T., Li, B., Farrell, E.J., and Sherman, D.J. (2009). Protocols for characterizing aeolian mass-flux profiles. *Aeolian Research* 1 (1–2): 19–26. doi: 10.1016/j.aeolia.2009.02.001.

Ellis, J.T. and Sherman, D.J. (2013). Wind-blown sand. In: *Treatise on Geomorphology,*

Volume 11. *Aeolian Geomorphology* (ed. J.F. Schroder, N. Lancaster, A.C.W. Baas and D.J. Sherman), 85–108. Rotterdam: Elsevier.

Ellis, J.T., Sherman, D.J., Farrell, E.J., and Li, B. (2012). Temporal and spatial variability of aeolian sand transport: implications for field measurements. *Aeolian Research* 3 (4): 379–387. doi: 10.1016/j.aeolia.2011.06.001.

Erpul, G., Gabriels, D., Cornelis, W.M. et al. (2009). Sand transport under increased lateral jetting of raindrops induced by wind. *Geomorphology* 104 (3–4): 191–202. doi: 10.1016/j.geomorph.2008.08.012.

Farrell, E.J. and Sherman, D.J. (2006). Process-scaling issues for aeolian transport modelling in field and wind tunnel experiments: roughness length and mass flux distributions. *Journal of Coastal Research Special Issue* 39: 384–389.

Fenton, L.K., Michaels, T.I., and Beyer, R.A. (2014). Inverse maximum gross bedform-normal transport 1: How to determine a dune-constructing wind regime using only imagery. In: Proceedings: 3rd International Planetary Dunes Workshop, Lowell Observatory, Flagstaff, AZ, June 12–15, 2012, *Icarus* 230(Special Issue): 5–14, doi:10.1016/j.icarus.2013.04.001.

Fryberger, S.G. and Dean, G. (1979). Dune forms and wind regime. In: *A Study of Global Sand Seas*, (ed. E.D. McKee), 137–169, United States Department of the Interior, U.S. Geological Survey, Professional Paper 1052.

Fryrear, D.W. (1985). Soil cover and wind erosion. *Transactions of the American Society of Agricultural Engineers* 28 (3): 781–784.

Fryrear, D.W. and Saleh, A. (1996). Wind erosion: field length. *Soil Science* 161 (6): 398–404. doi: 10.1097/00010694-199606000-00007.

Gares, P.A., Davidson-Arnott, R.G.D., Bauer, B.O. et al. (1996). Alongshore variations in aeolian sediment transport, sediment size and moisture at Carrick Finn Strand, Ireland. *Journal of Coastal Research* 12 (3): 673–682.

Gillette, D.A., Blifford, I.H., and Fenster, C.R. (1972). Measurements of aerosol size distributions and vertical fluxes on land subject to wind erosion. *Journal of Applied Meteorology* 11 (6): 977–987. doi: 10.1175/1520-0450(1972)011<0977:MOASDA>2.0.CO;2.

Gillette, D.A., Herbert, G., Stockton, P.H., and Owen, P.R. (1996). Causes of the fetch effect in wind erosion. *Earth Surface Processes and Landforms* 21 (7): 641–659. doi: 10.1002/(SICI)1096-9837(199607)21:7<641::AID-ESP662>3.0.CO;2-9.

Gillette, D.A., Ono, D., and Richmond, K. (2004). A combined modeling and measurement technique for estimating windblown dust emissions at Owens (dry) Lake, California. *Journal of Geophysical Research. Earth Surface* 109: F01003. doi: 10.1029/2003JF000025.

Gillette, D.A. and Stockton, P.H. (1986). Mass momentum and kinetic energy fluxes of saltating particles. In: *Aeolian Geomorphology, 17th Annual Binghampton Geomorphology Symposium* (ed. W.G. Nickling), 35–56. London: Allen and Unwin.

Gillette, D.A. and Stockton, P.H. (1989). The effect of nonerodible particles on wind erosion of erodible surfaces. *Journal of Geophysical Research: Atmospheres* 94 (D10): 12885–12893. doi: 10.1029/JD094iD10p12885.

Gillies, J.A. (2013). Dust emissions and transport – near surface (chapter 11.4). In: *Treatise on Geomorphology, Volume 11. Aeolian Geomorphology* (ed. J.F. Schroder, N. Lancaster, A.C.W. Baas and D.J. Sherman), 43–63. Rotterdam: Elsevier.

Gordon, M. and McKenna Neuman, C. (2011). A study of particle splash on developing ripple forms for two bed materials. *Geomorphology* 129 (1–2): 79–91. doi: 10.1016/j.geomorph.2011.01.015.

Greeley, R., Blumberg, D.G., and Williams, S.H. (1996). Field measurements of the flux and speed of wind-blown sand. *Sedimentology* 43 (1): 41–52. doi: 10.1111/j.1365-3091.1996.tb01458.x.

Hesp, P.A. (1981). The formation of shadow dunes. *Journal of Sedimentary Petrology*

51(1):101–112.doi:10.1306/212F7C1B-2B24-11D7-8648000102C1865D.

Hesp, P.A. and Hastings, K. (1998). Width, height and slope relationships and aerodynamic maintenance of barchans. *Geomorphology* 22 (2): 193–204. doi: 10.1016/S0169-555X(97)00070-6.

Ho, T., Valance, A., Dupont, P., and Ould El Moctar, A. (2013). Scaling laws in aeolian sand transport. *Physical Review Letters* 106: 094501. doi: 10.1103/PhysRevLett.106.094501.

Hsu, S. (1971). Wind stress criteria in eolian sand transport. *Journal of Geophysical Research* 76 (36): 8684–8686. doi: 10.1029/JC076i036p08684.

Hunt, J.C.R. and Carlotti, P. (2001). Statistical structure at the wall of the high Reynolds number turbulent boundary layer. *Flow, Turbulence and Combustion* 66 (4): 453–475. doi: 10.1023/A:1013519021030.

Hunt, J.C.R. and Morrison, J.F. (2000). Eddy structure in turbulent boundary layers. *European Journal of Mechanics – B/Fluids* 19 (5): 673–694. doi: 10.1016/S0997-7546(00)00129-1.

Iversen, J.D. and Rasmussen, K.R. (1994). The effect of surface slope on saltation threshold. *Sedimentology* 41 (4): 721–728. doi: 0.1111/j.1365-3091.1994.tb01419.x.

Jackson, N.L. and Nordstrom, K.F. (1998). Aeolian transport of sediment on a beach during and after rainfall, Wildwood, NJ, USA. *Geomorphology* 22 (2): 151–157.

Jackson, N.L., Sherman, D.J., Hesp, P.A. et al. (2006). Small-scale spatial variations in aeolian sediment transport on a fine-sand beach. *Journal of Coastal Research. Special Issue* 39: 379–383.

Jackson, R.G.I.I. (1976). Sedimentological and fluid dynamic implications of turbulent bursting phenomenon in geophysical flows. *Journal of Fluid Mechanics* 77 (3): 531–560. doi: 10.1017/S0022112076002243.

Kaimal, J.C. and Finnigan, J. (1994). *Atmospheric Boundary Layer Flows: Their Structure and Measurement*. New York: Oxford University Press.

Kawamura, R. (1951). Study on sand movement by wind. Tokyo: Institute of Science and Technology, Report 5 (3/4): 95–112. (in Japanese; translated as Report, HEL-2-8, (1964) Hydraulics Engineering Laboratory, University of California, Berkeley, CA).

King, J., Nickling, W.G., and Gillies, J.A. (2006). Aeolian shear stress ratio measurements within mesquite-dominated landscapes of the Chihuahuan Desert, New Mexico, USA. *Geomorphology* 82 (3–4): 229–244. doi: 10.1016/j.geomorph.2006.05.004.

King, J., Nickling, W.G., and Gillies, J.A. (2008). Investigations of the law-of-the-wall over sparse roughness elements. *Journal of Geophysical Research-Earth Surface* 113: F02S07. doi: 10.1029/2007JF000804.

Kjelgaard, J.F., Chandler, D.G., and Saxton, K.E. (2004). Evidence for direct suspension of loessial soils on the Columbia Plateau. *Earth Surface Processes and Landforms* 29 (2): 221–236. doi: 10.1002/esp.1028.

Kline, S.J., Reynolds, W.C., Schraub, F.A., and Runstadl, P.W. (1967). The structure of turbulent boundary layers. *Journal of Fluid Mechanics* 30 (4): 741–773. doi: 10.1017/S0022112067001740.

Kocurek, G., Ewing, R.C., and Mohrig, D. (2009). How do bedform patterns arise? New views on the role of bedform interactions within a set of boundary conditions. *Earth Surface Processes and Landforms* 35 (1): 51–63. doi: 10.1002/esp.1913.

Kok, J.F., Parteli, E.J.R., Michaels, T.I., and Karam, D.B. (2012). The physics of wind-blown sand and dust. *Reports on Progress in Physics* 75: 106901. doi: 10.1088/0034-4885/75/10/106901.

Kok, J.F. and Rennó, N.O. (2008). Electrostatics in wind-blown sand. *Physical Review Letters* 100: 014501. doi: 10.1103/PhysRevLett.100.014501.

Kok, J.F. and Rennó, N.O. (2009). A comprehensive numerical model of steady state saltation (COMSALT). *Journal of Geophysical Research: Atmospheres* 114: D17204. doi: 10.1029/2009JD011702.

Lancaster, N. and Baas, A. (1998). Influence of vegetation cover on sand transport by wind:

field studies at Owens Lake, California. *Earth Surface Processes and Landforms* 23 (1): 69–82. doi: 10.1002/(SICI)1096-9837 (199801)23:1<69::AID-ESP823>3.0.CO;2-G.

Lee, J.A. (1987). A field experiment on the role of small scale wind gustiness in aeolian sand transport. *Earth Surface Processes and Landforms* 12 (3): 331–335. doi: 10.1002/esp.3290120311.

Lee, Z.S. and Baas, A.C.W. (2015). Variable and conflicting shear stress estimates inside a boundary layer with sediment transport. *Earth Surface Processes and Landforms* 41 ((4): 435–445.

Leenders, J.K., van Boxel, J.H., and Sterk, G. (2005). Wind forces and related saltation transport. *Geomorphology* 71 (3–4): 357–372. doi: 10.1016/j.geomorph.2005.04.008.

Lettau, K. and Lettau, H.H. (1978). Experimental and micrometeorological field studies of dune migration. In: *Exploring the world's Driest Climate, Wind Forces and Related Saltation Transport*, 110–147. Madison: University of Wisconsin, Institute of Environmental Sciences.

Li, B., Granja, H.M., Farrell, E.J. et al. (2009). Aeolian saltation at Esposende Beach, Portugal. *Journal of Coastal Research* SI56: 327–331.

Liu, X. and Dong, Z. (2004). Experimental investigation of the concentration profile of a blowing sand cloud. *Geomorphology* 60 (3–4): 371–381. doi: 10.1016/j.geomorph.2003.08.009.

Marticorena, B. and Bergametti, G. (1995). Modeling the atmospheric dust cycle: 1. Design of a soil-derived dust emission scheme. *Journal of Geophysical Research: Atmospheres* 100 (D8): 16415–16430. doi: 10.1029/95JD00690.

Marusic, I. and Hutchins, N. (2008). Study of the log-layer structure in wall turbulence over a very large range of Reynolds number. *Flow Turbulence and Combustion* 81 (1): 115–130. doi: 10.1007/s10494-007-9116-0.

Marusic, I., Mathis, R., and Hutchins, N. (2010). Predictive model for wall-bounded turbulent flow. *Science* 329 (5988): 193–196. doi: 10.1126/science.1188765.

McEwan, I.K. (1993). Bagnold's kink: a physical feature of a wind velocity profile modified by blown sand? *Earth Surface Processes and Landforms* 18 (2): 145–156. doi: 10.1002/esp.3290180206.

McEwan, I.K. and Willetts, B.B. (1991). Numerical model of the saltation cloud. In: *Aeolian Grain Transport 1: Mechanics* (ed. O.E. Barndorff-Nielsen and B.B. Willets), 53–66. Vienna: Springer-Verlag Acta Mechanica Supplementum 1.

McEwan, I.K. and Willetts, B.B. (1993). Sand transport by wind: a review of the current conceptual model. In: *The Dynamics and Environmental Context of Aeolian Sedimentary Systems* (ed. K. Pye), 7–16. London: Geological Society of London, Special Publication 72.

McKenna Neuman, C. and Nickling, W.G. (1989). A theoretical and wind tunnel investigation of the effect of capillary water on the entrainment of sediment by wind. *Canadian Journal of Soil Science* 69 (1): 79–96. doi: 10.4141/cjss89-008.

McMenamin, R., Cassidy, R., and McCloskey, J. (2002). Self-organised criticality at the onset of aeolian sediment transport. *Journal of Coastal Research, Special Issue* 36: 498–505.

Nalpanis, P., Hunt, J.C.R., and Barrett, C.F. (1993). Saltating particles over flat beds. *Journal of Fluid Mechanics* 251: 661–685. doi: 10.1017/S0022112093003568.

Namikas, S.L. (2003). Field measurement and numerical modelling of aeolian mass flux distributions on a sandy beach. *Sedimentology* 50 (2): 303–326. doi: 10.1046/j.1365-3091.2003.00556.x.

Namikas, S.L. and Sherman, D.J. (1995). A review of the effects of surface moisture content on aeolian sand transport. In: *Desert Aeolian Processes* (ed. V.P. Tchakerian), 269–294. London: Chapman and Hall.

Nickling, W.G. and McKenna Neuman, C. (2009). Aeolian sediment transport. In: *Geomorphology of Desert Environments* (ed. A.J. Parsons and A.D. Abrahams), 517–555. New York: Springer.

Nickling, W.G., McKenna Neuman, C., and Lancaster, N. (2002). Grainfall processes in the lee of transverse dunes, Silver Peak, Nevada. *Sedimentology* 49 (1): 191–209. doi: 10.1046/j.1365-3091.2002.00443.x.

Nield, J.M., Wiggs, G.F.S., and Squirrell, R.S. (2011). Aeolian sand strip mobility and protodune development on a drying beach; examining surface moisture and surface roughness patterns measured by terrestrial laser scanning. *Earth Surface Processes and Landforms* 36 (4): 513–522. doi: 10.1002/esp.2071.

Nikuradse, J. (1933). Strömungsgesetze in rauhen Rohren. VDI-Forschungsheft, beilage zu Forschung auf dem Gebiete de Ingenieurswesens, Ausgabe B (Band 4).

Okin, G.S. (2008). A new model of wind erosion in the presence of vegetation. *Journal of Geophysical Research. Earth Surface* 113: F02S10. doi: 10.1029/2007 JF000758.

Okin, G.S. (2013). Linked aeolian-vegetation systems (Chapter 11.22). In: *Treatise on Geomorphology, Volume 11. Aeolian Geomorphology* (ed. J.F. Schroder, N. Lancaster, A.C.W. Baas and D.J. Sherman), 428–439. Rotterdam: Elsevier.

Okin, G.S. and Gillette, D.A. (2001). Distribution of vegetation in wind-dominated landscapes: implications for wind erosion modeling and landscape processes. *Journal of Geophysical Research: Atmospheres* 106 (D9): 9673–9684. doi: 10.1029/2001JD900052.

Owen, P.R. (1964). Saltation of uniform grains in air. *Journal of Fluid Mechanics* 20 (2): 225–242. doi: 10.1017/S0022112064001173.

Pelletier, J.D. (2009). Controls on the height and spacing of eolian ripples and transverse dunes: a numerical modeling investigation. *Geomorphology* 105 (3–4): 322–333. doi: 10.1016/j.geomorph.2008.10.010.

Rasmussen, K.R., Iversen, J.D., and Rautahemio, P. (1996). Saltation and wind-flow interaction in a variable slope wind tunnel. *Geomorphology* 17 (1–3): 19–28. doi: 10.1016/j.geomorph.2015.03. 041 0169-555X(95)00090-R.

Rasmussen, K.R. and Sørensen, M. (2008). Vertical variation of particle speed and flux density in aeolian saltation: measurement and modeling. *Journal of Geophysical Research. Earth Surface* 113: F02S12. doi: 10.1029/2007JF000774.

Rasmussen, K.R., Valance, A., and Merrison, J. (2015). Laboratory studies of aeolian sediment transport processes on planetary surfaces. *Geomorphology* 244: 74–94. doi: 10.1016/j.geomorph.2015.03.041.

Raupach, M.R. (1991). Saltation layers, vegetation canopies and roughness lengths. In: *Aeolian Grain Transport 1: Mechanics* (ed. O.E. Barndorff-Nielsen and B.B. Willets), 83–96. Vienna: Springer-Verlag.

Raupach, M.R., Gillette, D.A., and Leys, J.F. (1993). The effect of roughness elements on wind erosion threshold. *Journal of Geophysical Research: Atmospheres* 98 (D2): 3023–3029. doi: 10.1029/92JD01922.

Ravi, S., Zobeck, T.M., Over, T.M. et al. (2006). On the effect of moisture bonding forces in air-dry soils on threshold friction velocity of wind erosion. *Sedimentology* 53 (3): 597–609. doi: 10.1111/j.1365-3091.2006.00775.x.

Rice, M.A., Willetts, B.B., and McEwan, I.K. (1995). An experimental study of multiple grain-size ejecta produced by collisions of saltating grains with a flat bed. *Sedimentology* 42 (4): 695–706. doi: 10.1111/j.1365-3091. 1995.tb00401.x.

Rice, M.A., Willetts, B.B., and McEwan, I.K. (1996). Observations of collisions of saltating grains with a granular bed from high-speed cine-film. *Sedimentology* 43 (1): 21–31. doi: 10.1111/j.1365-3091.1996. tb01456.x.

Robinson, S.K. (1991). Coherent motions in the turbulent boundary layer. *Annual Review of Fluid Mechanics* 23: 601–639. doi: 10.1146/annurev.fl.23.010191.003125.

Rubin, D.M. and Hunter, R.E. (1987). Bedform alignment in directionally varying flows. *Science* 237 (4812): 276–278. doi: 10.1126/science.237.4812.276.

Rubin, D.M. and Ikeda, H. (1990). Flume experiments on the alignment of transverse, oblique and longitudinal dunes in directionally varying flows. *Sedimentology* 37 (4): 673–684. doi: 10.1111/j.1365-3091. 1990.tb00628.x.

Saffman, P.G. (1965). The lift on a small sphere in a slow shear flow. *Journal of Fluid Mechanics* 22 (2): 385–400. doi: 10.1017/ S0022112065000824.

Sakamoto-Arnold, C.M. (1981). Eolian features produced by the December 1977 windstorm, Southern San Joaquin Valley, California. *Journal of Geology* 89 (1): 129–137. doi: 10.1086/628568.

Sauermann, G., Kroy, K., and Herrmann, H.J. (2001). Continuum saltation model for sand dunes. *Physical Review E, Statistical, Nonlinear and Soft Matter Physics* 64: 031305. doi: 10.1103/PhysRevE.64.031305.

Schlichting, H. and Gersten, K. (2000). *Boundary-Layer Theory*. Berlin: Springer.

Shao, Y. and Li, A. (1999). Numerical modelling of saltation in the atmospheric surface layer. *Boundary-Layer Meteorology* 91 (2): 199–225. doi: 10.1023/A:1001816013475.

Shao, Y. and Raupach, M.R. (1992). The overshoot and equilibration of saltation. *Journal of Geophysical Research: Atmospheres* 97 (D18): 20559–20564. doi: 10.1029/92JD02011.

Shao, Y., Raupach, M.R., and Findlater, P.A. (1993). The effect of saltation bombardment on the entrainment of dust by wind. *Journal of Geophysical Research: Atmospheres* 98 (D7): 12719–12726. doi: 10.1029/93JD00396.

Sharp, R.P. (1963). Wind ripples. *The Journal of Geology* 71 (5): 617–636. doi: 10.1086/ 626936.

Sherman, D.J. and Bauer, B.O. (1993). Dynamics of beach-dune systems. *Progress in Physical Geography* 17 (4): 413–447. doi: 10.1177/030913339301700402.

Sherman, D.J. and Farrell, E.J. (2008). Aerodynamic roughness lengths over movable beds: comparison of wind tunnel and field data. *Journal of Geophysical Research. Earth Surface* 113: F02S08. doi: 10.1029/2007JF000784.

Sherman, D.J., Jackson, D.W.T., Namikas, S.L., and Wang, J. (1998). Wind-blown sand on beaches: an evaluation of models. *Geomorphology* 22 (2): 113–133. doi: 10.1016/S0169-555X(97)00062-7.

Sherman, D.J. and Li, B. (2012). Predicting aeolian sand transport rates: a reevaluation of models. *Aeolian Research* 3 (4): 371–378. doi: 10.1016/j.aeolia.2011.06.002.

Shields, A.S. (1936). Anwendung der Ähnlichkeitsmechanik und der Turbulenzforschung auf die Geschiebebewegung [Application of the similarity mechanics and turbulence research in the study of sand movement]. *Mitteilungen der Preussischen Versuchsanstalt für Wasserbau und Schiffbau, Berlin* 26.

Spies, P.J. and McEwan, I.K. (2000). Equilibration of saltation. *Earth Surface Processes and Landforms* 25 (4): 437–453. doi: 10.1002/(SICI)1096-9837(200004) 25:4<437::AID-ESP69>3.0.CO;2-5.

Spies, P.J., McEwan, I.K., and Butterfield, G.R. (2000). One-dimensional transitional behaviour in saltation. *Earth Surface Processes and Landforms* 25 (5): 505–518. doi: 10.1002/(SICI)1096-9837(200005) 25:5<505::AID-ESP78>3.0.CO;2-D.

Sterk, G., Jacobs, A.F.G., and Van Boxel, J.H. (1998). The effect of turbulent flow structures on saltation sand transport in the atmospheric boundary layer. *Earth Surface Processes and Landforms* 23 (10): 877–887. doi: 10.1002/(SICI)1096-9837(199810) 23:10<877::AID-ESP905>3.0.CO;2-R.

Stout, J.E. (1990). Wind erosion within a simple field. *Transactions of the American Society of Agricultural Engineers* 33 (5): 1597–1600.

Stout, J.E. (2004). A method for establishing the critical threshold for aeolian transport in the field. *Earth Surface Processes and Landforms* 29 (10): 1195–1207. doi: 10.1002/esp.1079.

Stout, J.E. and Zobeck, T.M. (1996). Establishing the threshold condition for soil movement in wind-eroding fields. In: Proceedings of International Conference on Air Pollution from Agricultural Operations, February 1996, Kansas City, Missouri.

Midwest Plan Service (MWPS C-3), Ames, Iowa, (ed. A. MacFarland, K. Curtit, and L. Jacobson), 65–72.

Stout, J.E. and Zobeck, T.M. (1997). Intermittent saltation. *Sedimentology* 44 (5): 959–970. doi: 10.1046/j.1365-3091.1997.d01-55.x.

Stull, R.B. (1988). *An Introduction to Boundary Layer Meteorology*. Dordrecht: Kluwer.

Sutton, S.L.F., McKenna Neuman, C., and Nickling, W. (2013). Avalanche grainflow on a simulated aeolian dune. *Journal of Geophysical Research. Earth Surface* 118 (F3): 1767–1776. doi: 10.1002/jgrf.20130.

Svasek, J.N. and Terwindt, J.H.J. (1974). Measurement of sand transport by wind on a natural beach. *Sedimentology* 21 (2): 311–322. doi: 10.1111/j.1365-3091.1974.tb02061.x.

Ungar, J.E. and Haff, P.K. (1987). Steady state saltation in air. *Sedimentology* 34 (2): 289–300. doi: 10.1111/j.1365-3091.1987.tb00778.x.

Walmsley, J.L. (1989). Internal boundary-layer height formulae – a comparison with atmospheric data. *Boundary-Layer Meteorology* 47 (1–4): 251–262. doi: 10.1007/BF00122332.

Wang, D., Wang, Y., Yang, B., and Zhang, W. (2008). Statistical analysis of sand grain/bed collision process recorded by high-speed digital camera. *Sedimentology* 55 (2): 461–470. doi: 10.1111/j.1365-3091.2007.00909.x.

Wasson, R.J. and Hyde, R. (1983). Factors determining desert dune type. *Nature* 304 (5924): 337–339. doi: 10.1038/304337a0.

Werner, B.T. and Gillespie, D.T. (1993). Fundamentally discrete stochastic model for wind ripple dynamics. *Physical Review Letters* 71 (19): 3230–3233. doi: 10.1103/PhysRevLett.71.3230.

Werner, B.T. and Kocurek, G. (1999). Bedform spacing from defect dynamics. *Geology* 27 (8): 727–730. doi: 10.1130/0091-7613(1999)027<0727:BSFDD>2.3.CO;2.

White, B.R. (1996). Laboratory simulation of aeolian sand transport and physical modeling of flow around dunes. *Annals of Arid Zone* 35 (3): 187–213.

White, B.R. and Schultz, J.C. (1977). Magnus effect in saltation. *Journal of Fluid Mechanics* 81 (3): 497–512. doi: 10.1017/S0022112077002183.

Wiggs, G.F.S. (2011). Sediment mobilisation by the wind. In: *Arid Zone Geomorphology: Process, Form and Change in Drylands*, 3e (ed. D.S.G. Thomas). Chichester: Wiley.

Wiggs, G.F.S., Baird, A.J., and Atherton, R.J. (2004). The dynamic effects of moisture on the entrainment and transport of sand by wind. *Geomorphology* 59 (1–4): 13–30. doi: 10.1016/j.geomorph.2003.09.002.

Wiggs, G.F.S. and Weaver, C.M. (2012). Turbulent flow structures and aeolian sediment transport over a barchan sand dune. *Geophysical Research Letters* 39: L05404. doi: 10.1029/2012GL050847.

Wilkinson, R.H. (1984). A method for evaluating statistical errors associated with logarithmic velocity profiles. *Geo-Marine Letters* 3 (49–52): L05404. doi: 10.1029/2012GL050847.

Wolfe, S.A. and Nickling, W.G. (1993). The protective role of sparse vegetation in wind erosion. *Progress in Physical Geography* 17 (1): 50–68. doi: 10.1177/030913339301700104.

Yang, Z., Yuan, W., Bin, Y., and Pan, J. (2009). Measurement of sand creep on a flat sand bed using a high-speed digital camera. *Sedimentology* 56 (6): 1705–1712. doi: 10.1111/j.1365-3091.2009.01053.x.

Yizhaq, H. (2004). A simple model of aeolian megaripples. *Physica A: Statistical Mechanics and its Applications* 338 (1–2): 211–217. doi: 10.1016/j.physa.2004.02.044.

Zender, C.S., Bian, H.S., and Newman, D. (2003). Mineral Dust Entrainment and Deposition (DEAD) model: description and 1990s dust climatology. *Journal of Geophysical Research-Atmospheres* 108 (D14): 4416. doi: 10.1029/2002JD002775.

3

Wind Erosion

Jasper Knight

University of the Witwatersrand, Johannesburg, South Africa

3.1 Introduction

'Wind erosion' is something of a misnomer, because the formation of wind-eroded landforms is not the work of the wind itself, but of the mineral grains (and in some cases ice crystals) that the wind carries. Nevertheless, this chapter uses the term 'wind erosion' to describe how landforms are shaped by wind-blown particles. The wind can erode both hard bedrock or loose surface materials, although the preservation, beyond a few days or months of wind-eroded features developed in loose materials as on dunes, is low (Chapter 6). Here, we discuss the effects of wind erosion on bedrock or loose boulder-covered surfaces. Discussion of these wind-eroded landforms is grounded in the historical context of research by such workers as Charles Keyes (Goudie 2012) and Ralph Bagnold (Thorne and Soar 1996). Definitions of terms used in this chapter are provided in Table 3.1.

The capacity of wind-blown sand to erode the surface of coherent materials depends on the properties of the wind, the grains it transports, and of the substrate. The most significant properties to consider are listed in Table 3.2. The relevant properties of the wind include its speed, its turbulence, the frictional force that it exerts on the land surface, and the nature of wind interactions with the surface macro- and micro-topography. Many of these properties and interactions are site-specific, are difficult to downscale effectively from regional climate or boundary layer models, and are subject to high spatial and temporal variability. There are also important feedbacks between the atmospheric boundary layer and a variety of land surface processes, which are challenges to wind erosion modelling, and which wind tunnel models cannot capture well (e.g. Burri et al. 2011; Tong and Huang 2012; McKenna Neuman et al. 2013).

3.2 The Processes of Wind Abrasion

Sand-sized particles are the most commonly observed and the most geomorphologically effective abraders, as shown both by field and experimental studies (Kuenen 1960; Suzuki and Takahashi 1981; Anderson 1986; Bridges et al. 2004). Silt particles or ice crystals may also be important in some environmental settings, but are generally considered to be less effective abraders, because of their smaller size and lower impact energy. The processes that result in the entrainment and transport of sand particles, and the different ways of calculating sediment threshold velocity, are described in Chapter 2. Here, we focus on those elements of particle transport that are of direct relevance to wind erosion. The impact threshold, which is the

Aeolian Geomorphology: A New Introduction, First Edition. Edited by Ian Livingstone and Andrew Warren.
© 2019 John Wiley & Sons Ltd. Published 2019 by John Wiley & Sons Ltd.

Table 3.1 Definition of commonly used terms related to wind processes.

Term	Definition
Deflation	The process by which wind removes dry, unconsolidated materials from regional-scale land surfaces, resulting in land surface downwearing and the formation of broad, low-relief, bowl-like depressions (deflation hollows)
Entrainment	The process by which loose sediments are incorporated within the windstream
Wind abrasion	The microscale process of mechanical downwearing of hard rock surfaces by a wind-blown abrader (i.e. a sand grain). Wind abrasion takes place through the interaction of boundary layer winds with microtopography of the rock surface
Wind erosion	The net effect of wind abrasion and deflation, thus wind erosion reflects net sediment removal from the land surface by wind

Source: Adapted after Whittow (2000).

Table 3.2 Key environmental properties that can influence the likelihood of wind erosion taking place, and the impacts of such erosion.

Properties of the wind	Properties of the substrate
Wind speed, direction, duration	Sediment availability
Wind turbulence	Grain size/shape
Wind temperature (air density)	Boundary layer (micro) topography
Saltation and suspension load	Vegetation cover Surface moisture Surface roughness

minimum velocity required to lift a grain into transport, is lower on bedrock or other hard land surfaces than on loose or unconsolidated surfaces where kinetic energy absorption/diffusion is higher. In this instance, instead of kinetic energy being transferred to the impacted grain, the energy is absorbed into the substrate. The higher wind speeds commonly found above bedrock surfaces and desert stone pavements can lead to higher-energy erosive environments, greater lift of saltating particles into the windstream, and higher potential sediment fluxes. It is no surprise, therefore, that evidence for wind abrasion is most common under such environmental conditions.

High wind speeds and sediment fluxes increase the height of the saltation curtain and the height at which maximum abrasion occurs. This is measured indirectly as the height of the peak kinetic energy flux, which is around 40–50 cm above the land surface (Anderson 1986). The precise height depends on wind speed and turbulence (which affects the size distribution of transported grains, and the nature of grain–grain interactions), land surface roughness, sediment supply, and the protrusion of obstacles into the windstream. The highest density of abraders, at the height of the peak kinetic energy flux, may undercut rock protrusions. Sediment transport, and thus the potential for wind abrasion, does occur above the height of the peak kinetic energy flux, but is generally restricted to <2 m above the surface.

3.2.1 Environmental Controls on the Wind Transport of Particles

The likelihood of particle transport by wind, and the erosional effects of such transport, are strongly related to a range of site-specific land surface factors which exhibit spatial and temporal variability at a number of scales (see Table 3.2). Apart from wind speed, air density is another significant control on the capacity for wind transport of sediment particles. Air density varies as a function of air temperature, colder air being denser than warmer air (Figure 3.1). This means that, for any given wind speed, colder and denser air can transport higher volumes of sediment and coarser grain sizes. As a consequence, wind transport and thus the effects of wind erosion are most clearly seen in cold, extra-glacial areas (Figure 3.2), many of which also experience strong, unidirectional katabatic winds (Box 3.1). Wind-eroded features can,

nonetheless, form under warm climates and may attain the sizes of those found in extra-glacial areas, but most low-latitude, warm climate locations have experienced many episodes of aridity and deflation at relatively slow rates throughout the Quaternary, thus retaining as much evidence of abrasion as colder locations (Fujioka and Chappell 2011).

This chapter describes the major land-forms developed as a consequence of wind erosion. These landforms are best manifested on bedrock or boulder-covered land surfaces, which retain more evidence for wind erosion. Different types of wind-eroded landforms are also more commonly found in these locations, being formed by similar suites of processes (Goudie 2008). The generally high preservation of these landforms has a high potential to reconstruct the boundary-layer climates and geomorphological processes of the past. The most common wind erosion landforms are now described in ascending order of size.

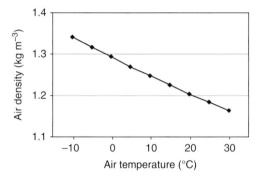

Figure 3.1 Illustration of the relationship between air density and temperature.

3.3 Ventifacts

Ventifacts are loose surface clasts (pebbles, cobbles or boulders) or upstanding bedrock protrusions that have been shaped by the

(a)

(b)

Figure 3.2 Wind-eroded features in Antarctica. (a) Polished and pitted boulder surfaces, Victoria Valley. (b) A wind-abraded boulder, Antarctic coast. *Source:* Photos: C. Bristow. See insert for colour representation of this figure.

Box 3.1 Katabatic Winds and Wind Erosion

Katabatic winds are generated by air flow down a thermal or topographic gradient as a result of seasonal or diurnal patterns of land surface heating and cooling. Typical examples include föhn or mistral winds, found in coastal settings where winds are generated by strong orographic relief. Katabatic winds flow at their strongest where there are significant atmospheric pressure gradients, and are thus best experienced in the unglaciated parts of east Antarctica where cold and arid air formed above the Antarctic ice sheet surface causes descending air masses and outward flow of this air towards and then away from ice sheet margins (Wendler et al. 1997; Parish and Cassano 2003; Nylen et al. 2004). The katabatic winds of east Antarctica are strong and unidirectional but the highest velocities occur in short gusts (Wendler et al. 1997; French and Guglielmin 1999). These winds are also strongest in the austral summer months and are largely directed by land surface topography.

Based on data from experimental abrasion sites, calculated free-stream wind speeds of katabatic winds can reach $<70\,\mathrm{m\,s^{-1}}$ ($250\,\mathrm{km\,h^{-1}}$) (Malin 1986) and are associated with very high rates of windblown snow transport (Scarchilli et al. 2010). Erosion by strong katabatic winds of snow or ice fields can form wind-parallel erosional features termed *zastrugi* or *sastrugi* (Mather 1962). The flux density and ice crystal size of windblown snow and ice decrease upwards in the windstream, similar to that of sediment particles. These snow and ice crystals are also important because they are potentially hard abraders that can shape rock surfaces. Under these conditions, ventifacts are commonly developed over short (decadal) timescales (Miotke 1982) and wind abrasion is considered to be the most effective geomorphic process in the region (French and Guglielmin 1999; Gillies et al. 2009). Larger-scale features such as honeycomb and cavernous (tafoni) weathering can reach $<30\,\mathrm{cm}$ in length, may form over 1000–4000 years, and may have had a polygenic origin, including by salt and thermal weathering as well as by wind abrasion (French and Guglielmin 1999) (Figure 3.2). A significant limitation on wind abrasion, however, is the availability of sediment grains or snow/ice crystals, thus regional aridity (Matsuoka et al. 1996).

---

bombardment by wind-blown particles. Ventifacts have been described from a range of mountain, desert, periglacial, and coastal environments worldwide (e.g. Wade 1910; Tremblay 1961; Sugden 1964; Selby 1977; Laity 1987, 1992; Bishop and Mildenhall 1994; Knight and Burningham 2001, 2003; Knight 2005, 2008; Mackay and Burn 2005) (Figure 3.3).

3.3.1 Morphology

The term *ventifact* was first used by Evans (1911, p. 335) to describe 'any wind-shaped stone'. His term was a replacement for the more specific German-language terms *Einkanter* (a rock or boulder surface with one abraded and one non-abraded face) and *Dreikanter* (a rock or boulder with three opposing, pyramidal sides, all of which have been abraded) (Wade 1910; Delo 1930). The most common morphological elements of a ventifact are one or more abraded faces, termed *facets*, which are sometimes separated by a sharp ridge-like *keel* that, where present, can usually be traced across the upper surface of the ventifact (Figure 3.4). Some ventifacts have only one dominant facet which therefore reflects very clearly the direction of the most geomorphologically-effective wind. In other cases, ventifacts with multiple facets have been interpreted as evidence for the clast having been moved by rotation or undercutting by the wind, rather than as evidence of changes in wind direction during the development of the ventifact (Kuenen 1928). The largest ventifacts reported are several metres horizontally and

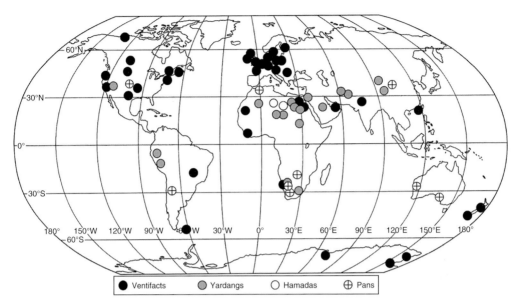

Figure 3.3 Global view of the main locations of different wind-eroded forms noted in the literature, showing the locations of ventifacts (Knight 2008), yardangs (Halimov and Fezer 1989; Goudie 1999), and hamadas and pans (Goudie 1989).

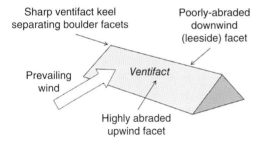

Figure 3.4 The main morphological elements of a ventifact, including facet and keel. *Source:* Adapted after Knight (2008).

up to 1 m in height (Mackay and Burn 2005; Knight 2008).

In detail, the wind-abraded facets of ventifacts are not uniform or smooth: they are frequently ornamented by smaller abrasion features (Figure 3.5). These smaller features are important because they reflect wind flow-substrate interactions more closely, and their morphologies are more strongly controlled by the properties of the substrate such as its mineralogy. These micro-features are best developed, or can be best identified, on fine-grained rocks that are the least likely to be affected by other weathering processes,

such as exfoliation or dissolution. Because the abrasion rate depends on wind speed and on the density of abrading particles within the windstream, as well as on the properties of the substrate, the morphology of ventifacts can vary substantially, both spatially and temporally. These morphologies evolve in a downwind direction as a result of patterns of the expansion or compression of the wind-stream as it interacts with the local micro-topography. Selby (1977) provided a numerical analysis of the dimensions of different micro-features and found that their length-to-width ratios could be used to discriminate between those formed by wind abrasion and those formed by weathering.

3.3.2 Wind Abrasion Microfeatures

Polishing takes place as a result of the abrasive action of quartz particles on fine-grained rock surfaces. The quartz of the abrading particles may also interact chemically with rock surfaces to produce a silica polish or coating, similar to desert varnish (Zerboni 2008). These coatings can preserve the underlying rock surfaces (Giorgetti and Baroni 2007).

Figure 3.5 Examples of ventifact microfeatures from north-west Ireland: (a) facets and keel, (b) pits developed on facets, (c, d) grooves. *Source:* Photos: J. Knight. See insert for colour representation of this figure.

Cosmogenic analysis of the polish can be used to date the formation of underlying wind-eroded surfaces (Ruszkiczay-Rudiger et al. 2011). Around 60% of the rocks in east Antarctica (Gillies et al. 2009) and north-west Ireland (Knight 2005) show polish.

Pits are enclosed, circular to elongated micro-depressions on the rock surface that are formed directly by the impact of an abrader (Várkonyi and Laity 2012). Pits pick out grain-scale variations in mineralogy, especially softer minerals, such as feldspar, or may exploit microscale variations in rock surface roughness, such as vesicles within basalt. The shape and size of pits vary from a few mm in diameter and depth, to 6 mm diameter and up to 0.5 cm in depth (Knight and Burningham 2003). Pits generally develop at a steep angle to the rock surface

and are elongated into the rock face, parallel to the direction of wind-flow. Steeper windward rock surfaces generally develop smaller and deeper pits, whereas on less steeply inclined surfaces, pits are more elongated and wider. They are therefore transitional to flutes and grooves.

Flutes and grooves are elongate depressions in the rock surface with scoop-like or scalloped edges, and smooth internal sides (Box 3.2). Many flutes and grooves have undulating or highly variable long profiles that are aligned in the direction of the wind (Selby 1977). The morphological distinction between flutes and grooves is unclear, and the terms have been used quite loosely. In both, small depressions developed on the upwind edge of the rock surface become shallower and broaden in a downwind direction, and have

---

Box 3.2 Formation and Significance of Helicoidal Vortices

Helicoidal (corkscrew-like) vortices are a likely candidate for the formation of wind-parallel erosional ridges and grooves across rock surfaces. Knight and Burningham (2003) described a series of wind-parallel grooves (3 m deep, 10–40 cm apart) developed up the face of a diabase boulder on the coast of Oregon, USA.

In Oregon, the grooves widen and deepen up the face, in a down-wind direction. These landforms are ascribed to wind flow within helicoidal vortices moving up the boulder face (Laity and Bridges 2009). In this example, airflow compression up the windward face of a rock surface results in increased windspeed and higher impact velocity of any transported particles, and thus higher rates of abrasion. Furthermore, interactions between wind dynamics and surface microtopography results in high flow vorticity. This can be manifested either parallel or perpendicular to wind direction, and result in lines of higher abrasion, forming pits or grooves (Whitney 1978). The repeated development and then shedding of wind vortices over rock surfaces represent a dynamic response of air pressure to microtopography (Tanner 1960; McKenna Neuman et al. 2013). This means that variations in microtopography and properties of the boundary-layer windflow are key controls on the development of such vortices.

---

clear to diffuse lateral margins. In general, neither flutes nor grooves have pits within them, but pits are common on the rock surfaces outside of these features. Flutes and grooves may also develop on oblique facets or downwind of pitted facets (Knight 2005).

3.3.3 Ventifact Evolution

Ventifacts form where loose sediments are present and where surface boulders or bedrock protrusions penetrate into the windstream, where vegetation is absent, and where the velocity of the prevailing wind is above the sediment threshold velocity. The processes and rates of the evolution of ventifacts have been examined both in field and laboratory studies in which blocks of different natural and synthetic materials have been exposed to abrasion (e.g. Schoewe 1932; Sharp 1964, 1980; Suzuki and Takahashi 1981; Bridges et al. 2004; Mackay and Burn 2005). These experiments show that time-averaged rates of wind abrasion are highly variable, both within and between materials of different types, but that all materials show similar patterns of downwearing, whereby the height of the ventifact decreases, the angle of the windward facet decreases, and the keel migrates downwind (Schoewe 1932).

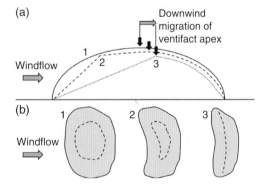

Figure 3.6 Diagram illustrating ventifact evolution over time. (a) Cross-sectional view showing changes in rock surface morphology from time periods 1–3, with development of the planar windward facet which decreases in angle and migrates downwind over time as abrasion takes place and the ventifact develops a more distinctive morphology and decreases in total volume. Abrasion moves the ventifact apex downwind over time and lowers the apex. (b) Plan view of the ventifact basal outline (solid line) and keel (dotted line), showing its development in time periods 1–3.

Laboratory studies suggest that only a 6% loss of mass from wind-abraded facets is required in order to form a ventifact (Kuenen 1960), but this value is somewhat arbitrary, because loss of mass can take place from all rock surfaces (Figure 3.6). Indeed, facets that have a very low value of overall mass loss may still show a range of well-developed micro-features.

Thus, there are problems in measuring and interpreting the attributes of ventifacts including their age, the depth of abrasion, the loss of rock mass, ambient wind speed, and sediment flux. This means that there are few good estimates of the timescales over which ventifacts form, or their abrasion rate. Estimates of timescales of ventifact formation vary considerably (Kuenen 1960), in part because the calculations of the timescale and abrasion rate are based on present-day climatic conditions (e.g. wind speed, sediment availability). This assumes that ventifacts are being actively formed today under contemporary conditions. However, ventifacts covered by lichens have also been observed (Knight 2005, 2008), and lichens can only colonise the rock surface if abrasion is virtually zero. In these cases, it is evident that the ventifacts are not being actively formed and thus observations of contemporary climate cannot be used to calculate their age or rate of abrasion.

In the light of these problems with the age of ventifacts and the role of climate in their formation, it is important to determine whether ventifacts form episodically (by individual storm events, say, of 10^{-4} years duration), cyclically (seasonally, of 10^{-2}–10^{-1} years duration) or during past cooler climate episodes (e.g. the Younger Dryas, Little Ice Age, of 10^{2}–10^{3} years duration) when winds were known to be stronger and thus more abrasive. Based on ventifact morphology, estimates of the ages of ventifacts observed in the field can be made. Some

ventifacts have been considered to be formed entirely under contemporary conditions (e.g. in Northern Ireland; Knight 2003), or reactivated episodically under these conditions (e.g. in north-west Ireland, northern Scotland, north-east England; Braley and Wilson 1997; Knight and Burningham 2001; Wilson et al. 2002), or are entirely relict features and not forming at present (e.g. in Denmark and central Scotland; Schlyter 1995; Christiansen 2004). Assumptions of ventifact age and depth of abrasion are required in order to calculate averaged abrasion rates (Table 3.3), but there are many uncertainties in both these measures (discussed by Greeley and Iversen 1985, pp. 131–133). Several experimental studies have examined the abrasion rates of different materials, but the results cannot be taken as representative of real rock surfaces under natural conditions. For example, using a lucite rod under field conditions, Sharp (1980) showed that abrasion was highly variable: in a 15-year experiment, 87% of the total abrasion took place during only 18% of the total time period, and inter-annual averaged abrasion rates varied by a factor of 15. This suggests that any short-term measurements of abrasion rates are not representative of the longer-term rates that are required to shape ventifacts.

3.4 Yardangs

The term yardang is a transliteration of a Turkic (central Asian) word to describe streamlined ridges, and was originally used

Table 3.3 Calculated averaged ventifact abrasion rates of 'natural' rock surfaces under field conditions.

Abrasion rate (mm yr^{-1})	Source	Comments
0.01–0.05	Knight (2003)	Basalt boulders on a jetty of known age, Northern Ireland
0.05–3.70	Malin (1986)	Various lithologies, Victoria Land, Antarctica
0.24–1.63	Knight and Burningham (2003)	Diabase boulders on a jetty of known age, Oregon, USA
10.40	Miotke (1982)	Dolerite, Victoria Land, Antarctica

to refer to landforms identified in the cold-desert Taklamakan region of north-west China (Blackwelder 1934; Goudie 1999). Yardangs are elongate, wind-parallel ridges that are developed in bedrock, particularly in sandstones, siltstones, and mudstones, where both physical and chemical weathering weaken rock surfaces and facilitate granular disintegration and exfoliation.

Yardangs can be distinguished from ventifacts on the basis of their size and relationship to wind direction: yardangs are much bigger and are aligned parallel rather than perpendicular to wind direction. Since the early twentieth century, yardangs have been identified in many arid and semi-arid locations worldwide (see Figure 3.3). Their common features include a continental (not coastal) setting; dry climatic conditions; sedimentary bedrock, particularly of softer rock types or of relatively young (Neogene) age; a mixture of exposed bedrock and loose surface sediments; and strong winds. Although fossil yardangs are known from the geologic record, Goudie (1999) argued that many contemporary yardangs developed as a consequence of increased aridity from the mid-Holocene onwards, exposing soft dried lake or river bed sediments to wind erosion (e.g. Gutiérrez-Elorza et al. 2002; Whitney et al. 2015).

Actively-forming yardangs have been described from China, Iran, and parts of the Sahara. In all these locations, strong, unidirectional winds and high sediment supply result in high rates of abrasion (Brookes 2001; Vincent and Kattan 2006; Al-Dousari et al. 2009). The formation of yardangs in these areas is probably also enhanced by strong thermal weathering, freeze–thaw action and salt crystal growth, which can detach sediment grains from the bedrock surface (Lancaster 1984; Goudie 1989). In the western Qaidam region of central China, Wang et al. (2011) showed that the rate of yardang abrasion was in the range 0.011–0.398 $mm\,yr^{-1}$ between 1986 and 2010, but that there was high inter-annual variability, probably related to the frequency of high winds. This rate of yardang abrasion is very similar to the rates of bedrock incision by wind across the same region, calculated using the cosmogenic ^{10}Be method (Rohrmann et al. 2013). Yardangs at Rogers Lake (Mojave Desert, USA) experience abrasion at the windward tip of $20\,mm\,yr^{-1}$ and on the lateral flanks of $0.5\,mm\,yr^{-1}$ (Laity 1994).

Although most studies have been concerned with yardangs that are undergoing active contemporary abrasion, examples of fossil yardangs have also been described (Tewes and Loope 1992; Inbar and Risso 2001; Vincent and Kattan 2006; Al-Dousari et al. 2009; Sebé et al. 2011). These relict (non-active) forms can be identified on the basis of their stratigraphic position and orientation relative to contemporary forms and wind regimes (Whitney et al. 2015).

3.4.1 Yardang Morphology and Evolution

Most yardangs are found in large fields where up to several hundred individuals, aligned parallel to one another, may be present (Brookes 2001). Individual yardangs are usually separated by a flat, sand-covered valley through which strong and unidirectional winds are funnelled. Over time, these valleys are widened by the abrasion of the flanks of the yardangs on either side (Al-Dousari et al. 2009; Sebé et al. 2011; Perkins et al. 2015) (Figure 3.7).

The processes of yardang development include wind abrasion by saltating particles, and deflation of loosened sediment grains from yardang slopes, which exposes the rock surface to further abrasion. In addition, yardangs can be affected by non-wind processes such as mass movement; episodic fluvial erosion around their margins; and subaerial thermal, freeze–thaw and salt weathering. Under strong, unidirectional winds, upstanding bedrock ridges and mesas are abraded on the upwind side and undercut by abrasion on yardang flanks. This results over time in a more streamlined and aerodynamic form. A length:

Figure 3.7 Examples of yardangs. (a) A field of yardangs from the Bodélé Depression, Chad, (b) a close-up of a single yardang from that field. *Source:* Photos: C. Bristow. (c) An individual 15 m-high yardang, from the Dunhuang Yardang National Geopark, China, (d) a group of parallel yardangs from the same field. *Source:* Photos: J. Bullard. See insert for colour representation of this figure.

width ratio of 4 : 1 is considered optimal with respect to the balance between skin-friction drag and pressure drag. In this abrasion process, the summit location migrates downwind. More complex forms, including cones and pyramids, can also develop (Halimov and Fezer 1989).

Most yardangs typically attain dimensions of around 30 m in height, >50 m in width and <5 km in length. Mega-yardangs up to 20 km long have also been identified (Goudie 2008). The outline morphology of individual yardangs is generally quite consistent, irrespective of their size, location, likely age or bedrock type. Yardangs are elongate in the direction of the most geomorphologically effective wind; the outline morphology of individual yardangs usually tapers downwind; morphology is axially symmetric; and the yardang summit is usually located near the upwind end with elevations decreasing downwind (Halimov and Fezer 1989). The slope of the windward end of the yardang decreases over time as a result of abrasion, whereas the downwind prow of the yardang becomes steeper over time or may become undercut at its base (forming a 're-entrant'; Wang et al. 2011), with an overhanging upper part caused by turbulence within the lowermost part of the saltation curtain (e.g. Lancaster 1984).

The reported dimensions of yardangs vary but, irrespective of substrate type, they generally attain similar length : width : height ratios of 10 : 2 : 1 (Halimov and Fezer 1989), which suggests that yardangs evolve as a function of wind interaction with bedrock and are less strongly controlled by environmental setting or bedrock type/structure. Ward and Greeley (1984) found a strong relationship (r^2 = 0.89) between yardang length and width. Wang et al. (2011) modelled the evolution of yardangs from the Qaidam basin, China. They showed that, although annually-averaged abrasion rates were generally high, there may be only 19–49 effective abrasion days per year, so that yardang evolution is highly episodic.

3.5 Hamadas and Stone Pavements

Hamadas are rocky, wind-eroded land surfaces that may be composed of bare bedrock that is covered by loose, angular and in places often large debris (pebbles to boulders) which has the same lithology as the underlying bedrock. Hamadas are common in the northern Sahara (Libya, Algeria) where they cover large areas and are thought to have developed by progressive deflation during arid phases of the Quaternary (Perego et al. 2011; Zerboni et al. 2011). Hamada surfaces have not been well described, but their macroscale morphology is commonly controlled by bedrock lithology and geologic structures. The angular boulders that commonly cover hamada surfaces may have been formed as a result of different weathering processes, in particular, thermal and salt weathering under arid, hot conditions, which shatter and loosen angular fragments from the underlying bedrock (Eppes et al. 2010). Alternatively, hamadas can develop through the deflation of fine sediments, allowing angular boulders to settle directly on to bedrock (Perego et al. 2011).

Stone pavements, also termed desert pavements, reg or gobi, are similar morphologically to hamadas in that their surfaces are covered by debris, but here the debris is of smaller size (pebbles), the clasts are often sub-rounded and form an interlocking surface cover of one clast thickness (Cooke 1970). These clasts armour the land surface and protect the underlying sediments from deflation or erosion by flashfloods (Williams and Zimbelman 1994; Dietze and Kleber 2012) (Figure 3.8). Below the surface cover there are usually unconsolidated sediments, not bedrock. Formation of stone pavements probably results from strong thermal, freeze–thaw and salt weathering, and strong winds that are able to carry small particles away.

Models of stone pavement evolution show that clasts comprising the pavements become clustered over time by erosion, thereby closing the gaps between clasts and reducing the ability of finer sediments to be blown away

(a)

(b)

Figure 3.8 Stone pavement (reg) surfaces from (a) Jordan. *Source:* Photo: J. Bullard, and (b) Simpson Desert, Australia. *Source:* Photo: C. Bristow. See insert for colour representation of this figure.

(Haff and Werner 1996). The net effect is to stabilise the land surface. Dating of these surfaces has shown that stone pavements and the micro-features developed thereon (rock varnish, polish, pits) can be in the range 10^3–10^7 years old (Wells et al. 1995; Quade 2001; Matmon et al. 2009). Stone pavements and desert landscapes can therefore develop and remain stable over very long time periods (McFadden et al. 1987; Fujioka and Chappell 2011). The ways in which stone pavements evolve, however, are not well understood. Apart from wind transport of fine sediments, other processes may also be significant in stone pavement formation, including rainsplash (Wainwright et al. 1995), sheetflow during episodic rainfall events (Williams and Zimbelman 1994; Dietz and Kleber 2012; Dietze et al. 2013), bioturbation by small desert animals (Haff and Werner 1996; Haff 2014), and burial through wind-blown sediment deposition and pedogenesis (Pelletier et al. 2007). Dong et al. (2002), using wind tunnel experiments, have shown that boundary layer roughness (and thus aerodynamic turbulence) are best developed when surface clast coverage is around 15% by area. As boundary layer turbulence is important in particle entrainment, this may mean that development of stone pavements by wind

erosion is most effective at around this value of 15% surface clast coverage. At values higher or lower than this, other processes may be more dominant.

3.6 Deflation Basins and Pans

Deflation basins are shallow and sometimes very large land surface depressions that, although they are ultimately polygenetic in origin, are likely to have experienced periods of wind erosion (Goudie 1989) (Figure 3.9a). Deflation basins are particularly common on sedimentary bedrock types where physical and chemical weathering processes can generate loose surface materials. The largest deflation basins include those in the Western Desert of Egypt, such as the Qattara Depression, which is around 300 km long, 145 km wide and attains a maximum depth of 134 m below sea level (Albritton et al. 1990). In the Western Desert, these basins are commonly deflated to the level of the water table, but their origin is more complex, with strong geologic control by the presence of suitable sedimentary strata (Aref et al. 2002). As such, there is a macroscale geologic control on deflation basin geometry and location (Laity 2008), including lines

(a)   (b)

Figure 3.9 Photos of (a) a deflation basin developed on a wind-eroded diatomite surface, Bodélé Depression, Chad. *Source:* Photo: C. Bristow, and (b) a pan in the linear dunefield between William Creek and Coober Pedy, South Australia. Note the linear dune in the background. *Source:* Photo: D. Nash. See insert for colour representation of this figure.

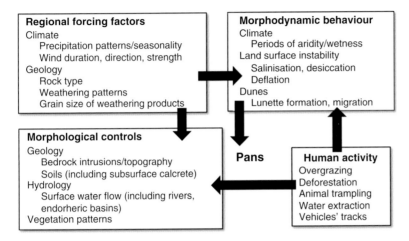

Figure 3.10 Model of pan evolution, illustrating the relationships between regional (geologic, climatic) processes and site-scale geomorphological processes. *Source:* Adapted after Goudie (1999).

of bedrock weakness and former river drainage systems (Goudie 1999). The main factors contributing to the evolution of such systems are described in Figure 3.10.

Climate change is also a significant control on the formation of deflation basins. For example, during wetter climate periods of the Quaternary, fine sediments and biogenic materials can be deposited in these depressions when they become water-filled (e.g. Telfer et al. 2009). During arid phases when the basins dry up, the loose sediments may become susceptible to deflation (Gutiérrez-Elorza et al. 2005). The resulting deflation may then be sediment-limited, ceasing when fine sediments are no longer available or when the groundwater table is reached. Such a polyphase history is likely for basins such as the Qattara Depression.

Pans are common in some semi-arid areas such as southern Africa, central USA, and southern and western Australia (see Figure 3.9b). They are shallow and generally small depressions that may contain water seasonally or ephemerally. Pans are also geomorphologically associated with sand dunes, sand sheets, and other features indicative of active wind transport (Goudie and Wells 1995). Pans can be distinguished from deflation basins based on their smaller size and the much more common presence of water. In geometry, pans are circular to elongate with smooth shorelines created by wind-driven waves. Winds can also form sand dune ridges and crescentic dunes (lunettes), in particular immediately downwind of pan margins when the pan is dry (Goudie 1989; Holmes et al. 2008).

3.7 Discussion

3.7.1 Wind Erosion and Boundary Layer Climates

Interactions between wind flow and land surface topography determine the volume and dynamics of wind-blown sediment transport and thus wind erosion capacity (Raupach and Finnigan 1997; Jackson et al. 2011; Webb et al. 2014). The relationship between wind erosion and boundary layer climates, however, remains poorly understood (Ping and Dong 2014). Wind tunnel experiments have been used to help inform this relationship, but these have focused mainly on the modes of sediment transport and rate fluxes rather than on the effects of wind abrasion (e.g. Li et al. 2008). Likewise, wind tunnel experiments using particle image velocimetry (PIV) have focused on grain–grain interactions in a free wind-stream, but not on grain interactions with the substrate (e.g. Yang et al. 2007). As a consequence, the detailed climatic conditions under which wind abrasion takes place at the microscale have not been well studied under laboratory conditions, or in instrumented field experiments (e.g. Pease and Gares 2013; Walker and Shugar 2013). Computational fluid dynamics models (e.g. Jackson et al. 2011) have been shown to be a useful approach to the parameterization of the conditions under which wind transport and abrasion can take place.

3.7.2 Reconstructing Past Wind Patterns from Wind-Eroded Features

Wind-eroded features, if they can be shown to be inactive, can be used to reconstruct the direction of the most geomorphologically-effective winds of the past. Relict wind erosion features are best identified in locations where past and present wind directions are very different, such as areas in North America and Europe that lay adjacent to the margins of the late Pleistocene ice sheets

(e.g. Rudberg 1968; Schlyter 1995; Christiansen and Svensson 1998; Christiansen 2004). For example, ventifacts are common across southern Sweden and Denmark (Schlyter 1995; Christiansen and Svensson 1998) and indicate regional easterly winds that were generated by a high-pressure cell over the former Scandinavian ice sheet (Isarin et al. 1997) (Figure 3.11).

3.8 Conclusions

Understanding wind erosion processes and products requires an interdisciplinary approach that considers the properties of both boundary-layer winds and of the substrate. The increasing availability and higher resolution of field equipment and remote sensing make it possible that such an interdisciplinary approach can be developed. These analytical tools can be used to monitor changes in sediment fluxes around, and geomorphic changes to, ventifacts and yardangs, and in evaluating spatial and temporal changes in the properties of hamadas and stone pavements. The use of high-resolution, multispectral remote sensing to map large areas of difficult or remote terrain is now also being developed (e.g. Perego et al. 2011). Finally, different cosmogenic radioisotopes (e.g. ^{3}He, ^{10}Be, ^{26}Al, ^{36}Cl) can be used individually or in combination to evaluate the age and in some cases the denudation history of wind-eroded land surfaces (e.g. Bierman and Caffee 2001).

There is great potential for developing future studies on the processes of and controls on wind erosion and their associated landforms; and this research direction can be usefully informed by both field and laboratory/experimental approaches that consider the dynamics of the boundary layer. Improved computing power and data quality in the analysis of wind erosion features from Mars and other planets also suggest ways in which similar studies could be deployed on Earth.

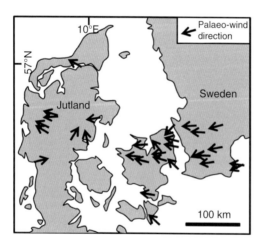

Figure 3.11 Spatial patterns of reconstructed palaeo-wind direction derived from ventifacts in Denmark and southern Sweden. *Source:* Adapted from Schlyter (1995) and Christiansen and Svensson (1998).

Acknowledgements

My field research on wind erosion is currently partly funded by the National Research Foundation (South Africa). I am particularly grateful to Charlie Bristow, Jo Bullard, and Dave Nash for generously sharing their field photos and for permission to use some of them in this chapter.

Further Reading

The key textbooks by Greeley and Iversen (1985) and Laity (2008) provide a good introduction to the range of processes and landforms associated with wind erosion.

References

Albritton, C.C., Brooks, J.E., Issawi, B., and Swedan, A. (1990). Origin of the Qattara depression, Egypt. *Geological Society of America Bulletin* 102 (7): 952–960. doi: 10.1130/0016-7606(1990)102<0952: OOTQDE>2.3.CO;2.

Al-Dousari, A.M., Al-Elaj, M., Al-Enezi, E., and Al-Shareeda, A. (2009). Origin and characteristics of yardangs in the Um Al-Rimam depressions (N Kuwait). *Geomorphology* 104 (3–4): 93–104. doi: 10.1016/j.geomorph.2008.05.010.

Anderson, R.S. (1986). Erosion profiles due to particles entrained by wind: application of an eolian sediment transport model. *Geological Society of America Bulletin* 97 (10): 1270–1278. doi: 10.1130/0016-7606 (1986)97<1270:EPDTPE>2.0.CO;2.

Aref, M.A.M., El-Khoriby, E., and Hamdan, M.A. (2002). The role of salt weathering in the origin of the Qattara depression, Western Desert, Egypt. *Geomorphology* 45 (3–4): 181–195. doi: 10.1016/S0169-555X(01) 00152-0.

Bierman, P.R. and Caffee, M. (2001). Slow rates of rock surface erosion and sediment production across the Namib Desert and escarpment, Southern Africa. *The American Journal of Science* 301 (4–5): 326–358. doi: 10.2475/ajs.301.4-5.326.

Bishop, D.G. and Mildenhall, D.C. (1994). The geological setting of ventifacts and wind-sculpted rocks at Mason Bay, Stewart Island, and their implications for late Quaternary paleoclimates. *New Zealand Journal of Geology and Geophysics* 37 (2): 169–180. doi: 10.1080/00288306.1994. 9514612.

Blackwelder, E. (1934). Yardangs. *Bulletin of the Geological Society of America* 45 (2): 159–166. doi: 10.1130/GSAB-45-159.

Braley, S.M. and Wilson, P. (1997). Ventifacts from the coast of Northumberland. *Proceedings of the Geologists' Association* 108 (2): 141–147. doi: 10.1016/S0016-7878(97)80036-3.

Bridges, N.T., Laity, J.E., Greeley, R. et al. (2004). Insights on rock abrasion and ventifact formation from laboratory and field analog studies with applications to Mars. *Planetary and Space Science* 52 (1–3): 199–213. doi: 10.1016/j.pss.2003.08.026.

Brookes, I.A. (2001). Aeolian erosional lineations in the Libyan Desert, Dakhla region, Egypt. *Geomorphology* 39 (3–4): 189–209. doi: 10.1016/S0169-555X(01) 00026-5.

Burri, K., Gromke, C., Lehning, M., and Graf, F. (2011). Aeolian sediment transport over vegetation canopies: a wind tunnel study with live plants. *Aeolian Research* 3 (2): 205–213. doi: 10.1016/j.aeolia.2011.01.003.

Christiansen, H.H. (2004). Wind polished boulders and bedrock in the Scottish highlands: evidence and implications of late Devensian wind activity. *Boreas* 33 (1): 82–94. doi: 10.1080/03009480310006998.

Christiansen, H.H. and Svensson, H. (1998). Wind-polished boulders as indicators of a late Weichselian wind regime in Denmark in relation to neighbouring areas. *Permafrost and Periglacial Processes* 9 (1): 1–21. doi: 10.1002/(SICI)1099-1530(199801/03) 9:1<1::AID-PPP271>3.0.CO;2-X.

Cooke, R.U. (1970). Stone pavements in deserts. *Annals of the Association of American Geographers* 60 (3): 560–577. doi: 10.1111/j.1467-8306.1970.tb00741.x.

Delo, D.M. (1930). Dreikanter in Wyoming and Montana. *Science* 72 (1876): 604. doi: 10.1126/science.72.1876.604.

Dietze, M., Groth, J., and Kleber, A. (2013). Alignment of stone-pavement clasts by unconcentrated overland flow: implications of numerical and physical modelling. *Earth Surface Processes and Landforms* 38 (11): 1234–1243. doi: 10.1002/esp.3365.

Dietze, M. and Kleber, A. (2012). Contribution of lateral processes to stone pavement formation in deserts inferred from clast orientation patterns. *Geomorphology* 139–140: 172–187. doi: 10.1016/ j.geomorph.2011.10.015.

Dong, Z., Xiaoping, L., and Wang, X. (2002). Aerodynamic roughness of gravel surfaces. *Geomorphology* 43 (1–2): 17–31. doi: 10.1016/S0169-555X(01)00097-6.

Eppes, M.C., McFadden, L.D., Wegmann, K.W., and Scuderi, L.A. (2010). Cracks in desert pavement rocks: Further insights into mechanical weathering by directional insolation. *Geomorphology* 123 (1–2): 97–108. doi: 10.1016/j.geomorph.2010.07.003.

Evans, J.W. (1911). Dreikanter. *The Geological Magazine, Decade V* 8 (6): 334–335. doi: 10.1017/S0016756800111483.

French, H.M. and Guglielmin, M. (1999). Observations on the ice-marginal, periglacial geomorphology of Terra Nova Bay, northern Victoria land, Antarctica. *Permafrost and Periglacial Processes* 10 (4): 331–347. doi: 10.1002/(SICI)1099-1530 (199910/12)10:4<331::AID-PPP328> 3.0.CO;2-A.

Fujioka, T. and Chappell, J. (2011). Desert landscape processes on a timescale of millions of years, probed by cosmogenic nuclides. *Aeolian Research* 3 (2): 157–164. doi: 10.1016/j.aeolia.2011.03.003.

Gillies, J.A., Nickling, W.G., and Tilson, M. (2009). Ventifacts and wind-abraded rock features in the Taylor Valley, Antarctica. *Geomorphology* 107 (3–4): 149–160. doi: 10.1016/j.geomorph.2008.12.007.

Giorgetti, G. and Baroni, C. (2007). High-resolution analysis of silica and sulphate-rich rock varnishes from Victoria Land (Antarctica). *European Journal of Mineralogy* 19 (3): 381–389. doi: 10.1127/0935-1221/ 2007/0019-1725.

Goudie, A.S. (1989). Wind erosion in deserts. *Proceedings of the Geologists' Association* 100 (1): 83–92. doi: 10.1146/annurev. earth.36.031207.124353.

Goudie, A.S. (1999). Wind erosional landforms: yardangs and pans. In: *Aeolian Environments, Sediments and Landforms* (ed. A.S. Goudie, I. Livingstone and S. Stokes), 167–180. Chichester: Wiley.

Goudie, A.S. (2008). The history and nature of wind erosion in deserts. *Annual Review of Earth and Planetary Sciences* 36: 97–119. doi: 10.1146/annurev.earth.36.031207. 124353.

Goudie, A.S. (2012). Charles Rollin Keyes and extravagant aeolation. *Aeolian Research* 4 (1): 51–53. doi: 10.1016/j.aeolia.2012.01.001.

Goudie, A.S. and Wells, G.L. (1995). The nature, distribution and formation of pans in arid zones. *Earth-Science Reviews* 38 (1): 1–69. doi: 10.1016/0012-8252(94)00066-6.

Greeley, R. and Iversen, J.D. (1985). *Wind as a Geological Process on Earth, Mars, Venus and Titan*. Cambridge: Cambridge University Press.

Gutiérrez-Elorza, M., Desir, G., and Gutiérrez-Santolalla, F. (2002). Yardangs in the semiarid central sector of the Ebro depression (NE Spain). *Geomorphology* 44 (1–2): 155–170. doi: 10.1016/S0169-555X(01)00151-9.

Gutiérrez-Elorza, M., Desir, G., Gutiérrez-Santolalla, F., and Marín, C. (2005). Origin and evolution of playas and blowouts in the semiarid zone of Tierra de Pinares (Duero

Basin, Spain). *Geomorphology* 72 (1–4): 177–192. doi: 10.1016/S0169-555X(01) 00151-9.

Haff, P.K. (2014). Biolevitation of pebbles on desert surfaces. *Granular Matter* 16 (2): 275–278. doi: 10.1007/s10035-013-0438-4.

Haff, P.K. and Werner, B.T. (1996). Dynamical processes on desert pavements and the healing of surficial disturbances. *Quaternary Research* 45 (1): 38–46. doi: 10.1006/qres.1996.0004.

Halimov, M. and Fezer, F. (1989). Eight yardang types in Central Asia. *Zeitschrift für Geomorphologie* 33 (2): 205–217.

Holmes, P.J., Bateman, M.D., Thomas, D.S.G. et al. (2008). A Holocene-late Pleistocene aeolian record from lunette dunes of the western free state panfield, South Africa. *The Holocene* 18 (8): 1193–1206. doi: 10.1177/0959683608095577.

Inbar, M. and Risso, C. (2001). Holocene yardangs in volcanic terrains in the southern Andes, Argentina. *Earth Surface Processes and Landforms* 26 (6): 657–666. doi: 10.1002/esp.207.

Isarin, R.F.B., Renssen, H., and Koster, E.A. (1997). Surface wind climate during the Younger Dryas in Europe as inferred from aeolian records and model simulations. *Palaeogeography, Palaeoclimatology, Palaeoecology* 134 (1–4): 127–148. doi: 10.1016/S0031-0182(96)00155-1.

Jackson, D.W.T., Beyers, J.H.M., Lynch, K. et al. (2011). Investigation of three-dimensional wind flow behaviour over coastal dune morphology under offshore winds using computational fluid dynamics (CFD) and ultrasonic anemometry. *Earth Surface Processes and Landforms* 36 (8): 1113–1124. doi: 10.1002/esp.2139.

Knight, J. (2003). A note on the formation of ventifacts at Castlerock, Northern Ireland coast. *Irish Journal of Earth Sciences* 21: 39–45.

Knight, J. (2005). Controls on the formation of coastal ventifacts. *Geomorphology* 64 (3–4): 243–253. doi: 10.1016/j.geomorph.2004.07.002.

Knight, J. (2008). The environmental significance of ventifacts: a critical review. *Earth-Science Reviews* 86 (1–4): 89–105. doi: 10.1016/j.earscirev.2007.08.003.

Knight, J. and Burningham, H. (2001). Formation of bedrock-cut ventifacts and late Holocene coastal zone evolution, County Donegal, Ireland. *The Journal of Geology* 109 (5): 647–660. doi: 10.1086/321959.

Knight, J. and Burningham, H. (2003). Recent ventifact development on the Central Oregon coast, Western USA. *Earth Surface Processes and Landforms* 28 (1): 87–98. doi: 10.1002/esp.432.

Kuenen, P.H. (1928). Experiments on the formation of wind-worn pebbles. *Leidsche Geologische Medellinger* 3 (1): 17–28.

Kuenen, P.H. (1960). Experimental abrasion 4. Eolian action. *The Journal of Geology* 68 (4): 427–449. doi: 10.1086/626675.

Laity, J.E. (1987). Topographic effects on ventifact development, Mojave Desert, California. *Physical Geography* 8 (2): 113–132. doi: 10.1080/02723646.1987.10642315.

Laity, J.E. (1992). Ventifact evidence for Holocene wind patterns in the east-central Mojave Desert. *Zeitschrift für Geomorphologie, N.F. Supplementband* 84: 73–88.

Laity, J.E. (1994). Landforms of aeolian erosion. In: *Geomorphology of Desert Environments* (ed. A.D. Abrahams and A.J. Parsons), 506–535. London: Chapman & Hall.

Laity, J.E. (2008). *Deserts and Desert Environments*. Chichester: Wiley-Blackwell.

Laity, J.E. and Bridges, N.T. (2009). Ventifacts on earth and Mars: analytical, field, and laboratory studies supporting sand abrasion and windward feature development. *Geomorphology* 105 (3–4): 202–217. doi: 10.1016/j.geomorph.2008.09.014.

Lancaster, N. (1984). Characteristics and occurrence of wind erosion features in the Namib Desert. *Earth Surface Processes and Landforms* 9 (5): 469–478. doi: 10.1002/esp.3290090507.

Li, Z., Feng, D., Wu, S. et al. (2008). Grain size and transport characteristics of non-uniform sand in aeolian saltation. *Geomorphology* 100 (3–4): 484–493. doi: 10.1016/j.geomorph.2008.01.016.

Ping, L. and Dong, Z. (2014). The status of research on the development and characteristics of mass-flux-density profiles above wind-eroded sediments: a literature review. *Environmental Earth Sciences* 71 (12): 5183–5194. doi: 10.1007/s12665-013-2921-y.

Mackay, J.R. and Burn, C.R. (2005). A long-term field study (1951–2003) of ventifacts formed by katabatic winds at Paulatuk, western Arctic coast, Canada. *Canadian Journal of Earth Sciences* 42 (9): 1615–1635. doi: 10.1139/e05-061.

Malin, M.C. (1986). Rates of geomorphic modification in ice-free areas southern Victoria Land, Antarctica. *Antarctic Journal of the United States* 20 (5): 18–21.

Mather, K.B. (1962). Further observations on sastrugi, snow dunes and the pattern of surface winds in Antarctica. *Polar Record* 11 (71): 158–171. doi: 10.1017/S0032247400052888.

Matmon, A., Simhai, O., Amit, R. et al. (2009). Desert pavement-coated surfaces in extreme deserts present the longest-lived landforms on Earth. *Geological Society of America Bulletin* 121 (5–6): 688–697. doi: 10.1130/B26422.1.

Matsuoka, N., Moriwaki, I., and Hirakawa, K. (1996). Field experiments on physical weathering and wind erosion in an Antarctic cold desert. *Earth Surface Processes and Landforms* 21 (8): 687–699. doi: 10.1002/(SICI)1096-9837(199608)21:8<687::AID-ESP614>3.0.CO;2-J.

McFadden, L.D., Wells, S.G., and Jercinovich, M.J. (1987). Influences of eolian and pedogenic processes on the origin and evolution of desert pavements. *Geology* 15 (6): 504–508. doi: 10.1130/0091-7613(1987)15<504:IOEAPP>2.0.CO;2.

McKenna Neuman, C., Sanderson, R.S., and Sutton, S. (2013). Vortex shedding and morphodynamic response of bed surfaces containing non-erodible roughness elements. *Geomorphology* 198: 45–56. doi: 10.1016/j.geomorph.2013.05.011.

Miotke, F.-D. (1982). Formation and rate of formation of ventifacts in Victoria Land, Antarctica. *Polar Geography and Geology* 6 (2): 98–113. doi: 10.1080/10889378209377158.

Nylen, T.H., Fountain, A.G., and Doran, P.T. (2004). Climatology of katabatic winds in the McMurdo dry valleys, southern Victoria Land, Antarctica. *Journal of Geophysical Research* 109: D03114. doi: 10.1029/2003 JD003937.

Parish, T.R. and Cassano, J.J. (2003). The role of katabatic winds on the Antarctic surface wind regime. *Monthly Weather Review* 131 (2): 317–333. doi: 10.1175/1520-0493 (2003)131<0317:TROKWO>2.0.CO;2.

Pease, P. and Gares, P. (2013). The influence of topography and approach angles on local deflections of airflow within a coastal blowout. *Earth Surface Processes and Landforms* 38 (10): 1160–1169. doi: 10.1002/esp.3407.

Pelletier, J.D., Cline, M., and DeLong, S.B. (2007). Desert pavement dynamics: numerical modeling and field-based calibration. *Earth Surface Processes and Landforms* 32: 1913–1927. doi: 10.1002/esp.1500.

Perego, A., Zerboni, A., and Cremaschi, M. (2011). Geomorphological map of the Messak Settafet and Mellet (central Sahara, SW Libya). *Journal of Maps* 2011: 464–475. doi: 10.4113/jom.2011.1207.

Perkins, J.P., Finnegan, N.J., and de Silva, S.L. (2015). Amplification of bedrock canyon incision by wind. *Nature Geoscience* 8: 305–310. doi: 10.1038/NGEO2381.

Quade, J. (2001). Desert pavements and associated rock varnish in the Mojave Desert: how old can they be? *Geology* 29 (9): 855–858. doi: 10.1130/0091-7613(2001)029 <0855:DPAARV>2.0.CO;2.

Raupach, M.R. and Finnigan, J.J. (1997). The influence of topography on meteorological variables and surface-atmosphere interactions. *Journal of Hydrology* 190: 182–213. doi: 10.1016/S0022-1694(96) 03127-7.

Rohrmann, A., Heermance, R., Kapp, P., and Cai, F. (2013). Wind as the primary driver of erosion in the Qaidam Basin, China. *Earth and Planetary Science Letters* 374: 1–10. doi: 10.1016/j.epsl.2013.03.011.

Rudberg, S. (1968). Wind erosion – preparation of maps showing the direction of eroding winds. *Biuletyn Peryglacjalny* 17: 181–193.

Ruszkiczay-Rudiger, Z., Braucher, R., Csillag, G. et al. (2011). Dating Pleistocene aeolian landforms in Hungary, Central Europe, using in situ produced cosmogenic[10]Be. *Quaternary Geochronology* 6 (6): 515–529. doi: 10.1016/j.quageo.2011.06.001.

Scarchilli, C., Frezzotti, M., Grigioni, P. et al. (2010). Extraordinary blowing snow transport events in East Antarctica. *Climate Dynamics* 34 (7–8): 1195–1206. doi: 10.1007/s00382-009-0601-0.

Schlyter, P. (1995). Ventifacts as palaeo-wind indicators in southern Scandinavia. *Permafrost and Periglacial Processes* 6 (3): 207–219. doi: 10.1002/ppp.3430060302.

Schoewe, W.H. (1932). Experiments on the formation of wind-faceted pebbles. *American Journal of Science*, 5th Series, Whole no. 224 24 (140): 111–134. doi: 10.2475/ajs.s5-24.140.111.

Sebé, K., Csillag, G., Ruszkiczay-Rüdiger, Z. et al. (2011). Wind erosion under cold climate: a Pleistocene periglacial mega-yardang system in Central Europe (western Pannonian Basin, Hungary). *Geomorphology* 134 (3–4): 470–482. doi: 10.1016/j. geomorph.2011.08.003.

Selby, M.J. (1977). Transverse erosional marks on ventifacts from Antarctica. *New Zealand Journal of Geology and Geophysics* 20 (5): 949–969. doi: 10.1080/00288306. 1977.10420690.

Sharp, R.P. (1964). Wind-driven sand in Coachella Valley, California. *Geological Society of America Bulletin* 75 (9): 785–804. doi: 10.1130/0016-7606(1980)91<724:WSICVC>2.0.CO;2.

Sharp, R.P. (1980). Wind-driven sand in Coachella Valley, California: further data. *Geological Society of America Bulletin* 91 (12): 724–730. doi: 10.1130/0016-7606(1966)77[1045:KDMDC]2.0.CO;2.

Sugden, W. (1964). Origin of faceted pebbles in some recent desert sediments of southern Iraq. *Sedimentology* 3 (1): 65–74. doi: 10.1111/j.1365-3091.1964. tb00276.x.

Suzuki, T. and Takahashi, K. (1981). An experimental study of wind abrasion. *Journal of Geology* 89 (1): 23–36; and Journal of Geology 89 (4): 509–522, doi:10.1086/628562.

Tanner, W.F. (1960). Helicoidal flow, a possible cause of meandering. *Journal of Geophysical Research* 65 (3): 993–995. doi: 10.1029/JZ065i003p00993.

Telfer, M.W., Thomas, D.S.G., Parker, A.G. et al. (2009). Optically stimulated luminescence (OSL) dating and palaeoenvironmental studies of pan (playa) sediment from Witpan, South Africa. *Palaeogeography, Palaeoclimatology, Palaeoecology* 273 (1–2): 50–60. doi: 10.1016/j.palaeo.2008.11.012.

Tewes, D.W. and Loope, D.B. (1992). Palaeo-yardangs: wind-scoured desert landforms at the Permo-Triassic unconformity. *Sedimentology* 39 (2): 251–261. doi: 10.1111/j.1365-3091.1992.tb01037.x.

Thorne, C.R., and Soar, P. (1996). R. A. Bagnold: a biography and extended bibliography. *Earth Surface Processes and Landforms* 21 (11): 987–991. doi: 10.1002/(SICI)1096-9837(199611)21:11<987::AID-ESP699>3.0.CO;2-D

Tong, D. and Huang, N. (2012). Numerical simulation of saltating particles in atmospheric boundary layer over flat bed and sand ripples. *Journal of Geophysical Research* 117: D16205. doi: 10.1029/2011JD017424.

Tremblay, L.P. (1961). Wind striations in northern Alberta and Saskatchewan, Canada. *Geological Society of America Bulletin* 72 (10): 1561–1564. doi: 10.1130/0016-7606(1961)72[1561:WSINAA]2.0.CO;2.

Várkonyi, P.L. and Laity, J.E. (2012). Formation of surface features on ventifacts: modeling the role of sand grains rebounding within cavities. *Geomorphology* 139/140: 220–229. doi: 10.1016/j.geomorph.2011.10.021.

Vincent, P. and Kattan, F. (2006). Yardangs on the Cambro-Ordovician Saq sandstones, north-West Saudi Arabia. *Zeitschrift für Geomorphologie* 50 (3): 305–320. doi: 10.1127/zfg/50/2006/305.

Wade, A. (1910). On the formation of dreikante in desert regions. *Geological Magazine Decade V* 7 (9): 394–398. doi: 10.1017/S0016756800135162.

Wainwright, J., Parsons, A.J., and Abrahams, A. (1995). A simulation study of the role of raindrop erosion in the formation of desert pavements. *Earth Surface Processes and Landforms* 20: 277–291. doi: 10.1002/esp.3290200308.

Walker, I.J. and Shugar, D.H. (2013). Secondary flow deflection in the lee of transverse dunes with implications for dune morphodynamics and migration. *Earth Surface Processes and Landforms* 38 (14): 1642–1654. doi: 10.1002/esp.3398.

Wang, Z., Wang, H., Niu, Q. et al. (2011). Abrasion of yardangs. *Physical Review E – Statistical, Nonlinear, and Soft Matter Physics* 84: 031304. doi: 10.1103/PhysRevE.84.031304.

Ward, A.W. and Greeley, R. (1984). Evolution of the yardangs at Rogers Lake, California. *Geological Society of America Bulletin* 95 (7): 829–837. doi: 10.1130/0016-7606(1984)95<829:EOTYAR>2.0.CO;2.

Webb, N.P., Okin, G.S., and Brown, S. (2014). The effect of roughness elements on wind erosion: the importance of surface shear stress distribution. *Journal of Geophysical Research-Atmospheres* 119 (10): 6066–6084. doi: 10.1002/2014JD021491.

Wells, S.G., McFadden, L.D., Poths, J., and Olinger, C.T. (1995). Cosmogenic ^{3}He surface-exposure dating of stone pavements: implications for landscape evolution in deserts. *Geology* 23 (7): 613–616. doi: 10.1130/0091-7613(1995)023<0613:CHSEDO>2.3.CO;2.

Wendler, G., Stearns, C., Weildner, G. et al. (1997). On the extraordinary katabatic winds of Adélie Land. *Journal of Geophysical Research* 102 (D4): 4463–4474. doi: 10.1029/96JD03438.

Whitney, M.I. (1978). The role of vorticity in developing lineation by wind erosion. *Geological Society of America Bulletin* 89 (1): 1–18. doi: 10.1130/0016-7606(1978)89<1:TROVID>2.0.CO;2.

Whitney, J.W., Breit, G.N., Buckingham, S.E. et al. (2015). Aeolian responses to climate variability during the past century on Mesquite Lake Playa, Mojave Desert. *Geomorphology* 230: 13–25. doi: 10.1016/j.geomorph.2014.10.024.

Whittow, J.B. (2000). *The Penguin Dictionary of Physical Geography*. London: Penguin.

Williams, S.H. and Zimbelman, J.R. (1994). Desert pavement evolution: an example of the role of sheetflood. *Journal of Geology* 102 (2): 243–248. doi: 10.1086/629666.

Wilson, P., Christiansen, H.H., and Ross, S.M. (2002). The geomorphological context and significance of wind-abraded gravels, boulders and outcrops from the coast of Scotland. *Scottish Geographical Journal* 118 (1): 41–57. doi: 10.1080/00369220218737135.

Yang, P., Dong, Z., Guangqiang, Q. et al. (2007). Height profile of the mean velocity of an aeolian saltating cloud: wind tunnel measurements by particle image velocimetry. *Geomorphology* 89 (3–4): 320–334. doi: 10.1016/j.geomorph.2006.12.012.

Zerboni, A. (2008). Holocene rock varnish on the Messak plateau (Libyan Sahara): chronology of weathering processes. *Geomorphology* 102 (3–4): 640–651. doi: 10.1016/j.geomorph.2008.06.010.

Zerboni, A., Trombino, L., and Cremaschi, M. (2011). Micromorphological approach to polycyclic pedogenesis on the Messak Settafet plateau (Central Sahara): formative processes and palaeoenvironmental significance. *Geomorphology* 125 (2): 319–335. doi: 10.1016/j.geomorph.2010.10.015.

4

Dust: Sources, Entrainment, Transport

Joanna Bullard and Matthew Baddock

Loughborough University, Loughborough, UK

4.1 Introduction

Some dust storms are dramatic events, over land or sea on Earth, or on the surfaces of planets and moons like Mars and Titan (see Chapter 11). On Earth, major events have substantial economic, social, and environmental effects. The 'Red Dawn' dust storm in Australia on 22–23 September 2009 covered an area of approximately 3000 km from north to south and 400 km from west to east (Figure 4.1). It moved across rural areas and major cities alike, and cost AUS$299 million in New South Wales alone (Leys et al. 2011; Tozer and Leys 2013). Not all dust events are as big: much smaller features such as dust devils may only affect an area of a few square metres and last for less than an hour (Balme and Greeley 2006).

The larger dusty events may grab the headlines, but their immediate geomorphological impact can be difficult to discern, because, although they move many thousands of tonnes of dust, it is taken from very large areas and deposited over even larger ones. It is, however, possible to identify landforms and landscapes that are formed by the transport of dust. Some are related to the repeated entrainment of dust, altering soil characteristics through the selective removal of sedimentary fractions and organic matter; others to its deposition. The largest dust 'events' often originate from the same place repeatedly, over very many years, denuding vast areas of topsoil, forming loess (see Chapter 5) or contributing to soils and peat bogs on land, or to lake and ocean bottom sediments.

Collectively, the amount of material in dust transport on Earth has been estimated, using models, to be approximately 2000 Tg yr^{-1}, of which 1500 Tg is deposited on land and 500 Tg in the oceans (Shao et al. 2011). Global dust emissions are approximately one-tenth of the amount of material delivered to the oceans by rivers, but their relative importance is regionally very variable (Table 4.1). Dust emissions are six times that of fluvial sediment flux in northern Africa and twice the fluvial load in the Middle East. In Australia, dust emissions and fluvial sediment delivery are of the same order of magnitude. In contrast, in some regions, such as Europe, Indonesia, and Malaysia dust emissions are negligible and almost all sediment is transported by rivers. When in transport, dust can affect the radiative properties of the atmosphere causing warming (via absorption) or cooling (via scattering) depending on the particle characteristics and position within the atmospheric column, and consequently has an impact on atmospheric dynamics (Alizadeh-Choobari et al. 2014).

Although dust has featured in academic papers since at least 1646, and been discussed by scientists such as Darwin (1846) and von

Aeolian Geomorphology: A New Introduction, First Edition. Edited by Ian Livingstone and Andrew Warren.
© 2019 John Wiley & Sons Ltd. Published 2019 by John Wiley & Sons Ltd.

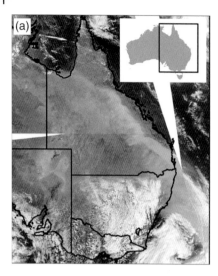 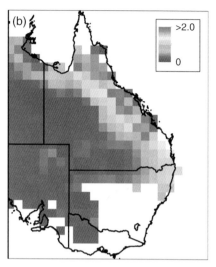

Figure 4.1 (a) Satellite image of 'Red Dawn' dust event; (b) MODIS Aerosol Optical Depth (AOD) of 'Red Dawn' dust event. See insert for colour representation of this figure.

Table 4.1 Comparison of the outputs of dust flux models (from Tanaka and Chiba 2006) and fluvial sediment flux (observations compiled by Milliman and Farnsworth 2011) for different regions.

Region	Dust flux Tg a^{-1}	Fluvial sediment flux Tg a^{-1}
North Africa	1087	168
South Africa	63	187
Middle East	221	106
North Asia	negligible	150
Central Asia	140	2700
East Asia	214	2690
North America	2	1900
South America	44	2250
Australia	106	160
Europe	negligible	850
Indonesia and Malaysia	negligible	6800
Other	111	1039
Total	1988	19000

Richthofen (1882), research into the sources of dust, its entrainment and transport, and its role in the global climate system has only gained momentum in the past 25 years (Stout et al. 2009). This reflects both a heightened

recognition of the role of dust in atmospheric, terrestrial, marine, and cryospheric systems, but also improvements in technology that make it easier to detect and measure dust using satellite remote sensing and in the field, and improved computing capabilities that enable more advanced models of global dust processes in the Earth system.

This chapter focuses on dust sediments, sources, entrainment, and transport. Chapter 5 looks at dust in the creation of landforms, and Chapter 12 at the management of accelerated wind erosion to control the emission of dust.

4.2 What Is Dust?

Dust can be defined in different ways. For many purposes, it is most useful to define it in terms of particle size: it typically comprises particles and aggregates < 100 μm in diameter, thus including clays, silts, and very fine sands. However, the feature that typically distinguishes dust from other fine aeolian sediments is its mode of transport: dust is transported by suspension (rather than by creep and saltation) over distances up to 20000 km from its source (Pye 1987). In a synthesis of measurements of dust deposition, Lawrence and Neff (2009) differentiated local, regional, and global dust deposits in terms of their distance from source, deposition rates, particle sizes, and mineral contents (Table 4.2). Long-distance transport causes dispersion of the sediments over extensive areas and means that dust deposits typically form a thin mantle over a pre-existing landscape as opposed to the bedforms that are characteristic of sands. Repeated deposits of dust can lead to the build-up of very deep layers of sediment that are highly erodible by wind and water to form distinctive landscapes such as the Chinese loess plateau (see Chapter 5).

The characteristics of dust vary according to the sediments themselves (e.g. geology) and the geomorphological processes operating on the land surfaces that are the source of the dust (e.g. glacial grinding typically produces more angular particles than fluvial comminution). Conditions in the source area fundamentally affect the composition of dust. In addition to quartz mineral particles, dust can include feldspars, calcite, and halite, and dust storms can transport organic material such as pollen, diatoms, and bacteria. The 'Red Dawn' dust storm (described earlier) contained pollen from over eight different plant groups including grasses, shrubs, and trees, and bacteria from seven different families including cyanobacteria (de Deckker et al. 2014). The characteristics of a dust plume can change during transport. Table 4.2 indicates that with increasing distance from source,

dust contains a higher proportion of clay and silt-sized particles, a lower proportion of quartz and higher proportions of feldspars and other minerals.

4.3 Production and Entrainment

The ease with which it is possible for a particle to be entrained by the wind depends on the interaction between meteorological variables and the cohesive, gravitational, and aerodynamic character of the dust itself. For particles < 10 μm in diameter, cohesive forces tend to dominate, and for particles > 300 μm in diameter, gravitational forces dominate; this means that most wind-transported sediments are in the 10–300 μm range. The mean grain size of particles of quartz in metamorphic and igneous rocks is 700 μm, which means, in turn, that particles need to be broken down before substantial aeolian sediment transport can occur (see also Chapter 1). Fine particles are produced by physical and chemical weathering as well as mechanical processes, such as grinding and abrasion during transport. Wright et al. (1998) conducted a series of experiments to try to determine the relative effectiveness of these different processes in the production of fine sediment (Table 4.3). The origin of loess particles is discussed in Chapter 5.

Different processes dominate in different regions. For example, under contemporary conditions, glacial grinding operates primarily at high latitudes and altitudes. Grinding is

Table 4.2 Characteristics of local, regional and global dust deposits (Lawrence and Neff 2009).

Deposition Class	Distance from source km	Deposition rate g m^{-2} yr^{-1} (mean)	Particle size (%			Mineral content (%)			
			<2 μm	2–50 μm	>50 μm	Quartz	Feldspars	Carbonates	Clay minerals
Local	0–10	50–500 (200)	20	50	30	50≥60	5–30	0–25	5–30
Regional	10–1000	1–50 (20)	25	60	15	30–50	5–35	0–25	5–40
Global	>1000	0–1 (0.4)	30	70	0	15–30	20–40	0–25	20–60

Table 4.3 A comparison of the relative effectiveness of silt-producing mechanisms (calculation of the theoretical maximum amount of silt produced from 1 kg of the original sample) (Wright et al. 1998).

Run type	Amount of fine (<63 μm) material produced (g kg^{-1})	Run duration
Glacial grinding	47.4	24 h
Aeolian abrasion	287	96 h
Fluvial tumbling: spheres	900	32 h
Salt weathering: Na$_2$SO$_4$	41.6	40 cycles
Frost weathering	0.44	360 cycles

capable of producing large amounts of fine silt-sized material that can be washed out from underneath the ice, deposited on glacial floodplains and then taken by strong ice-driven winds, which carry large, usually seasonal, dust storms (Crusius et al. 2011; Prospero et al. 2012) (Figure 4.2). In hot, dry deserts, at lower latitudes, several different processes – such as aeolian abrasion and weathering – take place and combine to produce large quantities of dust-sized material (Wright 2001). The wind itself abrades particles, physically breaking them down during transport through collision with other airborne particles or the ground surface. These processes can also remove the coatings of fine clays that adhere to sand grain surfaces (Bullard et al. 2004). Crouvi et al. (2010) suggested that extensive areas of loess downwind of dunefields in low latitudes – such as in North Africa and the Middle East – are the result of active aeolian abrasion of grains from sand seas (see Box 1.1 for the earlier history of this hypothesis).

The importance of saltation-impact in ejecting particles into suspension is described in Chapter 2. While this is arguably the most important mechanism of dust entrainment, direct entrainment of dust in the absence of saltation has also been observed (Loosmore and Hunt 2000). Sweeney and Mason (2013) suggested that the lack of coarse sand means that saltation impact is rare in loess landscapes. However, when unvegetated, loess can be deflated by direct entrainment of particles that occurs at a lower threshold friction velocity than for sandy soils.

The particle size of emitted dust depends on both the characteristics of the source area and the mechanism of dust entrainment and transport. Assuming they are present and exposed to the wind, particles in the range 10–100 μm are most likely to be ejected into suspension. Finer sediments are often harder to entrain because they are bound together by moisture, electrostatic forces or may be incorporated into physical or biological soil crusts. Ishizuka et al. (2008) found that as little as <1 mm rainfall caused a physical crust to form that reduced the emission of dust of 0.5–3 μm diameter because the fine particles were cemented within the crust. Higher wind speeds may enable more and finer dust to be entrained if emissions are driven by saltating sand (Chapter 2; Alfaro and Gomes 2001; Sow et al. 2009), which can also break down surface crusts (McKenna Neuman et al. 1996). Where soil particles are <10 μm in diameter, they may form aggregates; as the number of particles in an aggregate increases, its overall shape will become irregular and the aggregate less dense as voids are incorporated (Goossens 2005). These aggregates can be stable during transport at low wind velocities, but may be broken down to their constituent fine particles at higher wind speeds.

One of the factors contributing to the status of the Bodélé Depression as 'the dustiest place on Earth' is the character of its sediment. The Depression is located within the area of Palaeolake Mega-Chad and comprises a mosaic of different geomorphologies, each of which has a different erodibility due to surface type and the availability of loose material (Chappell et al. 2008). Diatomite formed on the bed of the lake aggregates into irregularly-shaped sand-sized (c.1000 μm diameter) pellets which can be blown to form barchan dunes. During transport these pellets can undergo abrasion which causes them to break down (Warren et al. 2007). The long-term

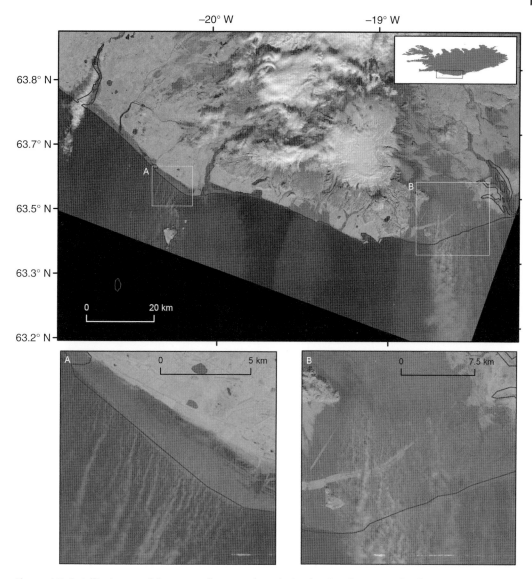

Figure 4.2 Satellite image of dust storm from southern Iceland, 5 October 2004. The dust sources are all proglacial floodplains; the plumes extend more than 500 km south over the Atlantic Ocean. See insert for colour representation of this figure.

rate of deflation in the Bodélé Depression is estimated at $2\,mm\,yr^{-1}$, but may be as high as $10\,mm\,yr^{-1}$ in small areas (Bristow et al. 2009).

4.4 Sources

Globally, and in contemporary conditions, most dust is derived from sub-tropical arid and semi-arid regions including North Africa,

the Middle East, central Asia, and central Australia. Middleton et al. (1986) used records from surface observations at World Meteorological Organisation (WMO) stations to identify key dust source regions. These records provide information about dust events in the form of dust codes (Table 4.4) and have the advantage that many stations have long records going back several decades, although there is some subjectivity in the way

Box 4.1 Satellite Remote Sensing of Dust

Major insights into the emission and transport of dust have been provided by views from satellites. The global coverage they offer, and the frequent and repeated monitoring from a range of sensors on different orbiting platforms, now allow regional and larger-scale studies. Space-borne sensors that detect discrete portions of the electromagnetic spectrum reveal the various properties of dust, and in turn enable studies of the way aerosols in the atmosphere interact with radiation (e.g. scattering and absorption). Because the role of aerosols is still one of the greatest uncertainties in climate science, the capability to detect and quantify dust is now a key feature in the design of satellite missions (e.g. the NASA Earth Observing System) (Colarco et al. 2010).

While impressive dust outbreaks can be visualised in the true colour scenes from satellites (Figures 4.1 and 4.2), one of the most useful contributions from satellites is the ability to quantify the loading of dust in the atmospheric. Over several years, the frequency of observation of dust loading above certain thresholds can be used to produce maps of dust intensity, and thereby identify prominent dust source locations and transport pathways. From these data, the relative magnitudes of different source regions can also be gauged. The Aerosol Index derived from ultraviolet channels of the Total Ozone Monitoring Spectrometer (TOMS) enabled the first global studies of dust sources using remote sensing, and validated the significance at the broadest scale of arid, internally draining basins as sources of dust (Prospero et al. 2002; Washington et al. 2003). Since the early 2000s, the Moderate Resolution Imaging Spectroradiometer (MODIS) has provided Aerosol Optical Depth (AOD) assessments (Figure 4.1) for which a recent 'Deep Blue' version has greatly improved the retrieval of information on dust over bright land surfaces. In providing more reliable data from desert regions, Deep Blue has proven to be an extremely useful development for assessment of dust sources (Baddock et al. 2016). For instance, Deep Blue was used by Ginoux et al. (2012) to achieve a global survey of dust source

locations at especially high resolution (0.1° of longitude). When combined with land use data sets, this study also provided an assessment of the anthropogenic contribution to the emission of dust. One of the highlight findings here is that in the pre-eminent source of dust in northern Africa, only 8% of emissions are anthropogenic and these are mostly from the Sahel.

Because the MODIS orbit is sun-synchronous, it provides a daily picture (effectively twice daily), with MODIS data coming from two satellites, Terra, and Aqua). In contrast, the geostationary nature of the Spinning Enhanced Visible and Infrared Imager (SEVIRI), on board the Meteosat Second Generation (MSG) platform, provides data at 15-minute resolution, and thus allows dust events to be tracked forwards and backwards in the infrared. The MSG's coverage over both northern Africa (Ashpole and Washington 2012; Schepanski et al. 2012) and southern Africa (Vickery et al. 2013) allows detailed insights into the behaviour of sources, including the precise timing of dust emission; this has helped to detect bias in other remote sensing datasets. One of the biggest values of dust-loading information from remote sensing sources is that it can be used to evaluate the predictions of dust models, and improve modelling efforts by 'nudging' the output of models to match the satellite record (Zhao et al. 2013).

Nonetheless, there are important limitations to data from remote sensing that need to be borne in mind when using them to understand dust activity. These limitations are sensor-specific, for instance, TOMS AI performance is less effective for dust restricted to relatively low altitude (e.g. haboob-type events) and the overpass times of MODIS may result in the detection of dust in transport rather than active emission from source (Schepanski et al. 2007; Bryant 2013). One of the benefits of MODIS is its spatial resolution, because it allows the pinpointing of the sources of dust but this is only possible from cloud-free images because emission occurring under cloud can not be observed (Lee et al. 2012).

Table 4.4 Dust codes recorded at staffed and automatic weather stations.

		Present weather reported from a manned weather station		
	Code	**Description**		
No duststorm or sandstorm at the station except for 09	05	Haze		
	06	Widespread dust in suspension, not raised by wind at or near the station at the time of observation		
	07	Dust or sand raised by wind at or near the station at the time of observation, but no well-developed dust whirl(s) or sand whirl(s), and no dust storm or sand storm seen; or in the case of ships, blowing spray at the station		
	08	Well-developed dust whirl(s) or sand whirl(s) seen at or near the station during the preceding hour or at the time of observation, but no dust storm or sandstorm		
	09	Dust storm or sandstorm within sight at the time of observation, or at the station during the preceding hour.		
Duststorm or sandstorm at station	30	Slight or moderate dust storm or sandstorm	Has decreased during the preceding hour	
	31		No appreciable change during the preceding hour	
	32		Has begun or has increased during the preceding hour	
	33	Severe dust storm or sandstorm	Has decreased during the preceding hour	
	34		No appreciable change during the preceding hour	
	35		Has begun or has increased during the preceding hour	

	Present weather reported from an automatic weather station
04	Haze or smoke, or dust in suspension in the air, visibility equal to, or greater than, 1 km
05	Haze or smoke, or dust in suspension in the air, visibility less than 1 km

that observers judge conditions and there can be changes in recording practices, both of which present challenges in using the record (O'Loingsigh et al. 2010, 2014). Meteorological codes can be used to generate indices that quantify the amount of dust storm activity at a location. The Dust Storm Index (DSI) used in Australia differentiates Severe Dust Storm Days (SDS; codes 33–35), Moderate Dust Storm Days (MDS; codes 30–32 and 98) and Local Dust Event Days (LDE; codes 07–09) and weights them as follows:

$$DSI = \sum_{i=1}^{n} \left[\left(5xSDS \right) + MDS + \left(0.05xLDE \right) \right]_i$$

where DSI is the Dust Storm Index at n stations, i is the ith value of n stations for i = 1–n. DSI can be used to map dust activity across Australia in space and through time (Figure 4.3). Other researchers have used surface visibility records from meteorological stations to examine trends in desert dust emissions (Mahowald et al. 2007).

Meteorological data provide indications of dustiness at a specific point, but the density of WMO stations is very variable. In Africa, which is the source of over 50% of the Earth's dust emissions, the density of meteorological stations is 1 in 26 000 km^2, eight times lower than the minimum density recommended by WMO. In addition, good spatial data and long-term trends require continuous, reliable data which are often not attainable for very sparsely populated regions. For better, reliable, global data, the most valuable source of information over the past 20 years has been satellite remote sensing (Box 4.1). Sensors such as TOMS have made it possible to identify areas where there are high concentrations of dust in the atmosphere and to use these to map potential dust sources.

(a)

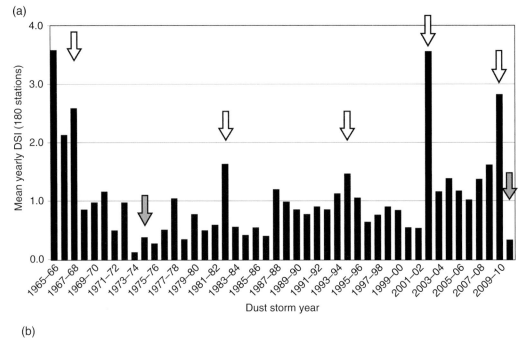

(b)

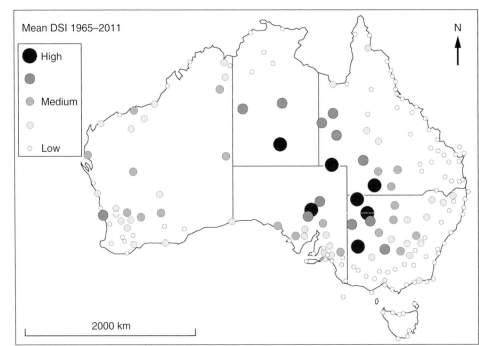

Figure 4.3 (a) Temporal trend in the dust storm index (*DSI*) for Australia from 1965–2011; (b) Spatial distribution of mean *DSI* values across Australia from 1965 to 2011. White arrows indicate periods of drought, grey arrows periods of high rainfall and flooding. *Source:* Redrawn from O'Loinsigh et al. (2014).

The key characteristics of dust sources at a global scale can be identified by combining ground-based meteorological data and satellite records. Most dust emission occurs from regions where the surface is erodible because of a lack of vegetation or other roughness elements, such as gravel, and where there are sufficiently strong winds to entrain particles. Prospero et al. (2002) suggested that all the major contemporary dust sources were associated with low points in the landscape, and located in arid regions receiving less than 250 mm rainfall per year. The most persistent dust source areas have plentiful supplies of available fine sediments, e.g. the Bodélé (see Section 4.3); others are supply-limited, but receive regular inputs of fine material during flood events, for example, the Lake Eyre Basin in Australia (Bullard et al. 2008). Earth's most prolific sources of dust may be the warm, arid regions, but the five million km^2 of cold arid regions (warmest month < 10 °C and < 250 mm yr^{-1} rainfall) can also be the source of substantial, seasonal dust storms (Figure 4.2; Bullard 2013). The largest global dust source region is the Sahara in North Africa, followed by Arabia and south-west and central Asia. Australia is the largest current dust source in the southern hemisphere, followed by southern Africa and South America (Figure 4.4).

It may be straightforward to identify the global-scale characteristics of dust sources, at the regional and local scale, but the magnitude, the frequency, and timing of dust emissions depend on smaller-scale geomorphology and landscape processes (Bullard et al. 2011). Dry lake beds are identified as key dust sources in global dust models, but they only account for approximately 1% of global desert surfaces. However, in the Mojave Desert and southern Great Basin of the south-west USA, playas and alluvial units produce almost the same quantity of dust per unit area. The greater surface area of alluvial deposits means that the amount of dust emitted from them is greater than for the playas (Reheis and Kihl 1995). Field measurements of surface erodibility in southern California show that, for the same wind speed, dust emissions are very variable, not only among different surface types, but also during different experimental runs on the same surface type (Sweeney et al. 2011, Figure 4.5). Attempts to

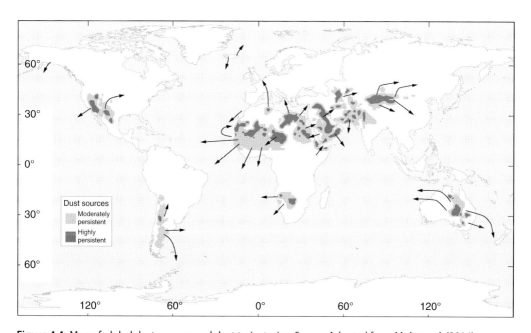

Figure 4.4 Map of global dust sources and dust trajectories. *Source:* Adapted from Muhs et al. (2014).

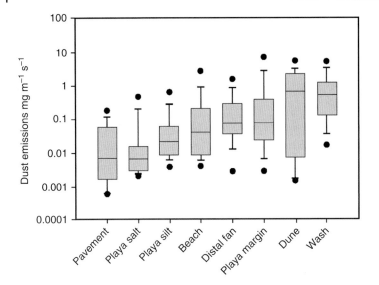

Figure 4.5 Variation in dust emissions from different land surfaces in southern California. The dots at the bottom and top of the diagrams represent the 5th and 95th percentiles, respectively, of measured values. The line in the boxes represents the median value. *Source:* Redrawn from Sweeney et al. (2011).

look at the relative importance of different geomorphic units as sources of dust are hampered not only by the different local conditions (such as microtopography, sediment size, and sorting), but also by the many ways in which landscapes have been classified by researchers, and in which dust flux is measured (Box 4.2). Bullard et al. (2011) proposed a classification of land surfaces for dust emission and applied it to parts of Australia, North Africa, and the USA (Table 4.5). Their results indicated that, whereas many ephemeral and dry lakes are sources of dust, as in the Chihuahuan Desert and in the Sahara (west of 17°E), in Australia, it is sand sheets and sand dunes that are the most frequent dust sources (during the period of study). Bullard et al. (2011) found very few dust storms to come from dunes in the western Sahara, but, in contrast, Crouvi et al. (2012) suggested that these were important dust sources in North Africa. The importance of dunes for dust production or as dust sources is likely to depend on dune type, such that active, migrating dunes are expected to be infrequent dust sources, but inactive dunes or those with stable cores are more likely to

produce dust. Other work has suggested that the margins of dunefields in the Sahara and China could be important dust sources, for example, where the interdunes receive inputs of fine material during periodic flooding (Ashpole and Washington 2013; Wanquan et al. 2013). This argument highlights the importance of the supply of sediment and its availability in the landscape. In areas without a large sediment store ('supply-limited'), the timing of dust storms can be linked to fresh inputs of sediment, for example, by flooding. There may be a time-lag between the supply of sediment and its availability caused by the need for moisture levels to drop, vegetation to die back or snow cover to melt before dust is emitted. In Namibia, dust emissions in the lower reaches of the Kuiseb River occur when strong, topographically funnelled winds follow periodic flooding which supplies fresh sediment (Vickery and Eckardt 2013). In the Copper River area of Alaska, where sediments are covered with snow until early summer, it is the coincidence of the maximum exposure of fine sediment at low river flow with strong down-valley winds in October and November that regularly triggers

Box 4.2 Field Observations and Measurements of Dust

There are many ways to sample wind-eroded sediment, including the fraction of dust-sized material that is moved primarily in suspension. Field measurements can quantify rates of dust emission from the surface and fluxes of horizontal transport, and such measurements have proven essential in improving predictive field-scale models of dust emission (Zobeck et al. 2003). Detailed field studies have also shed light on the processes of emission, such as Ono et al. (2011), who carried out dust process measurements on dry Mono Lake, California, and observed extreme variability in fine dust emission, relating this directly to changing surface conditions on the playa lake. As well as allowing estimations of flux, many techniques for the measurement of wind-eroded sediment result in the physical collection of entrained dust. This sample can be subsequently analysed for sedimentological and geochemical properties to further characterise the dust in terms of its particle size distribution and nutrient levels (e.g. Junran et al. 2007; Floyd and Gill 2011).

A key consideration in dust measurement is the accuracy of different sampling methods. Sampling efficiency, a measure of how well a given sampler collects sediment compared to the total amount in transport, varies for different samplers depending chiefly on wind-speed during measurement and by particle size. Yet the degree of efficiency must be known if accurate estimates of dust emission and transport are to be produced (Zobeck et al. 2003). Passive devices are one sampler type that rely on ambient wind conditions for the collection of sediment. Big Spring Number Eight (BSNE) (Box Figure 4.1a) and Modified Wilson and Cook (MWAC) (Box Figure 4.1b) are two commonly used passive samplers, popular due to their cheap cost and robust nature for remote deployment. At low wind speeds $(1-5\,m\,s^{-1})$ with sediment made up of 95% silt $(2-63\,\mu m)$, sampling efficiency was found to be 40% for BSNE and 80% for the MWAC (Goossens and Offer 2000). It is noteworthy, however, that the wind speeds during dust events when active dust emission is occurring are typically much greater than those assessed in this efficiency experiment.

Box Figure 4.1 Common dust measurement instruments are vertical arrays of (a) Big Spring Number Eight, (b) Modified Wilson and Cook traps, and (c) a DustTrak aerosol monitor deployed at 1.5 m above the surface. *Source:* Photograph of MWAC by Ted Zobeck/Scott Van Pelt.

The efficiency of *passive* samplers is limited by the fact that the path of fine particles in suspension is strongly controlled by the airstream, making the extent to which particles enter the device quite variable. Aeolian studies which focus on suspended particles commonly employ an *active* type of sampling apparatus, one which features a form of suction that draws a known air volume into the instrument. These instruments can estimate dust concentration by particles' scattering of light (e.g. a DustTrak aerosol monitor, (Box Figure 4.1c), or from the mass collected on filter papers (e.g. high volume samplers). Many active samplers are size-selective, for instance, sampling $<PM_{10}$, and most active samplers rarely sample the coarse dust fraction $>20\,\mu m$ (Goossens and Buck 2012) although filters can collect total suspended particles (TSP). When arranged in an array at different heights, the dust concentrations from active samplers can be used to calculate the vertical dust flux (mass per unit area per unit time) from an eroding land surface, as demonstrated by Zobeck and Van Pelt (2006), who deployed DustTraks at three heights above a dust-bearing agricultural field in West Texas.

Portable wind tunnels, designed to be deployed in field settings, have been commonly used in dust studies as they have the ability to apply a controlled erosive potential to a land surface, while quantifying dust emission by means of passive or active samplers (Van Pelt et al. 2013). By operating a tunnel on natural surfaces, the critical influence on the dust yield represented by *in situ* soil conditions such as crusting and surface roughness can be examined (Houser and Nickling 2001). The Portable In-Situ Wind Erosion Lab (PI-SWERL) is a battery-powered device which generates a shear stress on the surface within an enclosed cylinder by way of a blade rotating close to the ground, with a connected DustTrak to record PM_{10} concentration. A key benefit of the PI-SWERL is its portability which facilitates very many emission tests. This not only allows the variability in emission to be examined between different types of land surface, but also, within a single class of land surface, with the latter important for obtaining a true picture of spatial and temporal heterogeneity in emission (Sweeney et al. 2011; King et al. 2011) (see Figure 4.5).

While techniques and instruments that enable the characterisation of dust emission at field scales are well established, significant challenges remain for the successful linking of field observations, or their 'upscaling', to larger scales of enquiry, such as those considered by dust cycle models (Bullard 2010; Bryant 2013).

Table 4.5 Frequency of dust storms associated with different geomorphologies in three desert regions (Bullard et al. 2011).

Source	Frequency of dust storms (%)		
	Chihuahuan Desert, USA	Lake Eyre Basin, Australia	Sahara (west of 17°)
Ephemeral lake	29.5	11	3
Dry lake	18.4	0	63
Armoured, unincised high relief alluvial	0	0.6	0
Unarmoured, unincised high relief alluvial	20.7	0	0
Unarmoured, incised low relief alluvial	0	0	5
Unarmoured, unincised low relief alluvial	12	16.5	25
Stony surfaces	0.5	3.8	0
Sand sheet	1.8	5.7	2
Aeolian sand dunes	13.8	57.6	2
Low emission surfaces	3.2	4.9	0

spectacular Autumnal dust storms (Crusius et al. 2011).

Very many studies have concluded that overall the top three most important dust emitting surfaces are those associated with dry or ephemeral lakes, alluvial deposits (such as floodplains), and aeolian deposits. Surfaces least likely to emit dust are stony surfaces such as in desert pavements (e.g. Gillette et al. 1980; Bullard et al. 2011; Sweeney et al. 2011; Lee et al. 2012; Parajuli et al. 2014).

In addition to geomorphological and sedimentological characteristics, other factors can affect the emission of dust. An important one is land-cover (as by plants), which in turn can be influenced critically by agricultural practice (see Chapter 12). Cultivated croplands are particularly important dust sources in the USA (Lee et al. 2012) and in Australia pasture lands are significant dust sources. Ginoux et al. (2012) claimed that, globally, natural dust sources contributed 75% of emissions, and anthropogenic sources 25%. The balance is regionally very variable: in northern Africa, which is Earth's predominant source of dust, only 8% of emissions come from anthropogenically modified surfaces, whereas in Australia, 75% of dust is associated with anthropogenic sources.

Understanding the spatial and temporal variations in dust emissions at the global scale is primarily achieved using global dust models (Box 4.3). These suggest that, globally, annual dust emission has increased by 25% since the late nineteenth century, driven by both climatic change (accounting for 56% of the increase) and agricultural expansion (accounting for 40% of the increase) (Stanelle et al. 2014). Locally human-enhanced emission (the 'Anthropocene dust signal') has been recorded in lakes and soils (Neff et al. 2008; Marx et al. 2014). In the future, other changes to the quantity and distribution of dust sources may follow the retreat of ice sheets and glaciers (Bullard 2013). As ice retreats, it exposes new surfaces to wind erosion and the emission of dust, before the development of protective lag deposits or the colonisation of the surface by vegetation.

4.5 Dust Events and Weather Systems

There are several different types of dust event (see Table 4.4). The international meteorological definition of a dust storm is when visibility on the ground is less than 1000 m, and when the dust is being entrained from the surface within sight of the observer (Figure 4.6); this last criterion is frequently ignored, particularly by non-specialists, such that any event reducing visibility to this degree is often referred to as a dust storm. 'Dust haze' is produced by dust particles that have been lifted from the ground prior to the observation, whereas 'blowing dust' reduces visibility at eye level but not to less than 1000 m.

Dust events are associated with two main types of weather system. Large-scale dust events that last for several days are typically driven by pressure gradients associated with synoptic-weather systems. The most important of these are fronts in low-pressure systems with steep pressure gradients that are accompanied by strong winds. These systems are particularly important in the USA, North Africa, the Middle East, Central Asia, and Australia. The passage of a cold front displaces warm air which causes turbulence and the entrainment of dust. The September 2009 Australian dust storm described in Section 4.1 was associated with this type of weather system and is estimated to have transported 2.54 million tonnes of sediment (Leys et al. 2011); a similar Australian dust storm in October 2002 transported 3.35–4.85 million tonnes of material (Yiuchung et al. 2005).

Shorter-lived dust events are generally caused by convective atmospheric processes such as cold air downdrafts produced in thunderstorms (Figure 4.7). Within thunder clouds some air moves upwards driven by convection, and some air subsides due to the frictional drag of falling rain. The subsidence is enhanced as the rain below the cloud cools, becoming denser, creating downbursts of air. These downdrafts spread

Box 4.3 Numerical Dust Models

Modelling is the best tool for understanding the complex effect of mineral aerosol on climate and its interactions with other environmental systems. By representing all the stages of the dust cycle (emission, transport, and deposition), models give the best estimates of the spatial and temporal distribution of dust. This is especially so for the remote areas where most emission occurs, but where ground-observational data of dust are lacking.

Numerical models have been developed at different scales, depending on their intended application. Global dust models offer a perspective on long-term changes in atmospheric dust loading at the scale of the overall radiative budget, and thus of Earth's climate. However, the coarse degree-scale grid resolution of these models renders them too crude to represent the dust processes at the finer scales of sites of emission (Tegen and Schulz 2014).

Regional dust models with cell sizes < 50 km have therefore been developed, as for the Sahara (Todd et al. 2008; Haustein et al. 2012), and parts of Asia and Australia (Shao et al. 2007). These models offer detailed investigations of individual dust events and even in the provision of operational dust forecasts.

The accuracy at which the rate of emission is represented, and realistically quantified, is critical. The parameterisation of the dust emission process that is required in these models poses a major challenge, because emission is highly variable in space and time, and can occur at scales that are below the resolution of regional or global scale models. While most global dust models tend to agree on a total dust load around 1000–2000 Tg yr^{-1}, some of the key uncertainties and differences between different model estimates relate to the varying schemes used to predict emission (Shao et al. 2011).

In most models, dust flux from the surface is computed as a function of the wind speed at 10 m above ground level. This, in turn allows estimates of the threshold speed for the entrainment of dust (Ginoux et al. 2001), and the vertical dust. This allows the computation of the horizontal saltation flux which is needed to estimate the intensity of the sandblasting process. As the particle size distribution of surface sediments controls both entrainment (through threshold shear velocity) as well as the computed size distribution of the particles emitted, models must represent the particle size on the surface. However, these critical data must usually come from very coarse-scale maps of soil texture. The importance of such estimates is demonstrated by the improvements to earlier modelling outputs that were produced by explicit inclusion of 'preferential' source areas (Ginoux et al. 2001; Zender et al. 2003). When dust is aloft, the characteristics of the suspended load, and changes in the particle size distribution with distance and time from source, are modelled by numerical schemes that describe the transport process as driven by particle fallout resulting from both dry and wet deposition.

horizontally outwards from thunderstorms resulting in strong winds up to 50 m s^{-1} that cause dust storms along the storm-line (Goudie and Middleton 2006; see also Chapter 1). Dust storms formed in this way are known as haboobs and may account for up to 30% of dust production in the southern Arabian Peninsula (Miller et al. 2008) and up to 50% in North Africa. They are also common in the south-west USA; air quality measurements taken during a haboob near Phoenix, Arizona, in July 2011 clearly show the intense, short-lived nature of the event (Figure 4.8). During the peak of the dust storm, peak wind gusts reached 29 m s^{-1} and daily average PM$_{10}$ (particulate matter <10 μm diameter) near Phoenix reached 225 μg m^{-3} (the US Environmental Protection Agency (EPA) daily ambient air quality standard is 150 μg m^{-3} and the typical value for the Phoenix area is 30 μg m^{-3}) with hourly values exceeding 1900 μg m^{-3} (Raman et al. 2014).

Figure 4.6 Dust storm approaching the town of Bedourie in western Queensland, Australia, on 4 December 2014. *Source:* Photograph reproduced courtesy of Maggie Den Ronden. See insert for colour representation of this figure.

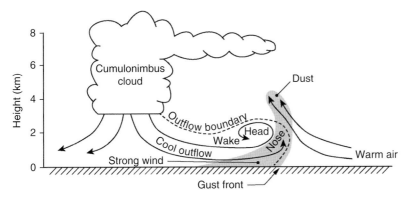

Figure 4.7 Cross-section schematic of a haboob. The leading edge of cool air is the outflow boundary. The rapidly moving and gusty cool air at the front causes dust (shaded) to be lofted high in the atmosphere. *Source:* After Warner (2004).

Dust devils have been widely reported on Earth and Mars and much recent research has been driven by the desire to understand the processes operating on Mars (Balme and Greeley 2006; see also Chapters 1 and 11). They form when rotating updrafts of air generate winds strong enough to entrain dust, causing the dust to rise. Dust devils are most frequently observed on hot, summer afternoons when strong surface heating by

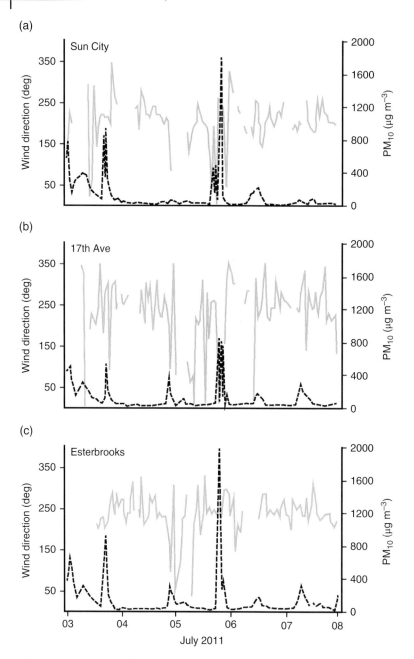

Figure 4.8 Time series of wind direction (grey lines) and PM_{10} concentrations (black dotted lines) from hourly observations at three locations near Phoenix, Arizona. The data clearly show the increased PM_{10} concentration associated with the passage of the haboob through the region on the evening of 5 July 2011. *Source:* Redrawn from Raman et al. (2014).

insolation combines with an unstable lower atmosphere. They can form over different surfaces, but are rare on well-vegetated areas with shrubs and trees. Unlike many other dust events, dust devils can leave clear, but often temporary, tracks across the landscape that reflect their size (width), direction of travel, and duration. On Mars, dust devil tracks are clearly visible (Chapter 11) but they are harder to see on Earth. Terrestrial dust devil tracks have been reported from China, Niger, and Peru and are visible where the vortices preferentially remove fine particles from the surface, changing the roughness and albedo of the surface over which they travel (Rossi and Marinangeli 2004; Reiss et al. 2011; Hesse 2012). Figure 4.9 shows dust tracks in the Ténéré Desert, Niger, where dunes and sand sheets are characterised by bimodal sands (Warren 1972). The fine sands are removed as the dust devil passes over the sediments; these tracks are a few tens of metres wide and up to 8.5 km long (Rossi and Marinangeli 2004). Dust devils are

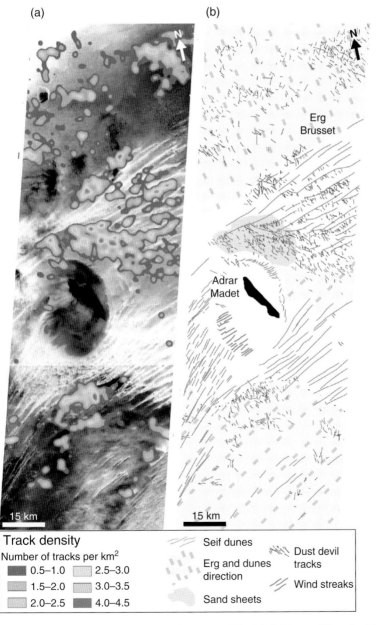

Figure 4.9 (a) Density of dust devil tracks in part of the Ténéré Desert, Niger, for 26 May 2001 shown over the ASTER green band mosaic. (b) Geomorphological map of 26 May 2001 ASTER coverage indicating dust devil tracks. *Source:* Redrawn from Rossi and Marinangeli (2004). See insert for colour representation of this figure.

typically much taller than they are wide; shorter, and narrower dust devils have a shorter duration than taller, broader ones. Sinclair (1965) suggested that dust devils with <3 m diameter would last less than 1 minute while those 15–30 m would last up to 5 minutes, but there are rare, larger, longer-lasting events. For example, dust devils lasting several hours and travelling tens of km have been reported from the south-west USA (Ives 1947). It is difficult to quantify the amount of dust transported in dust devils. Gillette and Sinclair (1990) estimated that they may account for two-thirds of fine air-borne dusts (<25 µm diameter) in the USA. Globally, Koch and Renno (2005) suggested that dry convective plumes and vortices together account for up to 35% of the aerosol budget but more recent research suggests a contribution of an order of magnitude less from these dust-lifting features (Jemmett-Smith et al. 2015).

4.6 Dust in Suspension

As dust is entrained from the surface, it forms a plume. Clean air, such as over oceans, away from dust transport pathways typically has a dust concentration of $<1 \mu g \, m^{-3}$. Most countries have air quality standards that indicate daily and annual values of PM_{10} that should not be exceeded. For example, the European Air Quality Framework Directive guidelines are that daily PM_{10} values should not exceed $50 \mu g \, m^{-3}$ on more than seven days per year and annual mean PM_{10} values must not exceed $20 \mu g \, m^{-3}$. The US Environmental Protection Agency standard indicates a mean daily PM_{10} concentration of $150 \mu g \, m^{-3}$ must not be exceeded more than once per year (averaged over three years). Close to source, the concentration of dust in the atmosphere is in the range $100–100000 \mu g \, m^{-3}$ (Goudie and Middleton 2006). This means that dust storms can cause an exceedance of air quality standards, and this can occur both proximally and distally to source when emissions are high. A dust concentration of $>500 \mu g \, m^{-3}$

is needed to achieve a reduction in visibility of <1 km, i.e. a WMO-defined dust storm (Ben Mohamed and Frangi 1986; Camino et al. 2015), but this concentration will vary depending on the sediment size characteristics. This is because finer particles have more influence on the impairment of visibility than they do on mass concentration, such that the reduction of visibility by a given dust concentration increases where fine particles dominate, as they do at greater distances from source (Baddock et al. 2014, Figure 4.10).

In Reykjavik, Iceland, dust storms account for approximately one third of all days when the PM_{10} health limit is exceeded. The closest sources of these dust storms are 100 km east of the city. Thorsteinsson et al. (2011) calculated that to create $200 \mu g \, m^{-3}$ of PM_{10} pollution in Reykjavík, about $35 \, g \, m^{-2} \, hr^{-1}$ (or 35 tonnes from each km^2 per hour) of material $\leq PM_{10}$ had to be blown away from the source. The most substantial global dust transport pathway is that from the world's largest dust source in North Africa, 9000 km west across the Atlantic Ocean (see Figure 4.4). In the summer months, dust transported along this pathway follows a more northerly trajectory towards the Caribbean. At Guadeloupe, aerosol concentration peaks from May to September and values $>50 \mu g \, m^{-3}$ occur on more than 20% of days in June and July; in contrast, in winter and spring, the pathway is located further south and dust is recorded at Cayenne where values $\geq 50 \mu g \, m^{-3}$ are recorded on 20–30% of days in March and April (Prospero et al. 2014). It has been estimated that $8–48 \, Tg \, a^{-1}$ $(8–50 \, kg \, ha \, a^{-1})$ of dust that has travelled along this east–west pathway is deposited in the Amazon Basin, providing an important source of nutrients to the soil (Hongbin et al. 2015).

Once dust is in suspension in the atmosphere, it is transported vertically and horizontally away from the source area. Most dust is deposited within a relatively short distance of the source – this is 'local dust' (Table 4.1) – but some particles are lifted high into the atmosphere and transported thousands of kilometres before being deposited.

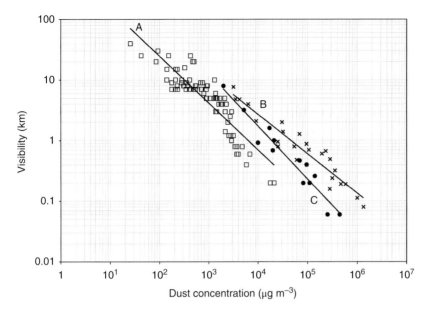

Figure 4.10 Three different relationships between dust concentration and visibility. Relationship A was determined for measurements taken 10–100 km from the dust source where fine particles dominate (Baddock et al. 2014); relationships B (Chepil and Woodruff 1957) and C (Patterson and Gillette 1977) were determined for measurements taken within 10 km of the dust source where there is a high proportion of coarse particles.

The length of time a particle remains in suspension, and hence the distance it can travel, depends on the characteristics of the air mass (e.g. wind velocity, turbulence), the rate at which the dust settles out of suspension (settling velocity), and the characteristics of the particles themselves. This means the characteristics of the dust plume change through space and time, typically decreasing in dust concentration and becoming finer with increasing distance travelled from the source. Using a kite-based sampling system, McGowan and Clark (2008) measured a 500 m vertical profile of PM_{10} concentrations through a regional dust haze in central Australia and found dust concentration decreased exponentially with height from $133 \, \mu g \, m^{-3}$ at the surface to $76 \, \mu g \, m^{-3}$ at 500 m height. Ansmann et al. (2012) described a Saharan dust plume that had travelled over central Europe where the highest dust concentrations of up to $500 \, \mu g \, m^{-3}$ were located at a height of 2 km; in this profile, the concentration of dust declined exponentially from the surface to approximately 500 m above it,

in the same way as described by McGowan and Clark (2008) but then increased from 500–2000 m before declining to near zero at 7 km height. As well as dust concentrations, particle size distributions can also change vertically through a plume. Chen and Fryrear (2002) measured a decrease in dust concentration from $1000 \, \mu g \, m^{-3}$ at the surface to $200 \, \mu g \, m^{-3}$ at 16 m height in a haboob over Texas; within this plume the concentration of particles $<10 \, \mu m$ diameter was largely invariant with height. The change in particle concentration was driven, primarily, by a decrease in the amount of coarser material as it settled out of suspension. Overall the total flux rate for the plume was $84\,960 \, kg \, km^{-2} \, h^{-1}$, of which approximately 20% was made up particles in the range $10–20 \, \mu m$ (Chen and Fryrear 2002).

If a dust storm has a single source of origin, samples of the plume collected downwind, or dust deposits, can have a distinctive characteristic that makes it possible to trace the dust to its source. However, many dust plumes that travel over land contain entrain material from more than one source, which can be differentiated by particle size or mineralogy. McTainsh et al. (1997) found that dust samples in Mali, West Africa, contained sediments from three different sources: distant (<5 μm), regional (20–40 μm), and local (50–70 μm). The latter two were primarily attributed to anthropogenic activities such as disturbance by livestock. Skonieczny et al. (2011) identified two key sources of a Saharan dust event in Senegal, West Africa, in March 2006, which differed in terms of grain size, composition, and mineralogy. Once dust travels beyond the land, over the oceans, there are no further opportunities to add

material to the plume and the suspended sediments become well mixed.

These transport processes clearly affect the characteristics of dust deposits. As discussed in Chapter 5, there is a relationship between the sedimentary characteristics of dust deposits, such as loess, and the distance the particles have travelled. One exception to this is where occasional, large grains (>63 μm) have been collected during dust storms and have been demonstrated to have travelled thousands of kilometres from source (Middleton et al. 2001; Menéndez et al. 2014). The mechanism by which these 'giant' dust particles have travelled such long distances is yet to be explained.

Further Reading

The literature on all aspects of dust, but particularly dust transport and the impact of dust in the Earth system has expanded rapidly since the mid-1990s. The book by Goudie and Middleton (2006) provides a useful overview of the role of dust in the global system. An excellent synthesis of the types of dust and rates of dust deposition in relation to distance from source can be

found in Lawrence and Neff (2009). Global dust emissions, aerosol impacts, and the geomorphological characteristics of dust sources at low latitudes have been reviewed by Shao et al. (2011), Goudie (2009), and Bullard et al. (2011) while the mechanisms driving dust emissions in glaciated and other high-latitude regions are discussed by Bullard (2013) and Bullard et al. (2016).

References

Alfaro, S.C. and Gomes, L. (2001). Modeling mineral aerosol production by wind erosion: emission intensities and aerosol size distributions in source areas. *Journal of Geophysical Research: Atmospheres* 106 (D16): 18075–18084. doi: 10.1029/2000 JD900339.

Alizadeh-Choobari, O., Sturman, A., and Zawar-Reza, P. (2014). A global satellite view of the seasonal distribution of mineral dust and its correlation with atmospheric circulation. *Dynamics of Atmospheres and Oceans* 68: 20–34. doi: 10.1016/j. dynatmoce.2014.07.002.

Ansmann, A., Seifert, P., Tesche, M., and Wandinger, U. (2012). Profiling of fine

and coarse particle mass: case studies of Saharan dust and Eyjafjallajokull/Grimsvotn volcanic plumes. *Atmospheric Chemistry and Physics* 12 (20): 9399–9415. doi: 10.5194/acp-12-9399-2012.

Ashpole, I. and Washington, R. (2012). An automated dust detection using SEVIRI: a multiyear climatology of summertime dustiness in the central and western Sahara. *Journal of Geophysical Research: Atmospheres* 117: D08202. doi: 10.1029/ 2011JD016845.

Ashpole, I. and Washington, R. (2013). A new high-resolution central and western Saharan summertime dust source map from automated satellite dust plume tracking. *Journal of Geophysical Research: Atmospheres* 118 (D13): 6981–6995. doi: 10.1002/jgrd.50554.

Baddock, M.C., Ginoux, P., Bullard, J.E., and Gill, T.E. (2016). Do MODIS-defined dust sources have a geomorphological signature? *Geophysical Research Letters* 43: 2606–2613. doi: 10.1002/2015GL067327.

Baddock, M.C., Strong, C.L., Leys, J.F. et al. (2014). A visibility and total suspended dust relationship. *Atmospheric Environment* 89: 329–336. doi: 10.1016/j.atmosenv.2014.02.038.

Balme, M. and Greeley, R. (2006). Dust devils on Earth and Mars. *Reviews of Geophysics* 44: RG3003. doi: 10.1029/2005RG000188.

Ben Mohamed, A. and Frangi, J.-P. (1986). Results from ground-based monitoring of spectral aerosol optical thickness and horizontal extinction: some specific characteristics of dusty Sahelian atmospheres. *Journal of Climate and Applied Meteorology* 25 (12): 1807–1815. doi: 10.1175/1520-0450(1986)025<1807: RFGBMO>2.0.CO;2.

Bristow, C.S., Drake, N., and Armitage, S. (2009). Deflation in the dustiest place on Earth: the Bodélé Depression, Chad. *Geomorphology* 105 (1–2): 50–58. doi: 10.1016/j.geomorph.2007.12.014.

Bryant, R.G. (2013). Recent advances in our understanding of dust source emission processes. *Progress in Physical Geography* 37 (3): 397–421. doi: 10.1177/0309133313479391.

Bullard, J.E. (2010). Bridging the gap between field data and global models: current strategies in aeolian research. *Earth Surface Processes and Landforms* 35 (4): 496–499. doi: 10.1002/esp.1958.

Bullard, J.E. (2013). Contemporary glacigenic inputs to the dust cycle. *Earth Surface Processes and Landforms* 38 (1): 71–89. doi: 10.1002/esp.3315.

Bullard, J.E., Baddock, M., McTainsh, G.H., and Leys, J. (2008). Sub-basin scale dust source geomorphology detected using MODIS. *Geophysical Research Letters* 35 (15): L15404. doi: 10.1029/2008GL033928.

Bullard, J.E., Baddock, M., Bradwell, T. et al. (2016). High-latitude dust in the Earth system. *Reviews of Geophysics* 54: 447–485. doi: 10.1002/2016RG000518.

Bullard, J.E., Harrison, S.P., Baddock, M.C. et al. (2011). Preferential dust sources: a geomorphological classification designed for use in global dust-cycle models. *Journal of Geophysical Research: Earth Surface* 116: F04034. doi: 10.1029/2011JF002061.

Bullard, J.E., McTainsh, G.H., and Pudmenzky, C. (2004). Aeolian abrasion and modes of fine particle production from natural red dune sands: an experimental study. *Sedimentology* 51 (5): 1103–1125. doi: 10.1111/j.1365-3091.2004.00662.x.

Camino, C., Cuevas, E., Basart, S. et al. (2015). An empirical equation to estimate mineral dust concentrations from visibility observations in Northern Africa. *Aeolian Research* 16: 55–58. doi: 10.1016/j.aeolia.2014.11.002.

Chappell, A., Warren, A., O'Donoghue, A. et al. (2008). The implications for dust emission modeling of spatial and vertical variations in horizontal dust flux and particle size in the Bodélé Depression, Northern Chad. *Journal of Geophysical Research: Atmospheres* 113: D04214. doi: 10.1029/2007JD009032.

Chen, W. and Fryrear, D.W. (2002). Sedimentary characteristics of a haboob dust storm. *Atmospheric Research* 61 (1): 75–85. doi: 10.1016/S0169-8095(01)00092-8 (Discussion: Le Roux, J.P. 2002. Journal of Sedimentary Research 72(3):441-442).

Chepil, W.S. and Woodruff, N.P. (1957). Sedimentary characteristics of dust storms. II. Visibility and dust concentration. *The American Journal of Science* 255 (2): 104–144. doi: 10.2475/ajs.255.2.104.

Colarco, P.R., da Silva, A., Chin, M., and Diehl, T. (2010). Online simulations of global aerosol distributions in the NASA GEOS-4 model and comparisons to satellite and ground-based aerosol optical depth. *Journal of Geophysical Research: Atmospheres* 115: D14207. doi: 10.1029/2009JD012820.

Crouvi, O., Amit, R., Enzel, Y., and Gillespie, A.R. (2010). Active sand seas and the formation of desert loess. *Quaternary Science Reviews* 29 (17–18): 2087–2098. doi: 10.1016/j.quascirev.2010.04.026.

Crouvi, O., Schepanski, K., Amit, R. et al. (2012). Multiple dust sources in the Sahara Desert: the importance of sand dunes. *Geophysical Research Letters* 39: L13401. doi: 10.1029/2012GL052145.

Crusius, J., Schroth, A.W., Gassó, S. et al. (2011). Glacial flour dust storms in the Gulf of Alaska: hydrologic and meteorological controls and their importance as a source of bioavailable iron. *Geophysical Research Letters* 38: L06602. doi: 10.1029/2010GL046573.

Darwin, C. (1846). An account of fine dust which often falls on vessels in the Atlantic Ocean. *Proceedings of the Geological Society* 2 (5, Part I): 26–30. doi: 10.1144/GSL.JGS.1846.002.01-02.09.

De Deckker, P., Munday, C.I., Brocks, J. et al. (2014). Characterisation of the major dust storm that traversed over eastern Australia in September 2009: a multidisciplinary approach. *Aeolian Research* 15: 133–149. doi: 10.1016/j.aeolia.2014.07.003.

Floyd, K.W. and Gill, T.E. (2011). The association of land cover with aeolian sediment production at Jornada Basin, New Mexico, USA. *Aeolian Research* 3: 55–66. doi: 10.1016/j.aeolia.2011.02.002.

Gillette, D.A., Adams, J., Endo, A. et al. (1980). Threshold velocities for input of soil particles into the air by desert soils. *Journal of Geophysical Research. Oceans and Atmospheres* 85 (C10): 5621–5630. doi: 10.1029/JC085iC10p05621.

Gillette, D.A. and Sinclair, P.C. (1990). Estimation of suspension of alkaline material by dust devils in the United-States. *Atmospheric Environment, A* 24 (5): 1135–1142. doi: 10.1016/0960-1686(90)90078-2.

Ginoux, P., Mian, C., Tegen, I. et al. (2001). Sources and distributions of dust aerosols simulated with the GOCART model. *Journal of Geophysical Research:*
Atmospheres 106 (D17): 20255–20274. doi: 10.1029/2000JD000053.

Ginoux, P., Prospero, J.M., Gill, T.E. et al. (2012). Global-scale attribution of anthropogenic and natural dust sources and their emission rates based on MODIS deep blue aerosol products. *Reviews of Geophysics* 50: RG3005. doi: 10.1029/2012RG000388.

Goossens, D. (2005). Quantification of the dry aeolian deposition of dust on horizontal surfaces: an experimental comparison of theory and measurements. *Sedimentology* 52 (4): 859–873. doi: 10.1111/j.1365-3091.2005.00719.x.

Goossens, D. and Buck, B.J. (2012). Can BSNE (big spring number eight) samplers be used to measure PM10, respirable dust, PM2.5 and PM1.0? *Aeolian Research* 5: 43–49. doi: 10.1016/j.aeolia.2012.03.002.

Goossens, D. and Offer, Z.Y. (2000). Wind tunnel and field calibration of six aeolian dust samplers. *Atmospheric Environment* 34 (7): 1043–1057. doi: 10.1016/S1352-2310(99)00376-3.

Goudie, A.S. (2009). Dust storms: recent developments. *Journal of Environmental Management* 90: 89–94.

Goudie, A.S. and Middleton, N.J. (2006). *Desert Dust in the Global System*. New York: Springer.

Haustein, K., Pérez, C., Baldasano, J.M. et al. (2012). Atmospheric dust modeling from meso to global scales with the online NMMB/BSC-dust model – part 2: experimental campaigns in Northern Africa. *Atmospheric Chemistry and Physics* 12 (6): 2933–2958. doi: 10.5194/acp-12-2933-2012.

Hesse, R. (2012). Short-lived and long-lived dust devil tracks in the coastal desert of Southern Peru. *Aeolian Research* 5: 101–106. doi: 10.1016/j.aeolia.2011.10.003.

Hongbin, Y., Mian, C., Tianle, Y. et al. (2015). The fertilizing role of African dust in the Amazon rainforest: a first multiyear assessment based on data from Cloud-Aerosol Lidar and Infrared Pathfinder Satellite Observations. *Geophysical*

Research Letters 42 (6): 1984–1991. doi: 10.1002/2015GL063040.

Houser, C.A. and Nickling, W.G. (2001). The emission and vertical flux of particulate matter <10μm from a disturbed clay-crusted surface. *Sedimentology* 48: 255–267. doi: 10.1046/j.1365-3091.2001.00359.x.

Ishizuka, M., Mikami, M., Leys, J. et al. (2008). Effects of soil moisture and dried rain droplet crust on saltation and dust emission. *Journal of Geophysical Research: Atmospheres* 113: D24212. doi: 10.1029/2008JD009955.

Ives, R.L. (1947). Behaviour of dust devils. *Bulletin of the American Meteorological Society* 28 (4): 168–174.

Jemmett-Smith, B.C., Marsham, J.H., Knippertz, P., and Gilkeson, C.A. (2015). Quantifying global dust devil occurrence from meteorological analyses. *Geophysical Research Letters* 42: 1275–1282. doi: 10.1002/2015GL063078.

Junran, L., Okin, G.S., Alvarez, L., and Epstein, H. (2007). Quantitative effects of vegetation cover on wind erosion and soil nutrient loss in a desert grassland of southern New Mexico, USA. *Biogeochemistry* 85: 317–332. doi: 10.1007/s10533-007-9142-y.

King, J., Etyemezian, V., Sweeney, M. et al. (2011). Dust emission variability at the Salton Sea, California, USA. *Aeolian Research* 3 (1): 67–79. doi: 10.1016/j.aeolia. 2011.03.005.

Koch, J. and Renno, N.O. (2005). The role of convective plumes and vortices on the global aerosol budget. *Geophysical Research Letters* 32: L18806. doi: 10.1029/2005 GL023420.

Lawrence, C.R. and Neff, J.C. (2009). The contemporary physical and chemical flux of aeolian dust: a synthesis of direct measurements of dust deposition. *Chemical Geology* 267 (1–2): 46–63. doi: 10.1016/j.chemgeo.2009.02.005.

Lee, J.A., Baddock, M.C., Mbuh, M.J., and Gill, T.E. (2012). Geomorphic and land cover characteristics of aeolian dust sources in West Texas and eastern New Mexico, USA. *Aeolian Research* 3 (4): 459–466. doi: 10.1016/j.aeolia.2011.08.001.

Leys, J.F., Heidenreich, S.K., Strong, C.L. et al. (2011). PM10 concentrations and mass transport during 'Red Dawn' – Sydney 23 September 2009. *Aeolian Research* 3 (3): 327–342. doi: 10.1016/j.aeolia.2011.06.003.

Loosmore, G.A. and Hunt, J.R. (2000). Dust resuspension without saltation. *Journal of Geophysical Research: Atmospheres* 105 (D16): 20663–20672. doi: 10.1029/2000JD900271.

Mahowald, N.M., Ballantine, J.A., Feddema, J., and Ramankutty, N. (2007). Global trends in visibility: implications for dust sources. *Atmospheric Chemistry and Physics* 7 (12): 3309–3339. doi: 10.5194/acp-7-3309-2007.

Marx, S.K., McGowan, H.A., Kamber, B.S. et al. (2014). Unprecedented wind erosion and perturbation of surface geochemistry marks the Anthropocene in Australia. *Journal of Geophysical Research. Earth Surface* 119 (1): 45–61. doi: 10.1002/2013 JF002948.

McGowan, H.A. and Clark, A. (2008). A vertical profile of PM10 dust concentrations measured during a regional dust event identified by MODIS Terra, Western Queensland, Australia. *Journal of Geophysical Research: Earth Surface* 113: F02S03. doi: 10.1029/2007JF000765.

McKenna Neuman, C., Maxwell, C.D., and Boulton, J.W. (1996). Wind transport on sand surfaces with photoautotrophic microorganisms. *Catena* 27 (3–4): 229–247. doi: 10.1016/0341-8162(96)00023-9.

McTainsh, G.H., Nickling, W.G., and Lynch, A.W. (1997). Dust deposition and particle size in Mali, West Africa. *Catena* 29 (3–4): 307–322. doi: 10.1016/S0341-8162(96) 00075-6.

Menéndez, M., Pérez-Chacón, E., Mangas, J. et al. (2014). Dust deposits on La Graciosa Island (Canary Islands, Spain): texture, mineralogy and a case study of recent dust plume transport. *Catena* 117: 133–144. doi: 10.1016/j.catena.2013.05.007.

Middleton, N.J., Betzer, P.R., and Bull, P.A. (2001). Long-range transport of 'giant' aeolian quartz grains: linkage with discrete sedimentary sources and implications for protective particle transfer. *Marine Geology* 177 (3–4): 411–417. doi: 10.1016/S0025-3227(01)00171-2.

Middleton, N.J., Goudie, A.S., and Wells, G.L. (1986). The frequency and source areas of dust storms. In: *Aeolian Geomorphology, 17th Annual Binghampton Geomorphology Symposium* (ed. W.G. Nickling), 237–259. London: Allen and Unwin.

Miller, S.D., Kuciauskas, A.P., Liu, M. et al. (2008). Haboob dust storms of the Southern Arabian Peninsula. *Journal of Geophysical Research: Atmospheres* 113: D01202. doi: 10.1029/2007JD008550.

Milliman, J.D. and Farnsworth, K.L. (2011). *River Discharge to the Coastal Ocean: A Global Synthesis*. Cambridge: Cambridge University Press.

Muhs, D., Prospero, J.P., Baddock, M.C., and Gill, T.E. (2014). Identifying sources of aeolian mineral dust: present and past. In: *Mineral Dust: A Key Player in the Earth System* (ed. P. Knippertz and J.-B.W. Stuut), 51–74. Cambridge: Cambridge University Press.

Neff, J.C., Ballantyne, A.P., Farmer, G.L. et al. (2008). Increasing eolian dust deposition in the Western United States linked to human activity. *Nature Geoscience* 1: 189–195. doi: 10.1038/ngeo133.

O'Loingsigh, T., McTainsh, G.H., Tapper, N.J., and Shinkfield, P. (2010). Lost in code: a critical analysis of using meteorological data for wind erosion monitoring. *Aeolian Research* 2: 49–57. doi: 10.1016/j.aeolia.2010.03.002.

O'Loingsigh, T., McTainsh, G.H., Tews, E.K. et al. (2014). The Dust Storm Index (DSI): a method for monitoring broadscale wind erosion using meteorological records. *Aeolian Research* 12: 29–40. doi: 10.1016/j.aeolia.2013.10.004.

Ono, D., Kiddoo, P., Howard, C. et al. (2011). Application of a combined measurement and modelling method to quantify windblown dust emissions from the exposed playa at Mono Lake, California. *Journal of the Air and Waste Management Association* 61 (10): 1036–1045.

Parajuli, S.P., Yang, Z.-L., and Kocurek, G. (2014). Mapping erodibility in dust source regions based on geomorphology, meteorology and remote sensing. *Journal of Geophysical Research. Earth Surface* 119 (9): 1977–1994. doi: 10.1002/2014JF003095.

Patterson, E.M. and Gillette, D.A. (1977). Measurements of visibility vs mass of airborne soil particles. *Atmospheric Environment* 11: 193–196.

Prospero, J.M., Bullard, J.E., and Hodgkins, R. (2012). High-latitude dust over the North Atlantic: inputs from Icelandic proglacial dust storms. *Science* 335 (6072): 1078–1082. doi: 10.1126/science.1217447.

Prospero, J.M., Collard, F.-X., Molinié, J., and Jeannot, A. (2014). Characterizing the annual cycle of African dust transport to the Caribbean Basin and South America and its impact on the environment and air quality. *Global Biogeochemical Cycles* 28 (7): 757–773. doi: 10.1002/2013GB004802.

Prospero, J.M., Ginoux, P., Torres, O. et al. (2002). Environmental characterization of global sources of atmospheric soil dust identified with the Nimbus-7 Total Ozone Mapping Spectrometer (TOMS) absorbing aerosol product. *Reviews of Geophysics* 40: 1002. doi: 10.1029/2000RG000095.

Pye, K. (1987). *Aeolian Dust and Dust Deposits*. London: Academic Press.

Raman, A., Arellano, A.F. Jnr., and Brost, J.J. (2014). Revisiting haboobs in the southwestern United States: an observational case study of the 5 July 2011 Phoenix dust storm. *Atmospheric Environment* 89: 179–188. doi: 10.1016/j.atmosenv.2014.02.026.

Reheis, M.C. and Kihl, R. (1995). Dust deposition in southern Nevada and California, 1984–1989: relations to climate, source area, and source lithology. *Journal of Geophysical Research: Atmospheres* 100 (D5): 8893–8918. doi: 10.1029/94JD03245.

Reiss, D., Raack, J., and Hiesinger, H. (2011). Bright dust devil tracks on Earth: implications for their formation on Mars. *Icarus* 211 (1): 917–920. doi: 10.1016/j.icarus.2010.09.009.

Rossi, P.A. and Marinangeli, L. (2004). The first terrestrial analogue to Martian dust devil tracks found in Ténéré Desert, Niger. *Geophysical Research Letters* 31: L06702. doi: 10.1029/2004GL019428.

Schepanski, K., Tegen, I., Laurent, B. et al. (2007). A new Saharan dust source activation frequency map derived from MSG-SEVIRI IR-channels. *Geophysical Research Letters* 34: L18803. doi: 10.1029/2007GL030168.

Schepanski, K., Tegen, I., and Macke, A. (2012). Comparison of satellite-based observations of Saharan dust source areas. *Remote Sensing of Environment* 123 (8): 90–97. doi: 10.1016/j.rse.2012.03.019.

Shao, Y., Leys, J.F., McTainsh, G.H., and Tews, K. (2007). Numerical simulation of the October 2002 dust event in Australia. *Journal of Geophysical Research: Atmospheres* 112: D08207. doi: 10.1029/2006JD007767.

Shao, Y., Wyrwoll, K.-H., Chappell, A. et al. (2011). Dust cycle: an emerging core theme in Earth system science. *Aeolian Research* 2 (4): 181–204. doi: 10.1016/j.aeolia.2011.02.001.

Sinclair, P.C. (1965). A microbarophone for dust devil pressure measurements. *Journal of Applied Meteorology* 4 (1): 116–121. doi: 10.1175/1520-0450(1965)004<0116:AMFDDP>2.0.CO;2.

Skonieczny, C., Bory, A., Bout-Roumazeilles, V. et al. (2011). The 7–13 March 2006 major Saharan outbreak: multiproxy characterization of mineral dust deposited on the West African margin. *Journal of Geophysical Research: Atmospheres* 116: D18210. doi: 10.1029/2011JD016173.

Sow, M., Alfaro, S.C., Rajot, J.L., and Marticorena, B. (2009). Size resolved dust emission fluxes measured in Niger during 3 dust storms of the AMMA experiment. *Atmospheric Chemistry and Physics* 9 (12): 3881–3891. doi: 10.5194/acp-9-3881-2009.

Stanelle, T., Bey, I., Raddatz, T. et al. (2014). Anthropogenically induced changes in twentieth century mineral dust burden and the associated impact on radiative forcing. *Journal of Geophysical Research: Atmospheres* 119 (23): 13526–13546. doi: 10.1002/2014JD022062.

Stout, J.E., Warren, A., and Gill, T.E. (2009). Publication trends in aeolian research: an analysis of the bibliography of Aeolian Research. *Geomorphology* 105 (1–2): 6–17. doi: 10.1016/j.geomorph.2008.02.015.

Sweeney, M.R. and Mason, J.A. (2013). Mechanisms of dust emission from Pleistocene loess deposits, Nebraska, USA. *Journal of Geophysical Research. Earth Surface* 118 (F3): 1460–1471. doi: 10.1002/jgrf.20101.

Sweeney, M.R., McDonald, E.V., and Etyemezian, V. (2011). Quantifying dust emissions from desert landforms, Eastern Mojave Desert, USA. *Geomorphology* 135 (1–2): 21–34. doi: 10.1016/j.geomorph.2011.07.022.

Tanaka, T.Y. and Chiba, M. (2006). A numerical study of the contributions of dust source regions to the global dust budget. *Global and Planetary Change* 52 (1–4): 88–104. doi: 10.1016/j.gloplacha.2006.02.002.

Tegen, I. and Schulz, M. (2014). Numerical dust models. In: *Mineral Dust: A Key Player in the Earth System* (ed. P. Knippertz and J.-B.W. Stuut), 201–222. Berlin: Springer.

Thorsteinsson, T., Gísladóttir, G., Bullard, J., and McTainsh, G. (2011). Dust storm contributions to airborne particulate matter in Reykjavik, Iceland. *Atmospheric Environment* 45: 5924–5933. doi: 10.1016/j.atmosenv.2011.05.023.

Todd, M.C., Bou Karam, D., Davazos, C. et al. (2008). Quantifying uncertainty in estimates of mineral dust flux: an intercomparison of model performance over the Bodélé Depression, northern Chad. *Journal of Geophysical Research* 113: D24107. doi: 10.1029/2008JD010476.

Tozer, P. and Leys, J. (2013). Dust storms: what do they really cost? *Rangeland Journal* 35: 131–142. doi: org/10.1071/RJ12085.

Van Pelt, R.S., Baddock, M.C., Zobeck, T.M. et al. (2013). Field wind tunnel testing of two silt loam soils on the North American

Central High Plains. *Aeolian Research* 10: 53–59. doi: 10.1016/j.aeolia.2012.10.009.

Vickery, K.J. and Eckardt, F.D. (2013). Dust emission controls on the lower Kuiseb River valley, Central Namib. *Aeolian Research* 10: 125–133. doi: 10.1016/j.aeolia.2013.02.006.

Vickery, K.J., Eckardt, F.D., and Bryant, R.G. (2013). A sub-basin scale dust plume source frequency inventory for Southern Africa, 2005–2008. *Geophysical Research Letters* 40 (19): 5274–5279. doi: org/10.1002/grl.50968.

Von Richthofen, F. (1882). On the mode of origin of the loess, decade II. *The Geological Magazine* 9 (7): 293–305. doi: 10.1017/S001675680017164X (also in Smalley, I.J. (ed), Loess; lithology and genesis. Benchmark papers in geology 26, (1975), Dowden, Hutchinson and Ross, Stroudsburg, PA, pp. 24–36.).

Wanquan, T., Haibing, W., and Xiaopeng, J. (2013). External supply of dust regulates dust emissions from sand deserts. *Catena* 110: 113–118. doi: 10.1016/j.catena.2013.05.014.

Warner, T.T. (2004). *Desert Meteorology*. Cambridge: Cambridge University Press.

Warren, A. (1972). Observations on dunes and bimodal sands in the Ténéré Desert. *Sedimentology* 19 (1): 37–44. doi: 10.1111/j.1365-3091.1972.tb00234.x.

Warren, A., Chappell, A., Todd, M.C. et al. (2007). Dust-raising in the dustiest place on Earth. *Geomorphology* 92 (1–2): 25–37. doi: 10.1016/j.geomorph.2007.02.007.

Washington, R., Todd, M., Middleton, N.J., and Goudie, A.S. (2003). Dust-storm source areas determined by the Total Ozone Monitoring Spectrometer and surface observations. *Annals of the Association of American Geographers* 93: 297–313. doi: 10.1111/1467-8306.9302003.

Wright, J.S. (2001). Making loess-sized quartz silt: data from laboratory simulations and implications for sediment transport pathways and the formation of 'desert' loess

deposits associated with the Sahara. In: Derbyshire, E. (ed.), Loess and palaeosols: characteristics, chronology and climate: a contribution to IGCP413. Proceedings, International Conference, Universität Bonn, March 25–April 01. *Quaternary International* 76–77: 7–19. doi:10.1016/S1040-6182(00)00085-9.

Wright, J.S., Smith, B.J., and Whalley, W.B. (1998). Mechanisms of loess-sized quartz silt production and their relative effectiveness: laboratory simulations. *Geomorphology* 23 (1): 231–256. doi: 10.1016/S0169-555X(97)00084-6.

Yiuchung, C., McTainsh, G.H., Leys, J. et al. (2005). Influence of the 23 October 2002 dust storm on the air quality of four Australian cities. *Water, Air, and Soil Pollution* 164 (1–4): 329–348. doi: 10.1007/s11270-005-4009-0.

Zender, C.S., Newman, D., and Torres, O. (2003). Spatial heterogeneity in aeolian erodibility: uniform, topographic, geomorphic, and hydrologic hypotheses. *Journal of Geophysical Research: Atmospheres* 108 (D17): 4543. doi: 10.1029/2002JD003039.

Zhao, C., Chen, S., Leung, L.R. et al. (2013). Uncertainty in modelling dust mass balance and radiative forcing from size parametrization. *Atmospheric Chemistry and Physics* 13: 10733–10753. doi: 10.5194/acp-13-10733-2013.

Zobeck, T.M., Sterk, G., Funk, R. et al. (2003). Measurement and data analysis methods for field-scale wind erosion studies and model validation. *Earth Surface Processes and Landforms* 28 (11): 1163–1188. doi: 10.1002/esp.1033.

Zobeck, T.M. and Van Pelt, R.S. (2006). Wind-induced dust generation and transport mechanics on a bare agricultural field. *Journal of Hazardous Materials* 132 (1): 26–38. doi: 10.1016/j.jhazmat.2005.11.090.

5

Loess

Helen M. Roberts

Aberystwyth University, Aberystwyth, UK

5.1   Introduction

Judged by the proportion of Earth's surface they cover, loesses are second to sand dunes. But, judged by their impact on human history, human well-being and on geo-science, loess is far out in front: loess soils were a vital ingredient in the productivity of some of the earliest arable farms, and continue to play an important role in present-day food and wine-production; and loess itself retains some of the best land-based records of the recent geological history, which can also be linked to records of far-travelled dust trapped in ice-cores and found in ocean-sediments.

5.2   Definitions of Loess and its Relationship to Dust

Loess is a fine-grained terrestrial sediment derived from wind-blown dust. The aeolian nature of loose yellow silt deposits termed 'Huangtu' (yellow earth) in China was recognised more than 2000 years ago (Liu et al. 1985). In Europe, 'loesch' was an easily worked yellow, limy soil in the language of the Upper Rhine Valley. In present usage, ('loess') the term was introduced to science in the early 1820s by Karl Caesar von Leonhard, and a discussion of loess appeared in volume 3 of Charles Lyell's 1833 classic

Principles of Geology (Smalley et al. 2001; Box 1.1). But it was not until 1882 that the theory of an aeolian origin for loess was advanced in Europe by von Richthofen (Zöller and Semmel 2001; see Box 5.1).

There have been many attempts to define loess (e.g. Smalley and Vita-Finzi 1968; Liu et al. 1985; Pécsi 1990; Pye 1987, 1995; Box 5.1). Pye's (1995) definition is the simplest: loess is a 'terrestrial clastic sediment, composed predominantly of silt-size particles, which is formed essentially by the accumulation of wind-blown dust'. Pécsi (1990) had already proposed that 'Loess [was] not just the accumulation of dust', arguing that, to be classified as loess, a deposit had to meet ten closely prescribed criteria for characteristics such as the structure, grain-size, mineralogy, chemical composition, porosity, and stability under dry, and wet conditions. It is true that most loess has undergone some degree of modification since the time of the deposition of the original material, but Pye's definition (Pye 1995) does not necessarily exclude the process behind Pécsi's criteria. Moreover, Pécsi's definition would exclude the massive accumulations of wind-blown silt found in Alaska and New Zealand, which are now widely acknowledged as loess (e.g. Muhs and Bettis Jr. 2003). All these definitions, however, recognise that 'loess' differs from 'dust', in having a sufficient thickness as

Aeolian Geomorphology: A New Introduction, First Edition. Edited by Ian Livingstone and Andrew Warren.
© 2019 John Wiley & Sons Ltd. Published 2019 by John Wiley & Sons Ltd.

Box 5.1 Loessification

'Loessification' essentially describes the transformation of non-loess ground into loess ground through weathering and soil formation processes (Smalley and Marković 2014). The term was originally coined by Berg (1916), who proposed that loess was formed entirely by in situ weathering of calcium-rich fine-grained sediments or bedrock, and it quickly became a very divisive term because Berg flatly and explicitly refused to acknowledge any role for aeolian processes. Russell (1944) also advocated a process of in situ loessification of previously deposited fine-grained alluvial sediments as critical to the formation of loess from the Lower Mississippi Valley, again adopting a position that was directly opposed to the aeolian theory of loess formation which was prevalent in North America at that time (Smalley et al. 2001). The aeolian theory and the theory of loessification can, of course, be somewhat reconciled if one acknowledges the critical role (now widely accepted) that aeolian processes play in the deposition of the silt which forms the basis of loess deposits, whilst also acknowledging the role of post-depositional soil forming and weathering processes which act on the silt deposits over time. The importance of this post-depositional development for the transformation of accumulated silt into loess was emphasised by Pécsi (1990) in his unambiguously titled paper 'Loess is not just the accumulation of dust'. However, while the exact nature, role, and significance of post-depositional processes of 'loessification' are still debated today, the common thread that soils scientists, geologists, sedimentologists, and geomorphologists all now agree upon is that loess is comprised of wind-blown dust. This definition, proposed by Pye (1984), does not preclude any post-depositional alteration of the silt which is deposited, but as Smalley et al. (2011) point out in their equally unambiguously titled paper, 'Loess is [almost totally formed by] the accumulation of dust', the accumulation of wind-blown silt is the one factor in the various processes described for 'loessification' that remains constant and critical for the formation of all deposits recognised as loess.

to be identifiable in the field as a distinct sedimentary unit.

Dust is also a significant contributor to some deep-sea sediments, but the term 'loess' is reserved exclusively for those deposits of wind-blown dust that have accumulated in terrestrial environments, and it probably includes the small number of terrestrial loesses that have survived rising sea levels and are now beneath sea level, as on the continental shelf of the East China Sea and the Yellow Sea (An et al. 2014).

The finer size of particles of dust, compared to that of sand, lends itself to potentially much longer distances of transport, but further distinctions can also be made. Thus, fine dusts (i.e. <10 μm and frequently <<2 μm diameter) can be transported across regional, continental, and even hemispheric boundaries; whereas coarse dusts (typically 10–50 μm), in general, have shorter transport distances.

Differences in size are generally accompanied by differences in composition: coarse dusts tend to be dominated by quartz and calcite; fine dusts tend to be dominated by clay minerals.

Most loess deposits consist predominantly of silt-sized particles but their content of sand and clay may increase or decrease their median and modal grain sizes. 'Typical' loess has a modal grain size of 30 μm (Pye 1995); its mineralogy is typically dominated by quartz (accounting for 50–70% of the composition), in addition to feldspars (5–30%), mica (5–10%), carbonates (0–30%), and clay minerals (10–15%) (Pye 1987). However, the geology of the source areas of a particular loess influences its mineral composition, which shows significant variation across the world, and even regionally (e.g. Muhs and Bettis Jr. 2003). Heavy minerals and volcanic glass are also frequently present (Pye 1995).

5.3 Distribution and Thickness of Loess Deposits

The differing levels of stringency in definitions of loess make it difficult to provide a definitive statement of its spatial extent and thickness. Primary air-fall loess (i.e. material that is wholly aeolian in origin) is estimated to cover around 5% of the land surface of the world (Pye 1987), while the areal extent of secondary loess (i.e. re-deposited loess or loess originating from non-aeolian processes) is estimated to be approximately 10% of the land surface (Liu et al. 1985). Detailed mapping and classification of primary versus secondary and other loess-like deposits can

pose difficult logistical challenges particularly in remote areas. For the Negev Desert of southern Israel, Crouvi et al. (2009) found that remote sensing was a valuable tool for mapping loess distribution at a regional scale, capable of identifying primary hilltop loess deposits (Figure 5.1). The distinction probably also applies in other parts of the world. Geographical Information Systems (GIS) can then be used to examine the significance and role of other environmental parameters such as rainfall, topography, and distance from potential sources, by coupling these parameters with the remotely sensed data regarding the distribution of loess types (Crouvi et al. 2009).

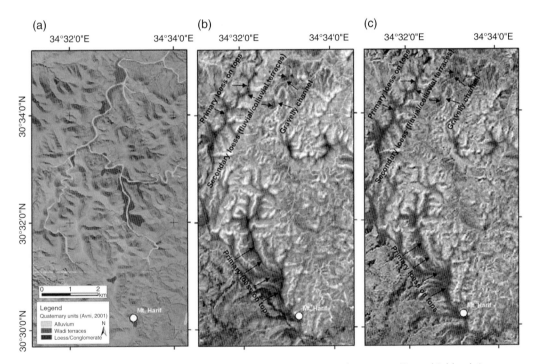

Figure 5.1 A map of loess as identified from remote sensing images in the western Negev highlands in Israel (Crouvi et al. 2009, Figure 5). The images show: (a) Quaternary units (alluvium, wadi terraces, loess/conglomerate) mapped by Avni (2001) at a scale of 1 : 50 000, projected on a rectified air photo; (b) an Advanced Spaceborne Thermal Emission and Reflection Radiometer (ASTER) image identifying primary loess by the dark red to dark purple shades; and (c) a Landsat Thematic Mapper (TM) image identifying primary loess by the brown, purple, red, and blue shades. Fluvial/colluvial loess terraces are in shades of green in ASTER (b) or blue in Landsat TM (c), depending on the clast- and vegetation-coverage, while channels with high gravel cover, or incised in bedrock, are shown in light purple in ASTER (b) or yellow to orange colours in Landsat TM (c). In both remote sensing images (b and c), north-facing slopes are greenish, probably because of the higher vegetation coverage compared to the south-facing slopes which are pink. *Source:* Crouvi et al. (2009). See insert for colour representation of this figure.

There are deposits of loess on all continents, except Antarctica (Figure 5.2). Much of the loess in the northern hemisphere is found in the mid-latitudes, fringing the former southernmost limits of major ice sheets in Europe, Russia, central Asia, the North American midcontinent and the Pacific north-west (Figure 5.2). There are particularly extensive mid-latitude loesses in China, but in this case there are no major adjacent ice sheets or glaciers. Loess is also found at high latitudes in Alaska and the Yukon. In the southern hemisphere, loess is found at low latitudes (from 23–38°S; Zárate 2003) in South America, and is also found on both South Island and North Island, New Zealand. In contrast, it is interesting to note that Australia preserves only very modest deposits of a clay-rich wind-blown sediment termed 'parna' (Butler 1956), which is being accepted by some as a clay-rich loess (e.g. Dare-Edwards 1984; Haberlah 2007). There are only limited occurrences of loess in the Middle East and Africa (Figure 5.2), in spite of the importance of contemporary dust there (Chapters 1 and 4).

Loess thicknesses worldwide are highly variable, spanning the range from a few cm in thickness to several hundreds of metres, but most deposits are typically less than 30 m. The thickest and spatially most extensive loess deposits are found in China, Central Asia, Central and Western Europe, North America, and Argentina. Loess deposits in Central Asia can exceed more than 200 m thickness, and the spatially extensive deposits of the Chinese Loess Plateau are typically of the order of 80–120 m in thickness, but can exceed 300 m in places; closer to the dust sources, in the western Chinese Loess Plateau, the loess deposits are exceptionally thick reaching up to 450 m. Loess deposits in Argentina and North America are typically 20–30 m thick (Pye 1987). Loess is found at sea level and at high altitudes; for example, although much of the loess in the Chinese Loess Plateau is found at an elevation of ~2000 m, loess is also found at 150 m below sea level in the Turpan Basin of XinJiang, north-western China, and at elevations in excess of 5000 m in the western Kunlun Shan, in the north-western Tibetan Plateau (An et al. 2014).

Loess deposits are one of the few geologic deposits to preserve a direct record of atmospheric circulation (Muhs and Bettis Jr. 2003), documenting wind directions and climates of the past. Reconstruction of these atmospheric circulation patterns is inferred from studying the thickness of the loess deposits, which declines with increasing distance from the source area, and also from trends in particle size and other data. For example, Figure 5.3 shows palaeowinds reconstructed for the Last Glacial period in the mid-continent of North America (~28 000–10 000 years ago) using the loess deposits from this central region. Loess is clearly at its thickest when it is close to the sources (as adjacent to the Nebraska Sand Hills, and on the eastern sides of the Mississippi River Valley, where thicknesses exceed 20 m and can reach ~40 m), and thins with distance from the source due to a reduction in sediment load. Particle size also tends to decrease away from the source as coarser sediment is winnowed out, and a decrease in carbonate content reflects syndepositional leaching where deposition rates are lower downwind (Muhs 2007). So, in the case of a north-to-south trending valley such as the Mississippi and Missouri River valleys (Figure 5.3), a decrease in loess thickness and particle size to the east of the river would suggest that the direction of the palaeowind that transported it was variously north-westerly, westerly, or south-westerly (Muhs 2007).

In the Chinese Loess Plateau, loess thickness and the grain size of the deposits decrease from the north-west to the south-east, suggesting that the sediments were chiefly transported by the winter monsoon (Liu et al. 1985 and Chapter 1), and derived from inland Asia (Porter 2001). Long-range transport took the finest-grained dusts far beyond the Loess Plateau to where they contributed to marine records from the North Pacific Ocean and to ice cores from Greenland (see Figure 5.10, redrawn from Porter 2001).

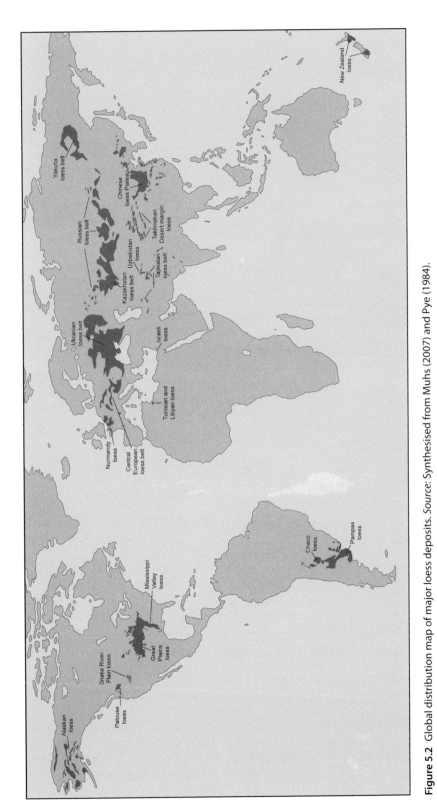

Figure 5.2 Global distribution map of major loess deposits. *Source:* Synthesised from Muhs (2007) and Pye (1984).

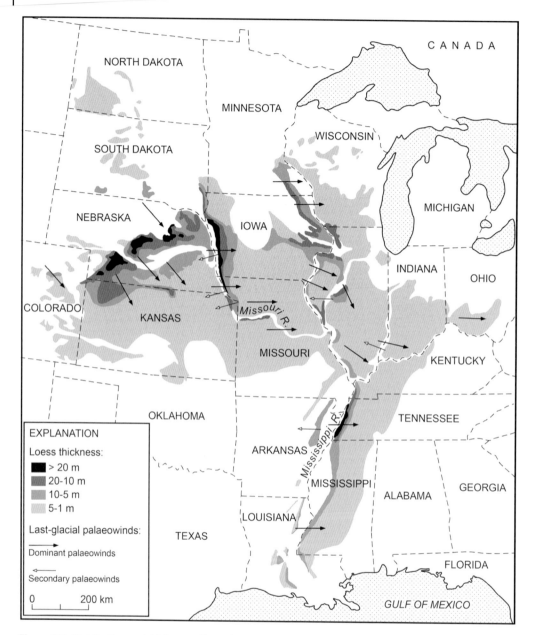

Figure 5.3 Distribution and thickness of loess from the mid-continent of North America. Reconstructed palaeowind directions based on loess thickness, particle size trends, and other data are also shown. *Source:* Redrawn from Muhs and Bettis Jr. (2003).

5.4 Loess Morphology

Loess tends to blanket the underlying topography, draping over a variety of landforms, including steep slopes and valley sides, terraces, alluvial fans, and pediments (Pye 1987).

The surface morphology of loess deposits can take a variety of forms depending upon the underlying topography, including rolling hills (as seen in Figure 5.4, which pictures the Palouse Loess of Washington State), plateaux, and steep-sided loess-covered valleys.

Figure 5.4 The gentle rolling loess hills of Washington State, north-western USA.

Following erosion, near-vertical cliffs or bluffs can be formed within loess deposits because of their cohesive strength, derived from their fine grain sizes and sufficiently strong, if weak, cementation. However, when saturated, loess deposits are prone to gullying, erosion, and collapse, forming vertical exposures. Perhaps one of the most famous exposures is the ~140 m thick section at Luochuan, in the central Chinese Loess Plateau (Figure 5.5), a relatively accessible site with loess spanning almost the whole of the Quaternary Period (the last 2.6 million years): this is now the focal-point of the Luochuan Loess National Geopark established in 2004.

The form and nature of loess deposits vary according to the proximity and nature of the source material, the terrain, and the conditions responsible for transport and deposition of the source material. For example, thick loess deposits are often found immediately downwind of adjacent silt-bearing rivers or glacio-fluvial meltwater channels, which were the source of dust (Figure 5.6a). These source-proximal loess deposits tend to be the coarsest, and grain-size and thickness rapidly decrease with increasing distance from the source, while the finer dust is transported and deposited much further afield. Where source material is of a greater range of grain-sizes, sediment-sorting processes can also give rise to aeolian sand and loess deposits being found close to each other, either as one contiguous deposit ranging from sand dunes, to sand sheets, sandy loess, and finally silty- and clayey-loess (Figure 5.6b), or as sand dunes and loess deposits separated by a clear zone of sediment bypassing (e.g. Figure 5.6c). The scenario shown in Figure 5.6c is of a situation in an arid zone with a rainfall gradient and hence also a gradient of vegetation density. These conditions can be found in a desert margin in which loess can accumulate in sparsely vegetated

Figure 5.5 Luochuan in the central Chinese Loess Plateau. See insert for colour representation of this figure.

areas some considerable distance downwind of the source of the dust in a dry lake bed, wadi or fan (Pye 1995). Lack of vegetation in the intermediate area means that most dust bypasses it (Figure 5.6c). In contrast, the scenario shown in Figure 5.6b occurs where there is sufficient moisture and hence vegetation to enable a gradual transition in grain-size (down-wind fining), with no zone of sediment bypassing, before the distal transition to loess.

Loess may also accumulate against topographic barriers that interrupt the flow of the wind, potentially leading to enhanced deposition on the windward side of the barrier (Figure 5.6d), but deposition on the summit and/or enhanced preferential deposition in the lee of the feature has also been observed. In the scenarios illustrated in Figures 5.6b–d, there is a single sediment source responsible for the proximal and distal accumulations of aeolian sand and loess. Muhs and Bettis Jr. (2003) considered a further scenario in which, in addition to a local

sediment source giving rise to the formation of proximal aeolian sand dunes, a distal sediment source also brings finer-grained sediments which are deposited over a wider area, forming loess deposits which have a distinctly different source to the aeolian sands, and may also have been deposited at very different times (Figure 5.6e). The role of vegetation in all these models of loess accumulation is critical, because vegetation disrupts the flow and reduces the wind-speed, hence reducing the re-entrainment of deposited dust, and it also provides a physical trapping mechanism for the dust.

5.5 The Generation of Loess Sediments

The mechanisms for the production of loess sediments have long been the subject of discussion, and have been something of an enigma, given the extraordinarily large quantities of dust that have been generated.

(a) Proximal loess accumulation

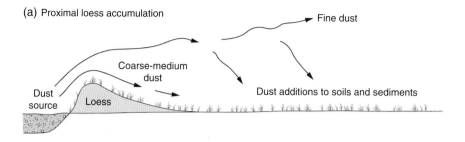

(b) Loess accumulation contiguous with sand sheet – sandy loess transition

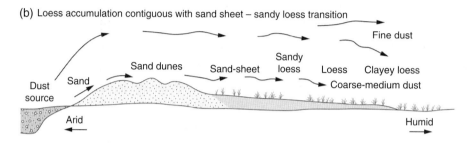

(c) Distal loess accumulation along a climatic gradient

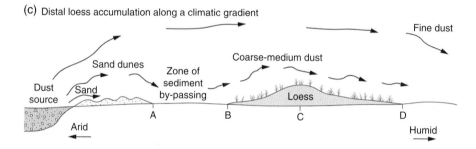

(d) Distal loess accumulation against a topographic obstacle

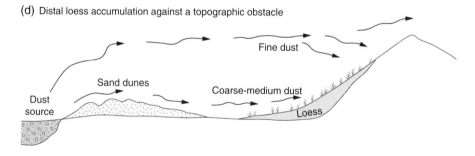

(e) Loess and sand accumulation from two separate sources

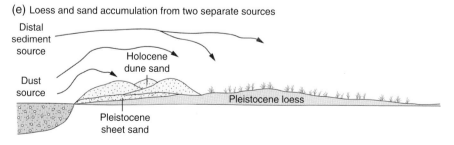

Figure 5.6 Schematic models to explain the formation of loess – and other related aeolian-deposits. *Source:* (a–d) are redrawn from Pye (1995); (e) is modified from Muhs and Bettis Jr. (2003).

Indeed, it is the extensive nature of loess deposits which gave rise to the widespread but erroneous view during the nineteenth century that loess was fluvial in origin. Today the aeolian origin of primary loess is widely accepted, but the mechanisms of the formation of loess are a matter of active debate. Two key models of loess sediment production are considered below: one accounts for production of silt-sized sediment in glacial settings; the other explores production in desert regions. In addition to these models, Muhs (2013) noted that volcanigenically produced fine silt and clay-sized tephra can also make a significant contribution to loess deposits, as in volcanically active regions such as South America, New Zealand, Alaska, Iceland, and Japan, and commented that silt-sized particles are also readily generated in areas with outcrops of fine-grained sedimentary rocks such as siltstone, mudstone, and shale (Muhs 2013).

5.5.1 The 'Glacial' Model

The classical model for the production of the silt that became loess involves glacial grinding followed by deposition in glacial outwash

channels and further comminution of grains by fluvial action, prior to deflation from sand and gravel bars exposed at times of low flow, aeolian transport and further reduction in size by abrasion, and the eventual deposition to form loess (Figure 5.7). Deposits thought to have been formed in this way have been termed 'glacial' loess. The observation that many major loess deposits were located at the fringes of former major ice-sheets (Figure 5.2), such as the Laurentide and Cordilleran Ice Sheets of North America, and the Fennoscandian Ice Sheet of Europe, seemed to add to the credibility of this model. There is certainly little question that silt is produced by glacial-grinding in the present day (Figure 5.8), and observations of contemporary glacierised catchments (i.e. which currently host a glacier) illustrate that glacial processes are extremely effective mechanisms of silt-generation even today. Sediment loads within rivers draining glacierised catchments are sometimes so great that the rivers appear a cloudy/milky-white colour, in dramatic contrast to rivers draining non-glacierised catchments which can often appear black because the waters are so clear. Studies by Hallet et al. (1996) in Alaska show

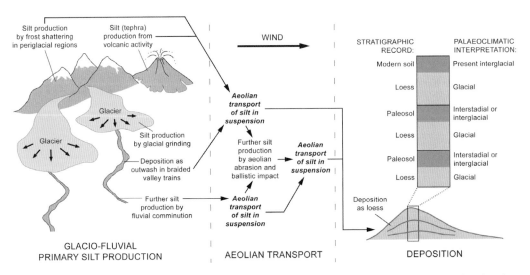

Figure 5.7 'Glacial' model of loess formation involving the production of silt-sized grains primarily by glacial grinding, prior to further transport and deposition. *Source:* Synthesised and modified from Muhs and Bettis Jr. (2003), Muhs (2007), and Muhs (2013). See insert for colour representation of this figure.

Figure 5.8 Dust in suspension on the Delta River floodplain, near Delta Junction, Alaska (July 2004). *Source:* Photograph courtesy of E.A. Bettis III. See insert for colour representation of this figure.

that sediment yields in the rivers draining glacierised basins are up to an order of magnitude higher than those draining non-glacierised basins.

5.5.2 The 'Desert' Model

While the 'glacial' model described above has been shown to be capable of generating the large quantities of silt-sized material suitable for the formation of loess deposits, not all loess deposits are located in areas where glacial-grinding is likely to have taken place. China is one of the key locations often cited when discussing this issue, and the origin and primary production mechanisms of Chinese loess are still very much a matter of debate. Although some researchers have argued that it is likely that processes of glacial-grinding were at work in the high mountainous areas fringing the desert basins of western China, others argue that glacial processes alone are unlikely to be capable of generating sufficient silt to account for the extensive loess deposits found in north-eastern China. Potentially clearer examples of the need for an alternative model of silt

production are provided by more modest loess deposits in hot arid or semi-arid areas which were never glaciated. Such so-called 'desert' loess requires non-glacigenic production mechanisms (Figure 5.9). In the 'desert' loess model, as applied in some parts of the world, frost shattering in adjacent high mountains would produce sediment initially, but a combination of physical and chemical mechanisms is then primarily responsible for the generation of silt-sized material. These mechanisms include the fluvial comminution, and aeolian abrasion and ballistic impacts that feature within the 'glacial' loess model, but additionally the 'desert' loess model includes processes such as salt weathering and chemical weathering of rocks, soils, and sediments (Figure 5.9).

To address the question of silt-production by non-glacigenic means, Crouvi et al. (2010) examined several small loess regions within subtropical deserts in Africa, the Middle East, and Arabia, each of which was found to be located downwind of a neighbouring sand sea at the time of loess deposition. Based on the mineralogy, timing of deposition, and the grain size of the sands and adjacent loess

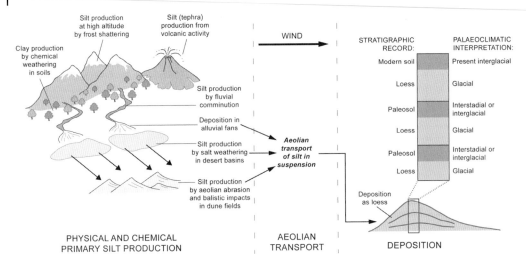

Figure 5.9 'Desert' model of loess formation involving the production of silt-sized grains primarily by various non-glacigenic processes, prior to further transport and deposition. *Source:* Synthesised and modified from Muhs and Bettis Jr. (2003), Muhs (2007), and Muhs (2013). See insert for colour representation of this figure.

(including the typical absence of coarse-silt sized grains), Crouvi et al. (2010) concluded that sand seas had played an important role in the production of 'desert' loess through aeolian abrasion of sand grains to produce silt-sized grains. Another area where there has been a debate about 'desert' loess is in south-eastern Tunisia (Dearing et al. 2001), which has also featured in a recent reiteration of the claim that dust can be created by the breakdown of sands in dunefields (Jerolmack and Brzinski III 2010).

Loess deposits in China may eventually be shown to derive from a mixture of glacigenic and non-glacigenic silt (see Figures 5.7 and 5.9 respectively). This has been shown to be the case for loess deposits of the mid-continent of North America, where loess derived from both non-glacigenic and glacigenic sources was found to contribute to the loess that accumulated at Loveland, Iowa, on the east bluff of the Missouri River throughout the last glacial period (Muhs et al. 2013), while west of the Missouri River at sites within the nearby Great Plains region of Nebraska and Colorado, significant deposition of solely non-glacigenically derived loess from the volcaniclastic White River Group siltstone took place during the last glacial

period (Muhs et al. 2008). 'Desert' loess shares with high latitude loesses the distinction that it accumulates more in the winter than the summer (see Section 5.5.3).

5.5.3 Seasonality and Continuity

There are many views about seasonality. In one, high latitude loesses share a characteristic with 'desert' loess in that they accumulate most during cool, glacial periods, their surface soils forming during wetter, interglacial times (see Figures 5.7 and 5.9, respectively). A second view is that the supply of silt to loess deposits fluctuates over time and does not cease during interglacial conditions, and that soil formation continues as the accumulation of silt continues: this is 'syndepositional' soil formation, being the continued formation of soil in accretionary conditions. This process is certainly observed in some accumulations of peri-desert loess, but the behaviour is less obvious in some high-latitude loess sequences. Another controversy exists as to whether loess accumulation in both high-latitude and peri-desert settings is continuous, even within any single glacial period. Some work has demonstrated that accumulation rates can fluctuate dramatically within a given loess

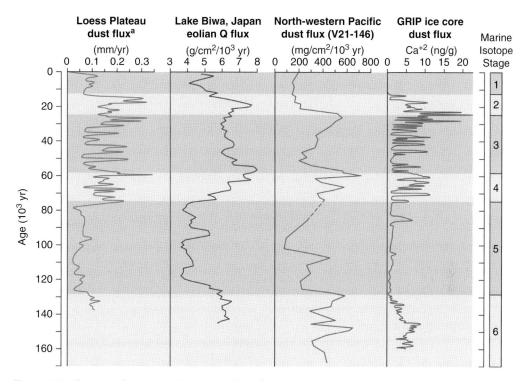

Figure 5.10 Changing dust accumulation rates ('dust flux') over the last glacial–interglacial cycle along a transect encompassing records from different depositional environments, ranging from (left-to-right) the Chinese Loess Plateau, a lacustrine core in Japan, a marine core from the Pacific Ocean, and an ice-core from the GRIP site on the Greenland Ice Sheet. *Source:* Redrawn from Porter (2001) which used data from Porter and An (unpublished), Xiao et al. (1997), Hovan et al. (1991), and GRIP Members (1993)

unit (e.g. Roberts et al. 2003). This work also suggests that loess profiles preserve hiatuses in the loess record (e.g. Stevens et al. 2006, 2007). Some loess sequences seem to be capable of preserving high-resolution records of relatively short-lived events (e.g. Heinrich events and Bond cycles are reported by Chen et al. (1997)). Nevertheless, there is a striking correspondence between the enhanced loess accumulation rates during glacial times and similar trends in the dust record within a variety of depositional archives e.g. lacustrine, marine, and ice-core records (see Figure 5.10).

5.6 Other Accumulations of Dust

Thus far, this chapter has focused on loess. The nature of aeolian dust, however, means that it has the potential to be transported over far greater distances than many other products of geomorphic change, and hence can be deposited in a wide variety of both source-proximal and source-distal settings. Aside from loess, the other key terrestrial setting for the accumulation of dust is lacustrine. Here, there is no reliance on trapping mechanisms such as the growth of vegetation because if wet conditions prevail in the lake basin, it will continue to trap dust and thus have the potential to preserve a long record of accumulation. However, for a meaningful record of dust accumulation to be constructed, any minerogenic aeolian dust input must be disentangled from other sources of sediment input, such as those from fluvial and colluvial processes. The task of distinguishing between aeolian dust and other minerogenic sediments is most easily achieved if the wind-transported dust is derived from a geochemically distinct source some distance from the local parent material from which fluvial and colluvial sediments were derived. Grain size can also be used to distinguish between sediments in situations where the sizes of the aeolian and fluvial sources are distinct.

Figure 5.10 shows the record of dust accumulation reconstructed for Lake Biwa, in Japan, over the last ~145 ka (Xiao et al. 1997). The 50 m-year record is constrained by various well-known tephra, and shows good agreement between the abundance of aeolian quartz over time (calculated as aeolian quartz flux in $g cm^{-2} 10^{-3}$ yr), and a corresponding increase in loess accumulation rate on the Chinese Loess Plateau (expressed as aeolian dust flux in $mm yr^{-1}$ for the classic loess site Luochuan) (see Figure 5.10), indicating that both records are capturing the same event in different depositional settings.

5.6.1 Dust on the Ocean Floor

Many ocean sediments preserve significant accumulations of dust. In some places, rising sea-levels have undoubtedly submerged terrestrial loess, although their fragility, when exposed to water and wave action, means that it is unlikely that any large or extensive drowned loess deposits have remained intact. But the potential of the marine record, extending to over 71% of the Earth's surface, offers a huge opportunity. It is true that marine sedimentary sequences are commonly bioturbated, which has reduced the resolution of the record. For marine and deep-sea settings located far from continental margins, aeolian dust makes a significant contribution to the sedimentary record, although it is likely to be derived from a diverse range of large-scale sources which become increasingly well-mixed with increasing distance of aeolian transport prior to deposition. Loess-like oceanic deposits close to coastlines and major estuaries appear to have come from a number of sources, including aeolian dust, and hence the challenge in their study is similar to that discussed for lacustrine settings. Beneath the oceans, as beneath lakes, the dust record must be disentangled from other sedimentary inputs including turbidites, volcanic ash, ice-rafted debris, and so on. Again, geochemistry (e.g. Olivarez et al. 1991) and mineralogy (e.g. Chester 1990; Hamann et al. 2009) have been used to decipher the dust contribution; grain size is most commonly used for this purpose; pollen, freshwater diatoms, and phytoliths can also yield useful data.

Aeolian dust, fortunately, has distinctive grain size characteristics. The end-member modelling approach (EMMA) proposed by Weltje (1997) and Prins and Weltje (1999) has been used to identify and quantify the aeolian sediment fraction in marine sediments. A study of sediment beneath the Arabian Sea identified three distinct end-members contributing to the sedimentary record, namely, proximal and distal dust and much finer fluvially-derived mud (Prins and Weltje 1999).

At longer timescales, the marine sedimentary record can demonstrate enhanced dust accumulation rates (recorded as dust flux in $mg cm^{-2} 10^{-3}$ yr) which correspond with records of glacial periods derived from terrestrial loess and lacustrine records. Figure 5.10 shows the dust flux determined for the last ~170 of a ~540 ka record derived from distal marine core V21–146 taken from the north-western Pacific Ocean, located to the east of the dust source areas in Asia (Hovan et al. 1991); dust flux here is shown to be greatest during glacial times, when loess accumulation in China and aeolian quartz flux in Lake Biwa, Japan, was also enhanced.

5.6.2 Ice Cores

Ice cores also retain a record of dust accumulation, usually derived from long-range transport. These records can be dated with much greater precision than marine and most terrestrial sequences, and because of the far-travelled nature of the dust preserved in the ice cores, the influence of local or regional sources is typically removed, leaving a signature that represents clear differences between large-scale glacial and interglacial conditions. Figure 5.10 shows the Ca^{+2} record (expressed in ng/g) as a proxy for dust, taken from the GRIP ice core (GRIP Members (Greenland Ice-core Project) 1993); the record shows a clear increase in dust flux during glacial periods. Although the absolute concentration of dust varies across the transect of sites shown in Figure 5.10 because of an increase in transport distance, the timing of these increases in dust flux shows good agreement across the various different dust deposits from source-distal ice-core records to the source-proximal terrestrial deposits of loess and lacustrine sediments, and the intermediate ocean records described earlier.

5.7 The Palaeoclimatic and Palaeoenvironmental Records from Loess and Dust

Unlike lacustrine or marine sediments that contain records of dust accumulation, few loess deposits preserve palaeoecological indicators such as pollen, phytoliths, diatoms, ostracods, foraminifera, or radiolaria, which have been used to reconstruct past environmental conditions in lacustrine and marine settings. However, fossil snail shells are commonly found in loess deposits and can be used for dating using radiocarbon techniques, as well as being used to reconstruct changing vegetation and climatic conditions over the time period of loess accumulation (e.g. Rousseau 2001).

The distribution, spatial extent, and changing particle-size and thickness of loess (and other accumulations of dust) reveal much about past ('palaeo') wind direction and wind strength, and likely source areas of dust, as discussed previously. In addition, the sequences of buff-coloured loess units and (typically red-brown coloured) intercalated palaeosols that are often clearly visible in the field (see Figure 5.5), bear witness to major changes from glacial to interglacial

conditions (see Figures 5.7 and 5.9). The palaeosols are not always visible to the naked eye, particularly in dry conditions in the field, but mineral magnetics, including magnetic susceptibility measurements, have been widely used to identify and characterise loess-palaeosol sequences (reviewed by Maher 2011). These records have been used to correlate loess-palaeosol sequences across different sites and regions (e.g. Kukla and An 1989), and also to link loess records to other proxies such as grain size and to oxygen isotope records from lacustrine and marine records (e.g. Heller and Liu 1984). In China, magnetic susceptibility measurements have been used to infer the strength of the warm, humid, East Asian summer monsoon. Mineral magnetics have also been used to yield quantitative reconstructions of palaeorainfall derived from palaeosols, although opinion is divided as to the validity of this approach (e.g. Maher et al. 1994; Porter et al. 2001), which requires site-specific consideration of the suitability of the record. The greatest success appears to have been achieved in 'suitable' soils described by Maher (2011) as accretionary, well-drained and well-buffered soils with weakly magnetic parent material, such as those found on the Chinese Loess Plateau.

Figure 5.10 shows that records of dust flux from all these materials and types of site demonstrate that dust accumulation has tended to increase during glacial periods and decline in interglacial times, regardless of how the dust has been generated (see Figure 5.9). Records of dust trapped within ice cores indicate that dust flux at high latitudes was enhanced by as much as 20 times during glacial compared to interglacial conditions, while marine and terrestrial records of dust flux from the last glacial maximum (LGM, ~21 ka) suggest that accumulation rates at the LGM were up to 10 times greater than those of the present day (Kohfeld and Harrison 2001). The increased dust flux recorded in these various dust archives during glacial periods could be due to a number of factors, including increased aridity, decreased intensity of the hydrologic cycle, strengthened and/or more persistent winds, decreased vegetation cover, and increased sediment supply and availability, including through exposure of continental shelves through lowered sea levels in glacial times (e.g. Mahowald et al. 1999; McGee et al. 2010).

Changes in dust flux recorded at a given site are often accompanied by changes in grain size, inferring a change in wind strength, and/or a change in the distance between the dust source and the area of deposition, and/or a change in the nature of the source material. Changes in grain size parameters (e.g. mean or median grain size, proportion of grains in coarse grain sizes [e.g. >40 μm], ratio of coarse to fine silt, or the ratio of coarse and medium silt to fine silt) have been widely interpreted as a proxy for changing wind strength and/or persistence. Across the Chinese Loess Plateau, these parameters have been widely used to infer the strength of the East Asian winter monsoon (e.g. Ding et al. 1994), with grain size increasing in loess units and hence winter monsoon intensity has been inferred to have increased during glacial periods. More recently, the distance from the dust source to the area of deposition has also been shown to be an important control on the grain size of Chinese loess on long (orbital) timescales (e.g. Yang and Ding 2008), reflecting changing desert margins which in turn are influenced by the East Asian summer monsoon which controls rainfall. Grain size records in China therefore serve as a proxy for changes in aridity (Yang and Ding 2014), with coarse grain sizes reflecting cool and dry conditions, while finer grain sizes indicate a shift to warmer and moister conditions. On millennial timescales, grain size changes in Chinese loess have been linked to abrupt climate events (such as Heinrich and Dansgaard/Oeschger events) triggered in the North Atlantic and polar regions which in turn bring about changes in the East Asian winter monsoon (e.g. Yang and Ding 2014).

Chinese loess/palaeosol sequences accumulated over the Quaternary period (i.e. during the last 2.6 million years), and frequently overlie a visually-striking, older red clay unit of mainly Neogene age. Much of this red clay was deposited between ~2.6 to 8 million years ago and is believed to be primarily aeolian in origin (e.g. Lu et al. 2001); however, in the western Chinese Loess Plateau the red clay deposits are much older, with basal ages of 22 million years (Ma) and even 25 Ma. There is significant controversy regarding the likely origin of these older red clay deposits, with fluvial/alluvial/slopewash processes potentially at work before ~eight million years ago (e.g. see review by Nie et al. 2016). Nevertheless, the Chinese loess/red clay sequences preserve a record of both global and regional climate history, and have been used to reconstruct changes in the Asian monsoon climate system and in aridification of inland Asia over the most recent tens of millions of years.

5.8 Dating Loess and Dust Deposits

At the most basic level, individual loess deposits offer a clear relative chronology of events based on the law of superposition and the visible alternation between loess units and intercalated palaeosols that may be preserved over numerous glacial–interglacial cycles. Rhythmic variability between loess and palaeosol sequences is also typically observed via proxies such as dust flux, grain size, magnetic susceptibility, and the degree of amino acid racemisation. The records from these individual loess deposits can be correlated to other terrestrial loess-palaeosol sequences and/or to records from more distal dust-bearing marine sediments or ice core records by making visual comparisons and wiggle-matching the proxy records (e.g. Figure 5.10). Alternatively, age-equivalent or isochronous stratigraphic markers such as tephra and palaeomagnetic reversals can be used to correlate between loess and/or other dust-bearing deposits. Thus, the coarse framework of loess-palaeosol sequences can be mapped onto that of marine and ice-core records. However, matching long sedimentary sequences at only a few key tie points and extrapolating between these tie-points (thereby assuming a constant accumulation rate, and a continuous sediment record with no hiatus between tie-points), and/or wiggle-matching proxy records between different sites/depositional settings or tuning one of the palaeoenvironmental proxies to orbital insolation, can lead to circular arguments or necessitate assumptions between records or proxies.

Instead, direct numerical or radiometric dating of the sediments is clearly preferable, ideally throughout the entire sedimentary sequence to enable records to be linked unambiguously and hence to explore (rather than necessarily assume) correlations. One technique for generating numerical chronologies for loess-palaeosol sequences is radiocarbon dating, which has typically been applied to humic acids found within the palaeosols. Such dating can be challenging where organic contents are low, or where the event being dated is the formation of the palaeosol which developed post-depositionally within the loess deposit, rather than dating the deposition of the silt that makes up the loess. Where suitable organic-rich material such as charcoal or particular species of gastropod shells (e.g. *Succineidae* shells (Pigati et al. 2013)) can be found within loess deposits, and are believed to be in situ, such macrofossils are usually preferable for radiocarbon dating rather than conducting bulk humic acid extractions on weakly-organic sedimentary units.

Recent years have also seen significant advances in the development and the use of luminescence dating (e.g. reviews by Roberts 2008, 2015; Rhodes 2011). Using luminescence techniques (Box 5.2), it is possible to

---

Box 5.2 Luminescence Dating of Sediments

The term 'luminescence dating' refers to a family of chronologic methods usually applied to the commonly occurring minerals quartz and feldspar. Luminescence dating techniques exploit a time-dependent signal that builds up in the mineral grains by exposure to naturally-occurring ionising radiation (principally from uranium, thorium, and potassium), recording the time elapsed since these mineral grains were last exposed to sunlight or to heating. For sedimentary deposits, the event being dated is the last exposure to sunlight, i.e. typically the time of deposition of the sediment.

Loess and dust deposits are ideally suited to luminescence dating because the fine-grained and aeolian nature of the deposits implies medium- to long-range transport distances, giving ample opportunity for the removal (or 'bleaching') of any previous luminescence signal during transport, prior to deposition. The broad chronostratigraphic framework offered by loess-palaeosol sequences spanning multiple glacial–interglacial cycles offered an ideal testing-ground for newly proposed luminescence methods, and hence loess deposits were the first terrestrial sediments to which luminescence dating techniques were systematically applied, tested, and developed (Roberts 2008). Consequently, some luminescence ages produced during certain developmental phases of luminescence dating methods are now considered to be questionable, whilst other luminescence ages and dating techniques endure and have been shown to be very robust. One of these is the Single Aliquot Regenerative dose (SAR) methods developed by Murray and Wintle (2000) for Optically Stimulated Luminescence (OSL) dating of quartz grains. New methods continue to develop and show great promise, e.g. new infrared stimulated luminescence (IRSL) methods arising from the work of Thomsen et al. (2008), using post-IR IRSL signals measured at elevated temperatures to reduce or even circumvent the issues of anomalous fading that had previously plagued much of the early dating work applied to polyminerals and feldspar grains.

date the mineral grains of quartz and/or feldspar that comprise sedimentary deposits, giving numerical chronologies that relate directly to the time of deposition when, for sediments in the natural environment, these grains were last exposed to sunlight prior to deposition and burial by the accumulation of further sediment grains. The term 'luminescence dating' actually encompasses a variety of methods applied chiefly to quartz and/or to feldspar grains that can be of different sizes, ranging from very-fine and fine-silts through to medium-sands. Feldspars and quartz are the first and second most abundant minerals on Earth, and thus unlike other numerical/radiometric techniques such as radiocarbon dating, it is rarely a challenge to find suitable, sufficient, in situ material for dating by luminescence. The numerical ages generated using luminescence techniques extend beyond the upper limit of radiocarbon dating, and have been used to provide independent chronologies for terrestrial loess-palaeosol sequences, lacustrine, and marine sediments, avoiding the potential circularity of tuning proxy records to orbital parameters, and instead permitting investigations of the synchronicity between loess/dust deposits (e.g. Lai and Wintle 2006; Stevens et al. 2006, 2007; Lai et al. 2007) and records of palaeoclimatic and palaeoenvironmental change.

The value of generating numerical chronologies rather than inferring the timescale from stratigraphic changes and/or from proxy data, with all of the inbuilt assumptions and circularity that potentially brings, is highlighted in recent work by Dong et al. (2015). Using optically-stimulated luminescence dating applied to three loess sections in a south-east to north-west transect across the Chinese Loess Plateau, they reveal that the Pleistocene/Holocene transition was defined by rapid changes in magnetic susceptibility that are not time-synchronous across the sites (occurring ~10.5, 8.5, 7.5 ka); this observation challenges the assumptions usually made about the synchronicity of such events across sites. Furthermore, Dong et al.

(2015) also note that these transitions are not synchronous with the 12.05 ka age of the Marine Isotope Stage 2/1 boundary to which such changes in magnetic susceptibility would normally have been automatically correlated if no independent numerical chronology had been available.

It is only through direct numerical dating of the sedimentary deposits that the quasi-continuous nature of loess-palaeosol sequences can be demonstrated. Hiatuses have been demonstrated within some loess deposits (e.g. Lu et al. 2006; Stevens et al. 2006, 2007; Buylaert et al. 2008; Yang and Ding 2014), while in other loess deposits, accumulation appears continuous at the resolution of the study conducted (e.g. Lai et al. 2007; Buylaert et al. 2008; Lai 2010; Sun et al. 2012; Yang and Ding 2014) and may not have ceased even during the formation of palaeosols. Luminescence dating of loess-palaeosol sequences has also revealed significant variability in the (mass) accumulation rates both within and between loess and palaeosol units (e.g. Roberts et al. 2003; Lu et al. 2007).

5.9 The Role of Dust in Climate Change

Enhanced rates of dust and loess accumulation have been noted as a response to glacial periods. The records are preserved in terrestrial loess-palaeosol sequences formed by both glacigenic and non-glacigenic (desert) processes, in lacustrine sediments, and also within distal marine and ice-core records (e.g. Figure 5.10, discussed previously). They probably reflect a shift to a dry, cold climate, strengthened, and persistent windy conditions, with abundant sediment supply and little vegetation cover. However, such increased dust flux during glacial times is not just a *product* of climate change: any increase in levels of atmospheric dust, regardless of how they are brought about, can also potentially act as an *agent* of climate change (Overpeck et al. 1996; Tegen et al. 1996; Harrison et al. 2001; Bar-Or et al. 2008).

As an agent of change, dust contributes to further climate change via feedback mechanisms, such as ocean–atmosphere forcings (e.g. through direct radiative forcing by mineral dust in the atmosphere), ice-sheet responses (e.g. through changes to albedo caused by dust deposition on ice-sheets) and bio-feedbacks (e.g. changes in land surface and ocean productivity caused by iron-fertilisation from dust added to nutrient-starved soils and ocean waters). For example, Roberts et al. (2003) noted unprecedented high mass accumulation rates compared to any other pre-Holocene locality worldwide, occurring between ~18–14 ka for the last glacial (Peoria) loess that accumulated in the North American mid-continent; this event was coincident with the presence of extra-limital cool-taxa snail species at the site (i.e. snails that are usually found in the much cooler localities far beyond this region), indicating that cold conditions persisted for several thousands of years after summer insolation exceeded present-day values. It was suggested that the unprecedented high dust flux during this time may have maintained a cooler-than-present climate in this region through dust-induced radiative forcing of climate, highlighting the need for Atmospheric General Circulation Models (AGCMs) to incorporate dust loading into regional scale models (Roberts et al. 2003).

Interest in dust flux and the role of dust in climate change is growing, and various attempts have been made to document and collate data regarding mass accumulation rates over key time periods (e.g. Kohfeld and Harrison 2001; Maher and Kohfeld 2009; Maher and Leedal 2014; Albani et al. 2015) prior to incorporation into regional or global atmospheric circulation models. In spite of these efforts to collect and collate data, there are still significantly large spatial and temporal gaps in the databases and our knowledge of dust flux remains inadequate for satisfactory testing of dynamic atmospheric global circulation models (AGCMs), although work focusing on either narrow time-slices, and/or regional studies of dust mass accumulation

rates has benefitted from such data synthesis exercises involving palaeodatasets (e.g. Albani et al. 2015). The record of ancient fluxes of dust preserved in loess, and other dust deposits, is important for the development and validation of regional and global dust and climate models, and can also improve our understanding of the impact of dust on climate. Global dust model work by Mahowald et al. (2006), for example, used mass accumulation rates determined from Last Glacial terrestrial loess deposits to include and constrain the size of glacigenic dust sources during the Last Glacial Maximum (LGM), to examine the repercussions of the inclusion of these glacigenic sources which, hitherto, typically had been omitted from previous models of LGM dust and climate.

5.10 Cultural, Economic, and Environmental Significance of Loess and Dust

Loess and dust influence the health and economy of a significant proportion of the world's population. Many loess soils are extremely fertile, and it is no coincidence that the key 'breadbaskets' that sustain so much agricultural activity are located on loess deposits. The great fertility and well-drained nature of loess deposits also make them particularly well-suited to the growth of vines for wine production. Indeed, some of the earliest farmers were those of the loess soils in China and these pioneers may have produced enough CO_2 to slow the cooling of the Earth, and hence, ironically, may have staved off being buried in a new layer of loess (Ruddiman et al. 2015).

In some regions, the addition of nutrient-rich dust may potentially play a significant role in enhancing otherwise relatively infertile soils; for example, Muhs et al. (2007) demonstrated the importance of inputs from African dust throughout much of the Quaternary period to the development of

(a)

(b)

Figure 5.11 Two examples of loess 'cave' dwellings in the Chinese Loess Plateau, carved into (a) loess exposures and (b) hillsides.

soils on Barbados. The addition of dust to the oceans can also affect productivity through the addition of iron and essential micronutrients, causing a growth in plankton, which in turn can potentially influence the climate by lowering the atmospheric concentration of carbon dioxide (e.g. Martínez-Garcia et al. 2014).

As well as playing an important role in the production of crops, loess also provides shelter in many parts of the Chinese Loess Plateau in the form of loess houses or 'caves' constructed either by carving dwelling-spaces into hillsides or by excavating an interior courtyard leading to dwelling-spaces (Figure 5.11). These loess dwellings maintain a consistent temperature throughout the year, and therefore benefit from being cool in the summer months and relatively warm in the winter months. In other loess regions which sustain vineyards, the stable temperature within loess deposits has also led to loess caves being excavated for the storage and maturation of wine. However, loess can also be very unstable when water is added and/or when earthquakes strike, and significant loss of life (e.g. >100 000 people in the 1920 Haiyuan earthquake, China) can be experienced due to the large-scale, catastrophic landslides that can be induced. Loess and dust also pose health concerns for human and animal health e.g. through

pneumoconiosis, or non-industrial silicosis, resulting from exposure to mineral dust concentrations of silica-rich particles of respirable size (<10 μm, and particularly <2 μm) (Derbyshire 2001).

5.11   Conclusion

Spatially-extensive loess and dust deposits shape large areas of our terrestrial landscape, and play an important role in the economy and health of many of the people living in these areas. Loess and its analogues beneath lakes and oceans, and dust trapped in ice cores, preserve records of past climate and environmental change. Dust also plays an active role in climate change through feedback mechanisms such as direct and indirect radiative forcing, changes to atmospheric chemistry, and changes to biogeochemical cycles in the oceans and on land through fertilisation of nutrient-starved waters and soils. There is still much debate regarding likely levels of dust in an increasingly warming world, and the nature of subsequent climate feedbacks in the future; however, even without climate change, there is clear potential for enhanced levels of dust in the future just on the strength of the inevitable increase in human population and the associated destabilisation of the landscape.

Further Reading

The single best general and comprehensive source of information on dust and loess remains Pye's book (1987), which gives an excellent and classic overview of the topic. The more recent book by Knippertz and Stuut (2014) addresses the characteristics of dust and dust deposits, but also considers the global dust cycle, and brings the subject to the present, when dust is viewed as an active agent of climate change, as well as providing a more passive record of past climate and environmental conditions. Various articles in the Encyclopaedia edited by Elias

(2013) cover loess, dust, and aeolian deposits from the perspective of the Quaternary records of past climate that they preserve, and an excellent overview of the 'geomorphology, stratigraphy, and paleoclimatic significance' of loess is provided by the open-access article by Muhs (2013). A detailed but approachable general review of luminescence dating is given by Rhodes (2011). Roberts (2008) charts the developments of luminescence dating as specifically applied to loess, from the earliest days of the technique through to the most recent developments (including Optically Stimulated Luminescence (OSL)) at the time of publication, while the Encyclopaedia article by Roberts (2015) includes some additional significant advances since that time. An insight into some of the characters and personalities involved in loess research of all kinds, as well as some of their key findings, is provided by *Loess Letter*, the newsletter of the International Union of Quaternary Research (INQUA) Loess Focus Group; this informal newsletter is playing an increasingly important role in archiving some of the classic loess texts, including some translated works that may otherwise go unread in the 'western' literature.

References

Albani, S., Mahowald, N.M., Winckler, G. et al. (2015). Twelve thousand years of dust: the Holocene global dust cycle constrained by natural archives. *Climate of the Past* 11: 869–903. doi: 10.5194/cp-11-869-2015.

An, Z., Sun, Y., Zhou, W. et al. (2014). Chinese loess and the East Asian Monsoon. In: *Late Cenozoic Climate Change in Asia: Loess, Monsoon and Monsoon-Arid Environment Evolution* (ed. Z.). An). Dordrecht: Springer. doi: 10.1007/978-94-007-7817-7_2.

Avni, Y. (2001). *Geological Map of Mt. Loz.* Jerusalem: Geological Survey of Israel.

Bar-Or, R., Erlick, C., and Gildor, H. (2008). The role of dust in glacial–interglacial cycles. *Quaternary Science Reviews* 27 (304): 201–208. doi: 10.1016/j.quascirev.2007.10.015.

Berg, L.S. (1916). O proiskhozhdenii lessa (the origin of loess). Izviestiia Imperatorskago russkago geograficheskago obshchestva 52(8): 579–647. (in Russian) (also in Smalley, I.J. (ed.), Loess; lithology and genesis. Benchmark papers in geology 26 (1975), Dowden, Hutchinson & Ross, Inc., Stroudsburg, PA, pp. 61–75, in English).

Butler, B.E. (1956). Parna – and aeolian clay. *Australian Journal of Science* 18 (5): 145–151.

Buylaert, J.-P., Murray, A.S., Vandenberghe, D. et al. (2008). Optical dating of Chinese loess using sand-sized quartz: establishing a time frame for late Pleistocene climate changes in the western part of the Chinese Loess Plateau. *Quaternary Geochronology* 3 (1–2): 99–113. doi: 10.1016/j.quageo.2007.05.003.

Chen, F., Bloemendal, J., Wang, J. et al. (1997). High-resolution multi-proxy climate records from Chinese loess: evidence for rapid climatic changes over the last 75 kyr. *Palaeogeography Palaeoclimatology Palaeoecology* 130 (1–4): 323–335. doi: 10.1016/S0031-0182(96).

Chester, R. (1990). The atmospheric transport of clay minerals to the world ocean. In: Farmer, V.C. and Tardy, Y. (eds), 9th International Clay Conference, Strasbourg, Germany, August 1989, Science Géologiques, Mémoires 88, pp. 23–32.

Crouvi, O., Amit, R., Enzel, Y., and Gillespie, A.R. (2010). Active sand seas and the formation of desert loess. *Quaternary Science Reviews* 29 (17–18): 2087–2098. doi: 10.1016/j.quascirev.2010.04.026.

Crouvi, O., Amit, R., Porat, N. et al. (2009). Significance of primary hilltop loess in reconstructing dust chronology, accretion rates, and sources: an example from the Negev Desert, Israel. *Journal of Geophysical Research: Earth Surface* 114: F02017. doi: 10.1029/2008JF001083.

Dare-Edwards, A.J. (1984). Aeolian clay deposits of South-Eastern Australia: parna or loessic clay? *Transactions of the Institute of British Geographers NS* 9 (3): 337–344. doi: 10.2307/622237.

Dearing, J.A., Livingstone, I.P., Bateman, M.D., and White, K. (2001). Palaeoclimate records from OIS 8.0–5.4 recorded in loess-palaeosol sequences on the Matmata Plateau, southern Tunisia, based on mineral magnetism and new luminescence dating.

In: Derbyshire, E. (ed.), Loess and palaeosols: characteristics, chronology and climate: a contribution to IGCP413. Proceedings, International Conference, Universität Bonn, March 25–April 1. *Quaternary International* 76–77: 43–56. doi: 10.1016/S1040-6182(00)00088-4.

Derbyshire, E. (2001). Geological hazards in loess terrain, with particular reference to the loess regions of China. In: Derbyshire, E. (ed.), Recent research on loess and palaeosols, pure and applied; selection of keynote addresses presented at "Loessfest '99". *Earth-Science Reviews* 54 (1–3): 231–260. doi: 10.1016/S0012-8252(01)00050-2.

Ding, Z., Yu, Z., Rutter, N.W., and Liu, T. (1994). Towards an orbital time scale for Chinese loess deposits. *Quaternary Science Reviews* 13 (1): 39–70. doi: 10.1016/S0277-3791(99)00017-7.

Dong, Y., Wu, N., Li, F. et al. (2015). Time-transgressive nature of the magnetic susceptibility record across the Chinese Loess Plateau at the Pleistocene/Holocene transition. *PLoS One* 10: e0133541. doi: 10.1371/journal.pone.0133541.

Elias, S.A. (2013). *Encyclopedia of Quaternary Science*, 2e. North-Holland: Elsevier.

GRIP Members (Greenland Ice-core Project) (1993). Climate instability during the last interglacial period recorded in the GRIP ice core. *Nature* 364: 203–207. doi: 10.1038/364203a0.

Haberlah, D. (2007). A call for Australian loess. *Area* 39 (2): 224–229. doi: 10.1111/j.1475-4762.2007.00730.x.

Hallet, B., Hunter, L., and Bogen, J. (1996). Rates of erosion and sediment evacuation by glaciers: a review of field data and their implications. *Global and Planetary Change* 12: 213–235. doi: 10.1016/0921-8181(95)00021-6.

Hamann, Y., Ehrmann, W., Schmiedl, G., and Kuhnt, T. (2009). Modern and late Quaternary clay mineral distribution in the area of the SE Mediterranean Sea. *Quaternary Research* 71 (3): 453–464. doi: 10.1016/j.yqres.2009.01.001.

Harrison, S.P., Kohfeld, K.E., Roelandt, C., and Claquin, T. (2001). The role of dust in climate changes today, at the last glacial maximum and in the future. *Earth-Science Reviews* 54 (1–3): 43–80. doi: 10.1016/S0012-8252(01)00041-1.

Heller, F. and Liu, T. (1984). Magnetism of Chinese loess deposits. *Geophysical Journal of the Royal Astronomical Society* 77: 125–141. doi: 10.1111/j.1365-246X.1984.tb01928.x.

Hovan, S.A., Rea, D.K., and Pisias, N.G. (1991). Late Pleistocene climate and oceanic variability recorded in Northwest Pacific sediments. *Paleoceanography* 6: 349–370. doi: 10.1029/91PA00559.

Jerolmack, D.J. and Brzinski, T.A. III (2010). Equivalence of abrupt grain-size transitions in alluvial rivers and eolian sand seas: a hypothesis. *Geology* 38 (8): 719–722. doi: 10.1130/G30922 (Erratum: 38(10): 886).

Knippertz, P. and Stuut, J.-B.W. (eds.) (2014). *Mineral Dust – a Key Player in the Earth System*. New York: Springer.

Kohfeld, K.E. and Harrison, S.P. (2001). DIRTMAP: the geological record of dust. In: Derbyshire, E. (ed.), Recent research on loess and palaeosols, pure and applied; selection of keynote addresses presented at "Loessfest '99". *Earth-Science Reviews* 54 (1–3): 81–114. doi: 10.1016/S0012-8252(01)00042-3.

Kukla, G. and An, Z. (1989). Loess stratigraphy in Central China. *Palaeogeography, Palaeoclimatology, Palaeoecology* 72: 203–225. doi: 10.1016/0031-0182(89)90143-0.

Lai, Z. (2010). Chronology and the upper dating limit for loess samples from Luochuan section in the Chinese Loess Plateau using quartz OSL SAR protocol. *Journal of Asian Earth Sciences* 37: 176–185. doi: 10.1016/j.jseaes.2009.08.003.

Lai, Z. and Wintle, A.G. (2006). Locating the boundary between the Pleistocene and the Holocene in Chinese loess using luminescence. *The Holocene* 16 (6): 893–899. doi: 10.1191/0959683606hol980rr.

Lai, Z., Wintle, A.G., and Thomas, D.S.G. (2007). Rates of dust deposition between

50 ka and 20 ka revealed by OSL dating at Yuanbao on the Chinese Loess Plateau. *Palaeogeography, Palaeoclimatology, Palaeoecology* 248: 431–439. doi: 10.1016/j.palaeo.2006.12.013.

Liu, T. et al. (1985). *Loess and the Environment*, 1e. Beijing: Ocean Press.

Lu, H., Stevens, T., Yi, S., and Sun, X. (2006). An erosional hiatus in Chinese loess sequences revealed by closely spaced optical dating. *Chinese Science Bulletin (Ke xue tong bao English)* 51 (18): 2253–2259. doi: 10.1007/s11434-006-2097-x.

Lu, H., Vandenberghe, J., and An, Z. (2001). Aeolian origin and palaeoclimatic implications of the 'Red Clay' (North China) as evidenced by grain-size distribution. *Journal of Quaternary Science* 16 (1): 89–97. doi: 10.1002/1099-1417(200101)16:1<89::AID-JQS578>3.0.CO;2-8.

Lu, Y., Wang, X.L., and Wintle, A.G. (2007). A new OSL chronology for dust accumulation in the last 130,000 yr for the Chinese Loess Plateau. *Quaternary Research* 67 (1): 152–160. doi: 10.1016/j.yqres.2006.08.003.

Maher, B.A. (2011). The magnetic properties of Quaternary aeolian dusts and sediments, and their palaeoclimatic significance. *Aeolian Research* 3 (2): 87–144. doi: 10.1016/j.aeolia.2011.01.005.

Maher, B.A. and Kohfeld, K.E. (2009). DIRTMAP Version 3. LGM, Late Holocene, and modern eolian fluxes from ice cores, marine sediments, marine sediment traps, and terrestrial deposits. http://www.lec.lancs.ac.uk/dirtmap3.

Maher, B.A. and Leedal, D.T. (2014). DIRTMAP: development of a web-based dust archive. *PAGES Magazine* 22: 90.

Maher, B.A., Thompson, R., and Zhou, L.P. (1994). Spatial and temporal reconstructions of changes in the Asian palaeomonsoon: a new mineral magnetic approach. *Earth and Planetary Science Letters* 125 (1–4): 461–471. doi: 10.1016/0012-821X(94)90232.

Mahowald, N.M., Kohfeld, K., Hansson, M. et al. (1999). Dust sources and deposition during the last glacial maximum and current climate: a comparison of model results with paleodata from ice cores and marine sediments. *Journal of Geophysical Research: Atmospheres* 104 (D13): 15895–15916. doi: 10.1029/1999JD900084.

Mahowald, N.M., Muhs, D.R., Levis, S. et al. (2006). Change in atmospheric mineral aerosols in response to climate: last glacial period, preindustrial, modern, and doubled carbon dioxide climates. *Journal of Geophysical Research: Atmospheres* 111: D10202. doi: 10.1029/2005JD006653.

Martínez-Garcia, A., Sigman, D.M., Ren, H. et al. (2014). Iron fertilisation of the Subantarctic Ocean during the last Ice Age. *Science* 343 (6177): 1347–1350. doi: 10.1126/science.1246848.

McGee, D., Broecker, W.S., and Winckler, G. (2010). Gustiness: the driver of glacial dustiness? *Quaternary Science Reviews* 29 (17–18): 2340–2350. doi: 10.1016/j.quascirev.2010.06.009.

Muhs, D.R. (2007). Loess deposits, origins and properties. In: *Encyclopedia of Quaternary Science*, vol. 2 (ed. S.A. Elias), 1405–1418. Amsterdam: Elsevier.

Muhs, D.R. (2013). The geologic records of dust in the Quaternary. *Aeolian Research* 9: 3–48. doi: 10.1016/j.aeolia.2012.08.001.

Muhs, D.R. and Bettis, E.A. Jr. (2003). Quaternary loess-paleosol sequences as examples of climate-driven sedimentary extremes. In: *Extreme Depositional Environments; Mega End Members in Geologic Time*, vol. 370 (ed. M.A. Chan and A.W. Archer), 53–74. Geological Society of America, Special Paper.

Muhs, D.R., Bettis, E.A. III, Aleinikoff, J.N. et al. (2008). Origin and paleoclimatic significance of late Quaternary loess in Nebraska; evidence from stratigraphy, chronology, sedimentology, and geochemistry. *Geological Society of America Bulletin* 120 (11–12): 1378–1407. doi: 10.1130/B26221.1.

Muhs, D.R., Bettis, E.A. III, Roberts, H.M. et al. (2013). Chronology and provenance of last-glacial (Peoria) loess in western Iowa and paleoclimatic implications. *Quaternary Research* 80 (3): 468–481. doi: 10.1016/j.yqres.2013.06.006.

Muhs, D.R., Budahn, J.R., Prospero, J.M., and Carey, S.N. (2007). Geochemical evidence for African dust inputs to soils of western Atlantic islands: Barbados, the Bahamas, and Florida. *Journal of Geophysical Research. Earth Surface* 112: F02009. doi: 10.1029/2005JF000445.

Murray, A.S. and Wintle, A.G. (2000). Luminescence dating of quartz using an improved single-aliquot regenerative-dose protocol. *Radiation Measurements* 32 (1): 57–73. doi: 10.1016/S1350-4487(99)00253-X.

Nie, J., Song, Y., and King, J.W. (2016). A review of recent advances in red-clay environmental magnetism and paleoclimate history on the Chinese Loess Plateau. *Frontiers in Earth Science* 4: 27. doi: 10.3389/feart.2016.00027.

Olivarez, A.M., Owen, R.M., and Rea, D.K. (1991). Geochemistry of eolian dust in Pacific pelagic sediments; implications for paleoclimatic interpretations. *Geochimica et Cosmochimica Acta* 55 (8): 2147–2158. doi: 10.1016/0016-7037(91)90093-K.

Overpeck, J.T., Rind, D., Lacis, A., and Healy, R. (1996). Possible role of dust-induced regional warming in abrupt climate change during the last glacial period. *Nature* 384: 447–449. doi: 10.1038/384447a0.

Pécsi, M. (1990). Loess is not just the accumulation of dust. *Quaternary International* 7–8: 1–21. doi: 10.1016/1040-6182(90)90034-2.

Pigati, J.S., McGeehin, J.P., Muhs, D.R., and Bettis, E.A. (2013). Radiocarbon dating late Quaternary loess deposits using small terrestrial gastropod shells. *Quaternary Science Reviews* 76: 114–128. doi: 10.1016/j.quascirev.2013.05.013.

Porter, S.C. (2001). Chinese loess record of monsoon climate during the last glacial-interglacial cycle. In: Derbyshire, E. (ed.), Recent research on loess and palaeosols, pure and applied; selection of keynote addresses presented at "Loessfest '99". *Earth-Science Reviews* 54 (1–3): 115–128. doi: 10.1016/S0012-8252(01)00043-5.

Porter, S.C., Hallett, B., Wu, X., and Zhisheng, A. (2001). Dependence of near-surface magnetic susceptibility on dust accumulation rate and precipitation on the Chinese Loess Plateau. *Quaternary Research* 55 (3): 271–283. doi: 10.1006/qres.2001.2224.

Prins, M.A. and Weltje, G.J. (1999). End-member modeling of siliclastic grain-size distributions: the late Quaternary record of eolian and fluvial sediment supply to the Arabian Sea and its paleoclimatic significance. *SEPM Special Publications* 63: 91–111. doi: 10.2110/pec.99.62.0091.

Pye, K. (1984). Loess. *Progress in Physical Geography* 8 (2): 176–217.

Pye, K. (1987). *Aeolian Dust and Dust Deposits*. London: Academic Press.

Pye, K. (1995). The nature, origin and accumulation of loess. *Quaternary Science Reviews* 14 (7–8): 653–667. doi: 10.1016/0277-3791(95)00047-X.

Rhodes, E.J. (2011). Optically stimulated luminescence dating of sediments over the past 200,000 years. *Annual Review of Earth and Planetary Sciences* 39: 461–488. doi: 10.1146/annurev-earth-040610-133425.

Roberts, H.M. (2008). The development and application of luminescence dating to loess deposits: a perspective on the past, present and future. *Boreas* 37 (4): 483–507. doi: 10.1111/j.1502-3885.2008.00057.x.

Roberts, H.M. (2015). Luminescence dating, loess. In: *Encyclopedia of Scientific Dating Methods* (ed. W.J. Rink and J. Thompson), 409–414. New York: Springer.

Roberts, H.M., Muhs, D.R., Wintle, A.G. et al. (2003). Unprecedented last-glacial mass accumulation rates determined by luminescence dating of loess from western Nebraska. *Quaternary Research* 59 (3): 411–419. doi: 10.1016/S0033-5894(03)00040-1.

Rousseau, D.D. (2001). Loess biostratigraphy: New advances and approaches in mollusk studies. In: Derbyshire, E. (ed.) Recent research on loess and palaeosols, pure and applied; selection of keynote addresses presented at "Loessfest '99". *Earth-Science Reviews* 54 (1–3): 157–171. doi: 10.1016/S0012-8252(01)00046-0.

Ruddiman, W.F., Fuller, D.Q., Kutzbach, J.E. et al. (2015). Late Holocene climate: natural or anthropogenic? *Reviews of Geophysics* 54 (1): 93–118. doi: 10.1002/2015RG000503.

Russell, R.J. (1944). Lower Mississippi valley loess. *Geological Society of America Bulletin* 55 (1): 1–40. doi: 10.1130/GSAB-55-1.

Smalley, I. and Marković, S.B. (2014). Loessification and hydroconsolidation: there is a connection. *Catena* 117: 94–99. doi: 10.1016/j.catena.2013.07.006.

Smalley, I.J., Jefferson, I.F., Dijkstra, T.A., and Derbyshire, E. (2001). Some major events in the development of the scientific study of loess. In: Derbyshire, E. (ed.), recent research on loess and palaeosols, pure and applied; selection of keynote addresses presented at "Loessfest '99". *Earth-Science Reviews* 54 (1–3): 5–18. doi: 10.1016/S0012-8252(01)00038-1.

Smalley, I.J., Marković, S.B., and Svircev, Z. (2011). Loess is [almost totally formed by] the accumulation of dust. *Quaternary International* 240 (1–2): 4–11. doi: 10.1016/j.quaint.2010.07.011.

Smalley, I.J. and Vita-Finzi, C. (1968). The formation of fine particles in sandy deserts and the nature of 'desert' loess. *Journal of Sedimentary Petrology* 38 (3): 766–774. doi: 10.1306/74D71A69-2B21-11D7-8648000102C1865D (also in Smalley, I.J. (ed), Loess; lithology and genesis. Benchmark papers in geology 26, (1975), Dowden, Hutchinson & Ross, Inc., Stroudsburg, PA, pp. 294–302.).

Stevens, T., Armitage, S.J., Lu, H., and Thomas, D.S.G. (2006). Sedimentation and diagenesis of Chinese loess: implications for the preservation of continuous, high-resolution climate records. *Geology* 34 (10): 849–852. doi: 10.1016/j.quascirev.2013.10.014.

Stevens, T., Thomas, D.S.G., Armitage, S.J. et al. (2007). Reinterpreting climate proxy records from late Quaternary Chinese loess: a detailed OSL investigation. *Earth-Science Reviews* 80 (1–2): 111–136. doi: 10.1016/j.earscirev.2006.09.001.

Sun, Y., Clemens, S.C., Morrill, C. et al. (2012). Influence of Atlantic meridional overturning circulation on the east Asian winter monsoon. *Nature Geoscience* 5: 46–49. doi: 10.1038/ngeo1326.

Tegen, I., Lacis, A.A., and Fung, I. (1996). The influence of mineral aerosols from disturbed soils on the global radiation budget. *Nature* 380: 419–422. doi: 10.1038/380419a0.

Thomsen, K.J., Murray, A.S., Jain, M., and Bøtter-Jensen, L. (2008). Laboratory fading rates of various luminescence signals from feldspar-rich sediment extracts. *Radiation Measurements* 43 (9–10): 1474–1486. doi: 10.1016/j.radmeas.2008.06.002.

Weltje, G. (1997). End-member modelling of compositional data: numerical statistical algorithms for solving the explicit mixing problem. *Journal of Mathematical Geology* 29 (4): 503–549. doi: 0.1007/BF02775085.

Xiao, J.L., Kumai, H., Yoshikawa, S. et al. (1997). Eolian quartz flux to Lake Biwa, Central Japan, over the past 145,000 years. *Quaternary Research* 48 (1): 48–57. doi: 10.1006/qres.1997.1893.

Yang, S. and Ding, Z. (2008). Advance-retreat history of the East-Asian summer monsoon rainfall belt over northern China during the last two glacial-interglacial cycles. *Earth and Planetary Science Letters* 274 (3–4): 499–510. doi: 10.1016/j.epsl.2008.08.001.

Yang, S. and Ding, Z. (2014). A 249 kyr stack of eight loess grain size records from northern China documenting millennial-scale climate variability. *Geochemistry, Geophysics, Geosystems* 15 (3): 798–814. doi: 0.1002/2013GC005113.

Zárate, M.A. (2003). Loess of southern South America. *Quaternary Science Reviews* 22 (118–119): 1987–2006. doi: 10.1016/S0277-3791(03)00165-3.

Zöller, L. and Semmel, A. (2001). 175 years of loess research in Germany – long records and 'unconformities'. In: E. Derbyshire (ed.), Recent research on loess and palaeosols, pure and applied; selection of keynote addresses presented at 'Loessfest '99'. *Earth-Science Reviews* 54 (1–3): 19–28. doi: 10.1016/S0012-8252(01)00039-3.

6

Desert Dunes: Form and Process

Giles Wiggs

University of Oxford, Oxford, UK

6.1 Introduction

Desert sand dunes have received considerable attention from geomorphologists in the last 100 years, with research developing from modest morphological description to highly intensive field monitoring of processes and complex systems modelling. While only about 20% of the desert surface is covered by wind-blown sand and, of this, only 60% is built into dunes, there are some arid regions that are dominated by active and dynamic dunes. Many semi-arid and sub-humid areas also comprise of dunes which are now stabilised by vegetation providing a sedimentary record of past environmental conditions (Chapter 7). In such areas, a knowledge of dune morphology and dynamics is important in understanding the response of these landscapes to global change (Thomas et al. 2005; Chapter 10), interpreting palaeo-environmental records of dune sediments (Leighton et al. 2013), and protecting against hazards due to sand dune encroachment over roads and railways (Al-Harthi 2002; Chapter 12).

Since the findings of Bagnold (1941) published in his classic book, *The Physics of Blown Sand and Desert Dunes*, our understanding of dune processes and dynamics has been greatly enhanced. A particular benefit that helped us develop our knowledge was the widespread availability and application of aerial photography and satellite imagery in the 1950s and 1960s. These enabled us to observe and interpret the interiors of sandy desert regions and allowed us to fully appreciate the diversity in dune form, pattern, and change over vast areas (Grove 1958; Breed and Grow 1979). This was followed in the 1980s and 1990s by a large number of field-based studies investigating windflow over dune forms (e.g. Tsoar 1983; Livingstone 1989; Lancaster et al. 1996). These studies often focused on the role of small-scale processes in dune dynamics and were inspired by an increasing awareness of the importance of flow-morphology feedback mechanisms on dune dynamics, and the impression of an equilibrium developing between them (e.g. Wiggs et al. 1996). Many of these field studies were directed towards measurement of the interaction between the flow field and dune morphology around individual dunes, and much was discovered concerning the application of our understanding of aeolian processes to sand dune dynamics (e.g. Lancaster et al. 1996; Frank and Kocurek 1996).

Field studies of single dunes necessarily offer data on dune morphology, processes and dynamics over very short scales in space and time when compared to those of dune and dunefield development (see Chapters 8 and 10 in this volume), and it has proved difficult to

Aeolian Geomorphology: A New Introduction, First Edition. Edited by Ian Livingstone and Andrew Warren.
© 2019 John Wiley & Sons Ltd. Published 2019 by John Wiley & Sons Ltd.

'up-scale' our understanding from such measurements to a complete comprehension of dune dynamics and interaction at the dunefield scale (Livingstone et al. 2007). However, more recent developments in modelling of dune dynamics have allowed us to hypothesise and test theories, often borne out of field experiments, at this larger dunefield scale. Major advances in computational fluid dynamics modelling (CFD) (Parsons et al. 2004a, b), complex systems modelling (Nield and Baas 2008), and analytical modelling (Hersen et al. 2004) have all contributed to our knowledge of dunefield dynamics. Indeed, the intellectual and practical exchange between field studies and modelling studies of dunes is continually growing with examples of field studies testing model outputs (e.g. Claudin et al. 2013; Ping et al. 2014) and models developing field-based hypotheses at the larger scale (Elbelrhiti et al. 2005).

Field-based dune studies are again proving popular in recent literature as test-beds for investigating the role of flow turbulence in dune dynamics (Baddock et al. 2011; Wiggs and Weaver 2012) and, given that sufficient time has elapsed since the single dune studies of the 1980s, there now exists an opportunity for modern re-surveys of dunes to provide data on dune movement and change over longer time periods (e.g. Livingstone 1993, 2003; Eckardt et al. 2013).

6.2 The Classification of Dunes

The classification of dunes can be problematic. It is a matter for debate with no single classification scheme successfully capturing the inherent complexities apparent in dune form or process of development. While there are many ways of classifying dunes, one of the more enduring is that of Livingstone and Warren (1996). This simple geomorphological classification (Figure 6.1) distinguishes between dunes that are 'anchored' (see Chapter 7) and those that are considered to be 'free' and formed only by aeolian action on loose sand. These 'free' dunes are then further divided by their shape into the major dune types that are evident in sandy deserts: transverse, linear, and star. Such a general classification could be considered over-simplistic in that many mobile dunes (classified as 'free' in Figure 6.1) can become stabilised (or 'anchored' in Figure 6.1) over time and/or space. Particularly problematic is any classification of dunes as 'coastal' because many of the dune types listed under 'free' can also occur in coastal regions and many of these are dynamic and not necessarily anchored by vegetation. However, within the wide range of complex environmental and climatic regimes within which dunes are found, simple geomorphological classifications, such as that shown in Figure 6.1, which differentiate

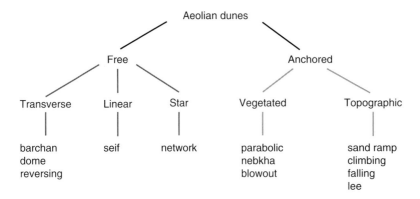

Figure 6.1 A simple geomorphological classification of dune types. *Source:* Adapted from Livingstone and Warren (1996).

between 'mobile' and 'stabilised' or 'semi-mobile' dune forms, are not unusual. With an appreciation of the necessary lack of sophistication in such schemes they can, nevertheless, provide a useful starting point for understanding dune genesis and diversity. In this regard, this chapter focuses on the forms and processes of 'free' dunes as identified in Figure 6.1 which are commonly found in desert regions.

A further important consideration is the differentiation between *compound* and *complex* dune forms. *Compound* dunes are those that are composed of a number of dunes of the same major dune type but at different scales superimposed onto each other (McKee 1979). An example of this is the multiple generations of superimposed linear dunes reported by Lancaster et al. (2002) in Mauritania (see Chapter 8). *Complex* dunes consist of the superimposition of dunes of a different type. A good illustration of this is

Figure 6.2 Complex dunes in the western part of Libya's Marzuq Sand Sea. Numerous smaller transverse and linear dune forms have developed on the slopes of massive mega-dunes. *Source:* NASA.

the existence of smaller transverse and linear features superimposed onto megadunes in Libya (see Figure 6.2), or transverse features emerging on the upper flanks of large linear dunes, as recognised in the Namib Desert by Livingstone et al. (2010).

6.2.1 Major Dune Types and Their Controls

The dune classification system of 'free' dunes in Figure 6.1 is based on the shape of the dune (transverse, linear, star), but dune shape is a manifestation of its constructing processes and this is amply demonstrated by Wasson and Hyde (1983), who differentiated between major dune types using the simple formative controls of equivalent sand thickness (EST, the depth of sand in a dunefield if all the dunes were flattened out) and wind directional variability (Figure 6.3). Here, wind directional variability is defined by the ratio of the calculated resultant drift potential (RDP) to drift potential (DP) where DP refers to the total amount of sand flux in any direction, and RDP is the sand flux in the primary directional vector (Fryberger and Dean 1979; section 2.7.5).

From Figure 6.3 it is clear that in areas with both low sand supply and low wind directional variability barchan dunes are formed. An increase in sand supply results in the coalescence of barchan dunes into transverse ridges, and an increase in wind directional variability results in the formation of linear dunes in bi-directional wind regimes. Star dunes are apparent where both sand supply and wind directional variability are at their greatest. Such a development in dune form can be realised at the dunefield scale from a coastal system, with low sand supply, and uni-directional winds, to an inland system where sand supply is greater and winds are more complex. A good example is the Namib sand sea which is characterised by northerly migrating barchans at the western coast, to extending linear dunes in the centre, and accumulating star dunes at the eastern edge

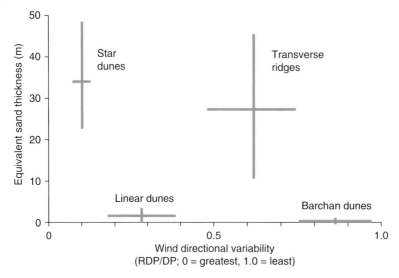

Figure 6.3 The major controls differentiating dune types, see text for details. *Source:* After Wasson and Hyde (1983).

where complex winds are responding to the emergence of the southern African plateau (Livingstone et al. 2010).

However, the simplicity of the classification system described in Figure 6.3 conceals the complex morpho-dynamics in both time and space which underpins sand dune development. For example, complex dunes have been widely reported in the literature with Livingstone et al. (2010) citing examples in the Namib of linear dunes with smaller transverse and barchan dunes superimposed on their upper slopes. Such occurrences are problematic for the Wasson and Hyde (1983) classification scheme because they represent distinctly different dune types co-existing in the same conditions of sand supply and wind regime. The reason that such complexity arises is that different sizes of dunes adjust to changing wind patterns at different rates, and this temporal dynamism is not allowed for in Figure 6.3. In the complex cases reported by Livingstone et al. (2010), the smaller barchan dunes on the flanks of the linear dune are able to respond to winds changing at the seasonal scale. In contrast, the larger linear dune has a much longer re-adjustment time in response to changing wind patterns and so it tends to remain in a state of flux, partially (but never fully) adjusting to the changing seasonal winds. It therefore builds a morphology that represents an integration of that partial geomorphological adjustment to a multitude of previous wind events.

This difference in the response and re-adjustment times for different scales of dune has been described as 'dune memory' (Warren and Kay 1987), such that large dunes 'remember' (or retain a geomorphological signature of) previous winds. This occurs because at any one time a sand dune (of whatever type) only responds to wind coming from a single direction. In other words, at any instant, every dune is acting like a proto-transverse dune and working towards generating a truly transverse form. The speed at which a dune may realise a transverse form relies particularly on the mass of sand that has to be eroded and deposited. For large dunes that are not already transverse in nature, the mass of sand that would need to be redistributed is large so that the wind direction alters well before a transverse form is achieved. From the point in time that the wind alters direction, the dune is again working towards generating a transverse form, but in response to a new wind direction. While the redistribution of sand in response to this new wind is taking place, some part of the dune form will remain as a consequence of the previous wind. This will be the case until the duration or strength of the new wind reaches a level such that traces of dune morphology consequent on the previous wind are eroded or buried. Large bodies of sand therefore tend to comprise a jigsaw of morphologies that represent a response of that dune to varying wind directions over time. In contrast, small bodies of sand re-adjust quickly to changing wind directions and so can generate a transverse morphology more readily.

A similar argument has been made by Werner and Kocurek (1997) concerning the coexistence of dunes of different orientations and sizes on Padre Island in Texas. They explain that the relative rates of adjustment of dunes might also be controlled by the continuity in the dune form, dependent upon the number of 'defects' (or terminations) in the dune pattern. Small, individual barchans display many terminations (i.e. a dunefield of small barchans will comprise a high density of dune horns) and so adjust rapidly to changing winds, in contrast to larger dune masses with fewer terminations per unit length, such as continuous transverse dune ridges.

6.3 Dune Dynamics

During the 1980s and 1990s, there was considerable field-based research investigating the dynamics of and aeolian processes operating on dune surfaces. The expectation in these early studies was that from understanding the smaller-scale operating processes on dunes we would be able to resolve behaviour and patterning at the larger dunefield scale. Much of this research was carried out on barchan and linear dune types as they present a relatively simple geomorphology (Livingstone 1988; Tsoar 1990; Lancaster et al. 1996; Wiggs et al. 1996). The approach used in these investigations often involved measuring wind speed profiles on windward dune slopes in order to determine shear velocity (u_*, Chapter 2). From such measurements of u_* it was anticipated that calculations could be made of sand transport rate (q) in a downwind direction (x), which could then be used to determine dune topographic change ($\Delta h/\Delta t$), whereby (Eq. 6.1):

$$\Delta u_{*x} \approx \Delta q_x \approx \Delta h/\Delta t \tag{6.1}$$

The hope was that this approach would allow the construction of simple models of dune processes, form and dynamics. Importantly, these field-based studies of airflow over dune surfaces disclosed the complexity evident in feedback mechanisms operating between dune form and flow. Specifically, they discerned a series of secondary flow regimes induced by the intrusion of dunes into the atmospheric boundary layer.

Figure 6.4 shows a theoretical schematic of airflow streamlines over a simple transverse sand dune identifying secondary flows on both the windward and the lee slopes. On the windward slope the intrusion of the dune into the boundary layer compresses the streamlines and thus accelerates the wind towards the crest of the dune. Downwind of the crest, a reverse flow eddy develops as the streamlines separate from the surface and expand towards the ground surface, creating a region of low wind velocity. Sand transport responds to these accelerating and decelerating flows. Figure 6.5 shows measurements of both wind velocity and sand transport rate upwind of and along the windward slope of a 10 m high barchan sand dune in Namibia (Weaver and Wiggs 2011). The acceleration of wind velocity on the windward slope is clearly seen with an increase in velocity of around 25% when compared to upwind values, and this is reflected in the more than doubling of the sand transport rate from the toe of the dune towards the crest and brink (from ~30 to ~75 g m^{-1} s^{-1}). This continuous increase in sand transport rate is a reflection of the increased wind shear eroding the

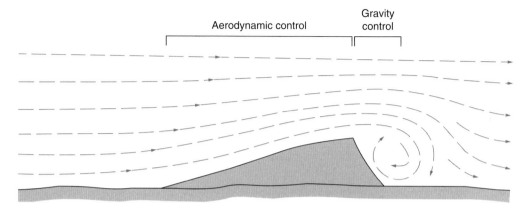

Figure 6.4 Streamlines over a simple transverse sand dune. Streamline compression on the windward slope leads to flow acceleration and the erosion of sand, while flow deceleration in the lee of the dune results in sand deposition through gravitational fallout.

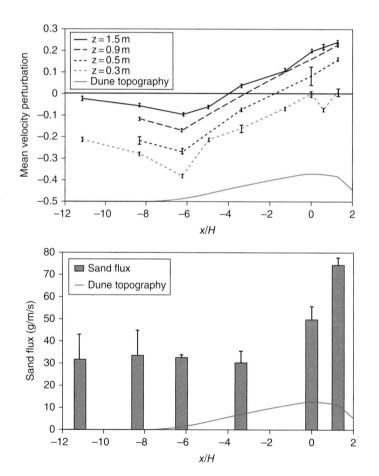

Figure 6.5 Perturbations in wind velocity (at different heights, z) and sand transport rate (flux) measured upwind of and along the windward slope of a 10 m high (max height = H) barchan sand dune. A wind velocity perturbation of 0 depicts equivalency to upwind velocity. Wind direction is from left to right. *Source:* After Weaver and Wiggs (2011).

surface of the windward slope as it accelerates up the dune. In contrast, deposition dominates on the lee slope as the driving force of the wind is dissipated and gravity pulls sand out of the airflow. This leeside grainfall normally occurs within ~ 1 m downwind of the dune brink (within one saltation leap length) and this zone of peak deposition of sand results in the formation of a growing 'bulge' on the lee-side. Eventually, further deposition of sand creates an over-steepening (beyond the critical angle of repose, ~34°) which results in a failure of the sand surface and an avalanche of sand towards the base of the lee side slope, known as the 'slip-face' (Anderson 1988; Nield et al. 2017).

In this way, a transverse dune migrates in a downwind direction as sand is eroded from the windward slope and deposited on the lee slope. The rate of downwind dune movement is therefore a consequence of the amount of sand being redistributed from the windward slope to the lee slope, and this is partially controlled by the degree of windward slope flow acceleration and subsequent lee-side deceleration. There is therefore a dynamic feedback mechanism operating between the dune topography, the degree of flow acceleration, the rate of sand transport, and the subsequent change in dune topography due to erosion of the windward slope. Rather than dunes merely responding to regional wind regimes (as intimated in Figure 6.3), they create secondary flows at a local scale which help control their own form and dynamics (see Figure 6.4). Such findings promoted ideas of the existence of 'equilibrium' dune forms in particular wind regimes and this encouraged a lot of further research in both the field and wind tunnels in order to elucidate the physical mechanisms controlling such equilibrium (e.g. Wiggs et al. 1996; McKenna-Neuman et al. 1997; Walker and Nickling 2002).

The field studies of dune dynamics described above provided good field data on the complexities of secondary windflow regimes, sand transport rates on dune surfaces, and dune topographic change. These data provided an important foundation for improving our understanding of the dynamics and processes operating on different dune types. However, they failed to provide a comprehensive understanding of dune evolution that allowed the identification of an 'equilibrium' dune form for a particular wind regime. Nor did the recorded data allow the construction of an enduring numerical model of windflow and dune dynamics or migration. There were three principal reasons for this.

First, there were issues relating to the measurement of shear velocity (u_*) on dune surfaces (see Chapter 2). It became clear that it was not possible to measure u_* on dune windward slopes where flow acceleration induced a curvature in the velocity profile. Such a curved profile prohibits the calculation of shear velocity which assumes a straight log-linear velocity profile (Mulligan 1988; Lancaster et al. 1996). Further, there were practical difficulties in measuring windflow within the 'inner-layer' which is the depth of flow (commonly <0.3 m above the dune surface) within which changes in wind stress were thought to be significant (Wiggs et al. 1996).

Secondly, questions arose as to the applicability of using measurements of average u_* in calculations of sand transport (q) on dune surfaces (Butterfield 1999). Rather, recent research has suggested that peaks in wind stress related to turbulence in the wind might provide a better solution for sand transport (Baddock et al. 2011; Weaver and Wiggs 2011; Wiggs and Weaver 2012; see Box 6.1 for further information about this].

Thirdly, the required upscaling of short-term measurements of windflow and sand transport parameters to larger-scale dune morphological change proved problematic. Whereas field measurements provide at-a-point data on the processes operating on a dune surface at any one time, as discussed above, dunes of different sizes (and also different elements of dunes such as surface ripples, slip face orientations, windward slope angles), respond to changing winds at different rates. It is therefore difficult to apply the findings from short-term field measurements to sand dunes that may only

Box 6.1 Turbulence, Sand Transport, and Dune Dynamics

Much of the research into windflow over sand dunes in the 1980s and 1990s focused on measuring mean values of wind velocity. These data were used to determine mean sand transport rates (q), in the belief that downwind changes in q lead to morphological change. It became clear, however, that time-averaged flow parameters did not necessarily provide reliable measures of the sand-transporting capacity of wind over dune surfaces. An example is the measurements of mean wind velocity shown in Figure 6.5. As the wind approaches the dune, it slows down because of the blockage presented by the dune body. If we consider the sand transport potential of the wind to be broadly related to mean windspeed, this region of low velocity at the toe of the dune should be characterised by sand deposition, leading to an upwind migration of the dune toe, behaviour not observed in the field and not apparent from the measurements of upwind sand flux in Figure 6.5 which maintain consistency.

At around this time, field sensors were beginning to be developed that could measure the high frequency perturbations in the windflow, such as ultrasonic anemometers (see Walker (2005) for a review; Box Figure 6.1) and sediment flux, such as grain impact sensors (see Baas (2004) and Chapter 2). Field researchers began to deploy these instruments in the field (e.g. Sterk et al. 1998; van Boxel et al. 2004; Leenders et al. 2005) and, allied with similar research in fluvial environments, collected data that suggested that sediment transport and entrainment do not just respond to time-averaged flow conditions; rather they involve a complex response to elements in turbulent windflow (also discussed in Chapter 2). Field measurements by Weaver and Wiggs (2011) and Baddock et al. (2011)

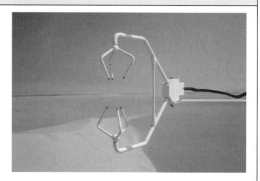

Box Figure 6.1 A sonic anemometer measuring 3D instantaneous windflow at the crest of a barchan dune in Namibia. *Source:* Giles Wiggs.

confirmed previous wind tunnel research that had shown that the low velocity zone upwind of and at the toe of dunes (see Figure 6.5) coincides with a region of highly turbulent flow showing peak values of instantaneous wind velocity of the order of less than one second. They suggest that the high energy turbulent peaks in the flow maintain sediment transport through the low velocity region. Further, Wiggs and Weaver (2012) found that distinct types of turbulent structure dominated sand transport in different regions over a barchan dune. They discovered that windflow in the toe region was dominated by the turbulent bursting phenomena, involving sweeps and ejections of high velocity wind. Turbulent structures that involved peaks in horizontal windspeed (called outward interactions) dominated the flow at the crest of the dune. These turbulent wind structures were found to account for up to 95% of the sand transport on the dune. We do not yet have a complete understanding of the influence of wind turbulence on sand transport and dune dynamics, but recent research represents a good foundation from which to explore these complex interactions.

be partially evolved towards an equilibrium with the measured windflow.

It has become clear that understanding the complete morphodynamics of sand dunes is

a complex and multiscale problem that requires a multiscale approach (Thomas and Wiggs 2008). Nevertheless, short-term field studies over the last 15 years have made huge advances in uncovering the relationships between wind turbulence, sand transport, and dune morphology (Livingstone et al. 2007; Baddock et al. 2011; Weaver and Wiggs 2011). Rather than being seen as an end in themselves, the understandings gained from these field studies are now being used in the development of larger-scale numerical or complex systems models of dune dynamics. Further, such field data can be used to test the output of these models and the theoretical assumptions that lie behind them (e.g. Claudin et al. 2013).

6.4 Dune Morphology

6.4.1 Barchan Dunes and Transverse Ridges

Barchan dunes represent the classic dune morphology (although they cover relatively little area of the sandy deserts), displaying a crescentic shape with the crest at the centre-line aligned perpendicular to the prevailing wind direction and horns directed downwind and parallel to it (Figures 6.6 and 6.7). Figure 6.3 shows that they have a *migrating* dynamic, emerging in environments with a relatively small sand supply and a strongly uni-directional wind regime. As such, they commonly occur in (but are not limited to) coastal dune fields. Good examples can be found on the Skeleton Coast of Namibia (Livingstone et al. 2010), in the Pampa La Joya of Peru (Gay 1999), and on the western Coast of Morocco (Elbelrhiti et al. 2005). Where sand supply is more plentiful, the horns of barchan dunes link up to form transverse ridges. Together, barchan dunes and transverse ridges make up about 10% of desert dunes globally (Breed and Grow 1979).

The windward slopes of barchan dunes are often in the range of 10–14° in response to the aerodynamic control on their erosion. In contrast, downwind slip-faces reflect the gravity control in this region as sand is deposited downwind of the brink resulting in a lee-slope at or close to the angle of repose of dry sand (32–34°, Nield et al. 2017). While their morphological form may differ substantially with varying degrees of separation in crest and brink (Figure 6.6), they display a fairly limited range in descriptive statistics. Hesp and Hastings (1998) found that the maximum height of barchans is of the order of 1/10th of both the length (toe to brink) and width (horn to horn).

In response to their formative uni-directional winds, barchans and transverse ridges are the most mobile of dunes, resulting in hazards to human activity where they may migrate over roads, railways and other installations (Dong et al. 2004b). Barchans range in height from a minimum of *c.* 1 m (Elbelrhiti et al. 2005) to over 50 m with a distinct inverse relationship evident within individual dunefields between dune height and migration rate (Figure 6.8). This can be explained by the differential in mass of sand between small and large dunes. Given that barchan dunes move by re-distributing sand from their windward slopes to their lee slopes (see Section 6.3) it follows that large dunes, with a greater mass of sand, take longer to reconstitute and are thus slower in migration. Recent measurements of barchan migration on the Skeleton Coast

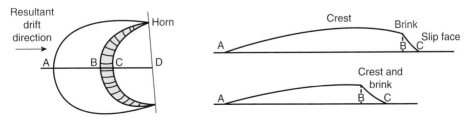

Figure 6.6 The morphology of barchan sand dunes (Hesp and Hastings 1998).

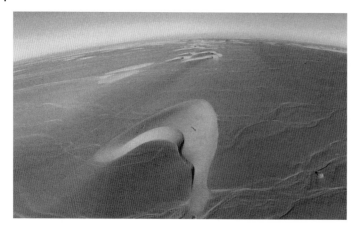

Figure 6.7 A 6 m high barchan dune on the Skeleton Coast of Namibia. The horns of the barchan are directed downwind and parallel to the prevailing wind direction and the lighter tone of the picture downwind of the horns depicts the sand streaming away from the dune. Barchan dunes often exist in 'trains' downwind of a specific sand source and in this picture upwind barchans can be seen in the background closer to the original sand source of the dry Huab river mouth. *Source:* Jerome Mayaud.

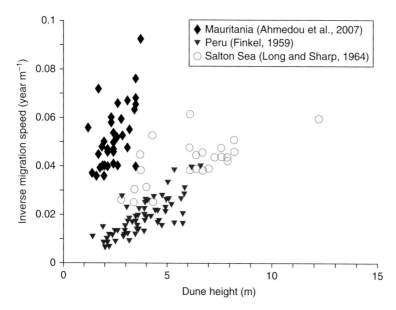

Figure 6.8 The relationship between barchan dune height and migration rate measured in Mauritania, Peru, and the Salton Sea (Ahmedou et al. 2007). Note inverse y-axis.

of Namibia reveal that a 9.7 m high dune worked on by Weaver and Wiggs (2011) had moved 130 m in nine years (14.4 m yr^{-1}). In contrast, nearby but smaller 3.3 –4.5 m high dunes (originally measured by Baddock et al. 2007) moved approximately 570 m in 13 years (43.8 m yr^{-1}).

The relationship between height and migration rate also helps to explain the crescentic morphology of barchan dunes. The largest mass of sand in a barchan is in a slice through its centre-line parallel to the prevailing wind direction (along the line A-B-C in Figure 6.6). This part of a dune therefore moves relatively slowly. In contrast, the flanks of a barchan are lower in height and contain less mass of sand meaning that they move downwind at a faster rate than the centre-line. The differing height between the centre-line of barchan dunes and their horns also has an impact on their respective sand budgets. On the centre-line, the height of the dune is such that all the sand eroded from the windward slope and swept over the brink is deposited on the upper part of the slip face. At the centre-line the dune therefore acts as a perfect sand trap. At the horns, however, the low dune height allows some of the saltating sand to leap beyond the dune mass and so evade the slip face. Indeed, at the outer edge of the horns, the dune mass is so low that no slip face develops. It is only at the horns, therefore, that barchan dunes can discharge sand downwind. The preferential sand transport corridors downwind of barchan horns often provide a sand source for the development of downwind dunes. In this way, barchan dunes within a dune field are often off-set from one another with the horn of one dune feeding sand onto the windward slope of a downwind dune (see Figure 6.7).

6.4.2   Linear Dunes

Linear dunes (sometimes referred to as *longitudinal* dunes) are the most common dune type in desert interiors, covering large areas of southern Africa, Australia, the Sahara, Arabia, and China. Their morphology is a repeated pattern of sinuous ridges that can reach up to 200 m in height and extend across a landscape for hundreds of kilometres (Figure 6.9). Their large mass makes them relatively stable features and this stability encourages the growth of vegetation, even in arid climates. This vegetation is usually on the lower flanks of the dunes, where wind speeds and sand transport rates are generally lower allowing the establishment and growth of plants. The stability of linear dunes, combined with their large mass, means they are able to persist in the landscape even through major climatic fluctuations. Linear dunes can therefore provide a valuable data resource for palaeo-environmental reconstruction in deserts (e.g. Stone and Thomas 2008; Chapter 10). The current vegetated (and stable) status of many linear dunefields such as the Kalahari Desert in southern Africa (Wiggs et al. 1995) and the Simpson and Strzelecki Deserts of eastern Australia (Hesse and Simpson 2006) may be threatened by human activities and climate change.

When actively dynamic in arid or hyper-arid deserts, most linear dunes tend to a sharp-crested morphology such as found in the Namib (Livingstone 2003) and Sinai (Tsoar 1983). In semi-arid environments, part-vegetated linear dunes experience little sand transport and little movement. In these circumstances most dunes have rounder and more muted crestlines, as in the Kalahari desert (Bullard et al. 1995). While it was long thought that such vegetated linear dunes were relics from previous (drier) climates, it is now recognised that these dunes tend to episodic activity in response to varying levels of vegetative cover through time (Wiggs et al. 1994, 1995; Hesse and Simpson 2006). They therefore show a 'pulsed' geomorphic response to environmental changes such as drought, grazing pressure and fire, dependent on the degree of erosivity (wind stress) at the time.

The dynamics of linear sand dunes have long been the subject of debate (see Livingstone 1988 and Tsoar 1990). Bagnold (1953) tentatively suggested that they were formed as a result of pairs of longitudinal helical roll vortices sweeping sand into a sinuous ridge. However, field-based studies of airflow over linear dunes in the 1980s (Tsoar 1983, 1989; Livingstone 1989) provided robust evidence that linear dunes developed in response to bi-directional (often seasonal) wind regimes (Figure 6.3). Some of these studies noted the particular importance of secondary flow regimes in the growth and dynamics of linear dunes. As with the flows over barchans described above, linear dunes induce an

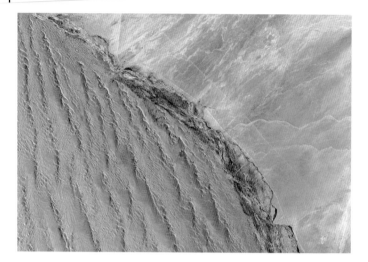

Figure 6.9 A satellite image of the northern edge of the Namib dunefield. The linear dunes (visible towards the bottom left of the picture) can reach over 100 m high and extend 400 km from the southern edge of the dunefield to the Kuiseb River in the north. Here, the advancing fronts of the dunes fall into the dry Kuiseb canyon as they migrate north. Every few years the Kuiseb river floods and the sand from the dunes is washed west into the South Atlantic Ocean, the gravel plain to the north of the Kuiseb canyon is therefore free of dunes. *Source:* NASA.

acceleration in wind towards their crest on the windward slope, and a reverse-flow eddy on their lee side. However, because linear dunes develop in regions with a seasonally bi-directional wind regime the flow approaching the windward slope often does so at an angle to the crest. This angle of approach induces a lee-side eddy downwind of the brink that is deflected and extended in an along-dune direction (i.e. more parallel to the crest, Figure 6.10).

In any one seasonal wind, this deflection of the lee-side eddy results in the along-dune transport of sand, parallel to the dune orientation, and in juxtaposition to the prevailing wind direction. In the following season, when the wind comes from an alternative direction, it erodes the former lee slope, and deposits sand on the former windward slope. In both wind directions, the development of the deflected lee-side eddy causes an along-dune deposition of sand, resulting in an *extending* dune form in a direction controlled by the combination of the two seasonally alternating winds (Figure 6.11; Tsoar 1983; Livingstone 1989). The secondary flows on linear dunes (both the accelerating windward

flow, and deflected lee-side flow) operate at an angle to the local dune slope and so sometimes generate secondary transverse dune forms (Livingstone 2003). These may take the form of barchan or transverse ridge morphologies on the flanks of linear dunes, so creating a multi-scale and complex dune form (Figure 6.12).

The bi-directional wind regimes in which linear dunes are formed are rarely symmetrical and so it has been suggested that over long periods of time linear sand dunes might also move laterally (perpendicular to the crestal vector). Evidence for such long-term movement has been presented by Bristow et al. (2005) in their investigation of the internal structure of a large linear dune in the Namib using ground-penetrating radar. The dip slopes of the internal sedimentary structures provided evidence of an overall east-west migration of the dune, in contrast to its overall south–north extension. Such lateral movement operates at a temporal scale which is too large to be observed from measurements of contemporary linear dune dynamics (Livingstone et al. 2007) and so the debate concerning the possibility of

Figure 6.10 The along-dune deflection of the lee-side eddy on linear sand dunes as a result of the approach angle of the wind on the windward slope (Livingstone 1988).

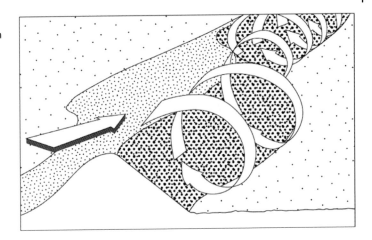

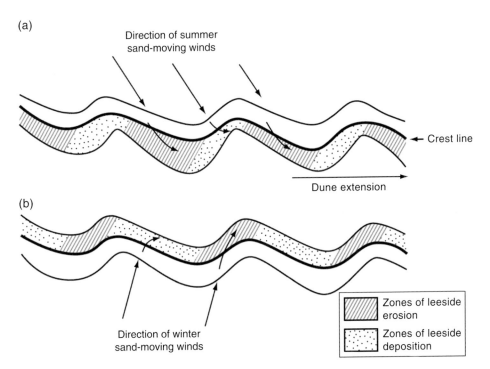

(a)

Direction of summer sand-moving winds

← Crest line

Dune extension

(b)

Direction of winter sand-moving winds

Zones of leeside erosion

Zones of leeside deposition

Figure 6.11 A seasonal bi-directional wind regime impacting on a linear sand dune. The along-dune deflection of the lee-side eddy, in each of the seasonal winds, results in the development and extension (from left to right in the case here) of the dune in a direction which is the vector of the two dominant wind regimes (Tsoar 1983).

lateral migration in linear dune dynamics continues.

6.4.3 Star Dunes

Star dunes are massive bodies of sand that develop in regions with multi-directional wind flows. These characteristics make their dynamics complex and both theoretically and practically difficult to interpret. While there is a good body of work describing the morphology of star dunes, there is a scarcity of work dedicated to their dynamics. The regions of plentiful sand supply and multi-directional winds in which star dunes occur (Figure 6.3) allow an *accumulating* dynamic

Figure 6.12 Transverse dunes on the flanks of a large linear dune in the northern Namib sand sea near Sossusvlei. *Source:* Giles Wiggs. See insert for colour representation of this figure.

with significant vertical growth (Tsoar et al. 2004). In such regions there is insufficient wind-directional uniformity to provide a significant net outward-transport of sand and, once formed, the slip faces of star dunes become very effective sediment traps. This results in the accumulation of a considerable sand mass such that heights of star dunes have been reported at over 300 m in the Namib desert (Lancaster 1989), northern Sahara (Mainguet and Chemin 1984) and China (Dong et al. 2004a).

With their large mass, star dunes have a considerable 'dune memory' (Warren and Kay 1987), such that, despite the frequently changing and complex wind regimes in which the dunes develop, their form never comes to an equilibrium with its seasonal wind regime. Rather, the dune extends small 'arms' in response to each formative wind regime (Figure 6.13) and it is these which give rise to the 'star' in their title. Given that star dunes continually accumulate sand and height, there has been some debate as to what controls their maximum size. Andreotti et al. (2009) have argued that a maximum cap on the

height of large dunes is provided by the atmospheric boundary layer which extends to a depth of between 1 and 2 km above the desert surface. Given that the depth of the atmospheric boundary layer is partly controlled by temperature lapse rates, which increase with continentality, this might help to account for the fact that most mega-scale star dunes are found in the interior of deserts.

6.5 Dune Orientation and Alignment

The alignment of transverse, linear, and star dunes as a result of varying seasonal wind regimes has been the subject of some debate. It has been considered that dunes align themselves either parallel (in the case of linear dunes) or transverse (in the case of transverse dunes) to the *net* resultant wind directional vector (Fryberger and Dean 1979). However, experimental studies by Rubin and Hunter (1987) showed that dunes align themselves such that the sand transport across the crest is maximised. This is called

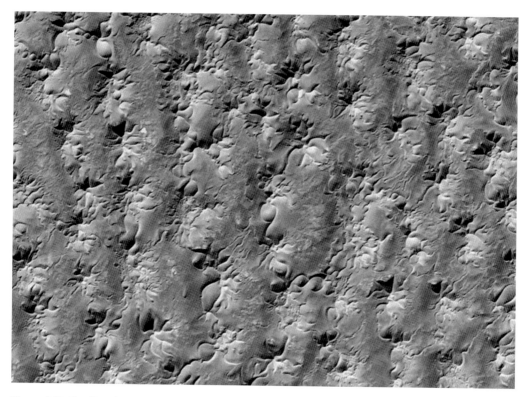

Figure 6.13 Satellite photograph of star dunes in Algeria showing the characteristic arm extensions resulting from the complex multi-directional wind regime in which this dune type develops. *Source:* NASA.

the *gross bedform-normal rule* and it takes into account that every sand-transporting wind (from any direction) across the dune contributes to its *gross* development. This is in contrast to the concept of *net* dune development (where opposing winds act to cancel each other out) when using the resultant directional vector to determine dune alignment. Using the gross bedform-normal approach, it is both the angular variability in the direction of the major winds and the ratio of the sand-transporting capacity between those winds that determine the eventual dune alignment. In a bi-modal wind regime, dunes can be classified according to this sand transport ratio and also the 'divergence' angle between the bedform trend and the computed angle of the resultant wind vector for the two winds (Kocurek and Ewing 2005). Generally, bedforms are considered to be linear where the divergence angle is between 0 and 15°, oblique at angles between 15 and

75°, and transverse at angles from 75 to 90° (Hunter et al. 1983).

Using dunes in Namibia, and the SW USA, Lancaster (1991) resolved each transport vector into bedform-normal and bedform-parallel components to determine the predicted dune alignment using the gross bedform-normal rule. He found close agreement between actual and predicted dune orientations confirming the underlying foundation of the gross bedform-normal rule that dune crests align as perpendicular as possible to all sand-transporting winds that comprise the overall wind regime. More recently, Fenton et al. (2014) investigated dune orientations in the Great Sand Dunes of Colorado to test the rule in reverse, i.e. predicting the orientation and magnitude of sediment transporting winds from the observed alignment of sand dunes. Such an approach could be useful in determining wind patterns from bedform orientation on other planets. The gross bedform-normal rule has also been tested experimentally at the landscape-scale in the Tengger Desert in Inner Mongolia by Ping et al. (2014). In a novel study Ping et al. mechanically levelled 16 ha of the dunefield using a bulldozer and monitored the re-development of the dunes over nearly four years. They found that over time the emerging dune trend was orientated in agreement with that predicted from the gross bedform-normal rule and at a divergence angle of 50°, classifying the dunes as oblique.

6.6 Dune Interactions and Equilibrium

While our understanding of the dynamics of desert dunes has been considerably advanced by small-scale, field-based studies of individual dunes, it has also been hampered by our concentration on dunes at this scale. As discussed above, it has proved difficult to up-scale these studies to the time- and length-scales at which larger sand dunes operate. Similarly, our reliance on single-dune studies has precluded a significant appreciation of the potential interactions that might operate between dunes in dunefields, which is where most dunes exist. While there have been a small number of field studies that have attempted to investigate relationships between dunes by measuring dune spacing and airflow interactions (e.g. Baddock et al. 2007) or monitoring geomorphological interactions over time (Gay 1999), the majority of research progress in recent years has been accomplished by various modelling techniques. These have included flow models such as CFD (e.g. Parsons et al. 2004a, b), but most of the progress in understanding larger-scale interactions has come from analytical and complex systems models.

6.6.1 Complex Systems Models

The complex systems approach to understanding dune interactions disregards the small-scale processes that have been the

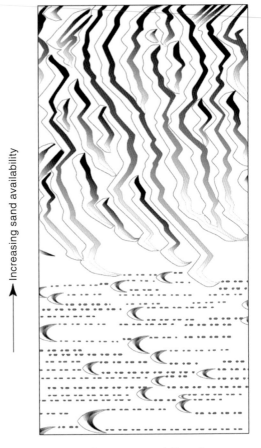

Increasing sand availability

Figure 6.14 Complex systems modelling of transverse and linear dune patterns with an increasing sand supply from bottom to top. Wind is from left to right (Baas 2007).

focus of field-based studies, instead treating the system at the scale of interest (i.e. the dune and dunefield). It is thus decoupled from the reductionist complications of modelling processes at the grain scale. Werner (1995) championed this approach and developed a model in a cellular grid based on 'slabs' of sand whose dynamics were governed by applying three simple rules: (i) a probability distribution for entrainment and deposition; (ii) an increased probability of deposition if a slab of sand descends on another slab of sand; and (iii) the development of an avalanche if the deposition of sand slabs creates an angle greater than the angle of repose of sand (32–34°).

Models like this create very realistic patterns of both linear and transverse dune systems, depending on the choice of boundary conditions and wind regime (Figure 6.14). Further, the models realistically replicate the merging and calving (or splitting) of dunes as the patterns develop. The model has been further developed by Momiji et al. (2002) and Baas (2002), who incorporated elements of flow and dynamics learned from field-based studies, and Baas and Nield (2007, 2010) have since included the impact and dynamics of vegetation on dune surfaces.

The basis of complex systems modelling is that dunes show self-organisation in patterns, whereby the evolution of dune pattern responds to the number of defects (terminations) within a dunefield (Werner and Kocurek 1997; Kocurek and Ewing 2005), the orientation of dunes relative to the mean

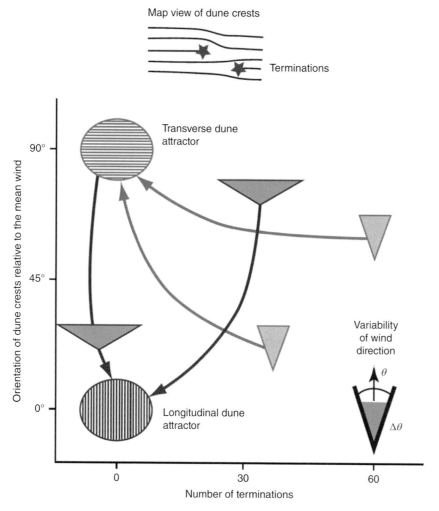

Figure 6.15 Dunefield self-organization: the emergence of linear and transverse attractor states toward which dunes evolve in complex systems modelling. Reproduced from Anderson (1996). The x-axis refers to the density of 'terminations' evident in a dunefield, sometimes called the 'defect density'. This is related to the level of continuity of dunes within a dunefield. For example, a dunefield comprising of small barchan dunes would include many horns where each dune terminated. This represents a high defect density. In contrast, a dunefield comprised of continuous transverse ridges would display very few dune terminations and so have a low defect density.

wind, and the wind directional variability. Anderson (1996) noted that the outcome of dune self-organisation was the emergence of two attractor states (linear and transverse) towards which the system evolves independent of initial conditions (Figure 6.15).

6.6.2 Analytical Models

Analytical models of dune dynamics use approximations of solutions for airflow over dunes to determine sand transport rates on dune surfaces and dune morphological change (Andreotti et al. 2002; Schwämmle and Herrmann 2003). The development of analytical models in recent years has proven particularly fruitful for investigating issues of larger scale dune dynamics and interaction.

For example, by employing a constant sand supply at the upwind edge of their modelling domain, Hersen et al. (2004) noted that over time small dunes tended to shrink and disappear, while large dunes grew larger (Figure 6.16). This indicates that single barchan dunes are perhaps not in equilibrium with their windflow, contrasting with the considerations driving the majority of field-based process studies in the 1980s and 1990s (Section 6.3). Hersen et al. (2004) explained this contrasting dynamic between different dune sizes in terms of the sand budgets evident for different sized dunes.

They observed that the amount of sand influx onto the windward slope of a barchan dune is proportional to the width of the dune (i.e. wider dunes receive more sand). The outflow of sand is proportional to the width of the horns, which field studies have shown are the only parts of a barchan that allow sand to be discharged. Hersen et al. (2004) recognised that small dunes tend towards relatively larger horn widths in comparison to the overall width of the dune, whereas larger dunes tended to have proportionally smaller horn widths. The ratio of horn width to barchan dune width therefore decreases with increasing dune width, so the output sand flux is not linearly proportional to dune width. This results in a negative net sand flux budget for small dunes (which shrink and disappear over time) and a positive net sand flux budget for large dunes (which grow over time).

As dunefields do not generally show a domination of large dunes, it is apparent that there must be mechanisms that operate at the dunefield scale which effectively transfer sand from one dune to another, counteracting the contrasting sand flux budgets for different-sized individual dunes. Such mechanisms might include merging by collision, dividing, and splitting (Katsuki et al. 2011) and also lateral transfer of sand by variation in the direction of sand-transporting winds (Hersen and Douady 2005). Hersen et al.

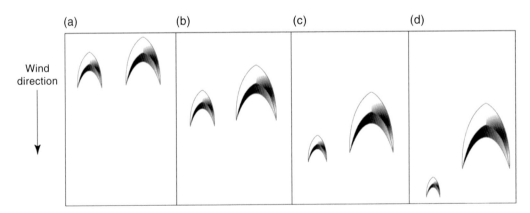

(a)   (b)   (c)   (d)

Wind direction

Figure 6.16 The disequilibrium of a single dune as shown by analytical modelling. With a constant upwind supply of sand small dunes shrink and disappear over time (a–d) while large dunes get larger (Hersen et al. 2004).

(2004) therefore questioned whether single dunes should be considered as having the potential for equilibrium in their dynamics. They suggested instead that we should be considering the possibility of equilibrium at the *dunefield* scale.

6.7 Conclusion

Our understanding of desert dunes and their dynamics has improved enormously over the last few decades. The field-based experimental studies in the latter part of the twentieth century provided a robust dataset from which the earlier classification of dunes was extended to an understanding of feedbacks between dune form and windflow, and the importance of secondary flows and turbulence. Although these studies were limited in their scale, they inspired the development of a variety of modelling approaches which have now begun to consider the interactions between dunes at a larger scale. The collaboration now evident between fieldworkers and modellers should allow us to answer significant questions concerning the equilibrium of dunes and dunefields, the role of sediment transfers between dunes, dynamic interactions between dunes, and the long-term development of dune landscapes.

Further Reading

A good general review on dune dynamics which summarises some of the key points of this chapter is provided by Livingstone et al. (2007). For advanced reading on field measurements of flow and turbulence over individual dunes, the reader is directed to Weaver and Wiggs (2011), while a larger-scale view of the interaction between windflow and dunefield dynamics is delivered by Jerolmack et al. (2012). Our understanding of the evolution of dunefields has been stimulated by the development of complex systems models and Baas (2007) offers a good introduction to this topic. Recent advances in our understanding of the processes of dune dynamics and interaction have come from analytical modelling that is sometimes combined with field measurements. The investigations of Hersen et al. (2004) provide a useful primer in this context.

References

Ahmedou, D.O., Mahfoudh, A.O., Dupont, P. et al. (2007). Barchan dune mobility in Mauritania related to dune and interdune sand fluxes. *Journal of Geophysical Research* 112: F02016. doi: 10.1029/2006JF00050000.

Al-Harthi, A.A. (2002). Geohazard assessment of sand dunes between Jeddah and Al-Lith, Western Saudi Arabia. *Environmental Geology* 42 (2): 360–369. doi: 10.1007/s00254-001-0501-z.

Anderson, R.S. (1988). The pattern of grainfall deposition in the lee of aeolian dunes. *Sedimentology* 35: 175–188. doi: 10.1111/j.1365-3091.1988.tb00943.x.

Anderson, R.S. (1996). The attraction of sand dunes. *Nature* 379 (6560): 24–25. doi: 10.1038/379024a0.

Andreotti, B., Claudin, P., and Douady, S. (2002). Selection of dune shapes and velocities. Part 1: dynamics of sand, wind and barchans. *European Physical Journal B* 28: 321–329. doi: 10.1140/epjb/e2002-00236-4.

Andreotti, B., Fourière, A., Ould-Kaddour, F. et al. (2009). Giant aeolian dune size determined by the average depth of the atmospheric boundary layer. *Nature* 457 (7233): 1120–1123. doi: 10.1038/nature07787.

Baas, A.C.W. (2002). Chaos, fractals and self-organisation in coastal geomorphology: simulating dune landscapes in vegetated environments. *Geomorphology* 48 (1–3): 309–328. doi: 10.1016/S0169-555X(02) 00187-3.

Baas, A.C.W. (2004). Evaluation of saltation flux impact responders (Safires) for measuring instantaneous aeolian sand transport. *Geomorphology* 59 (1–4): 99–118. doi: 10.1016/j.geomorph.2003.09.009.

Baas, A.C.W. (2007). Complex systems in aeolian geomorphology. *Geomorphology* 91 (3–4): 311–331. doi: 10.1016/j.geomorph. 2007.04.012.

Baas, A.C.W. and Nield, J.M. (2007). Modelling vegetated dune landscapes. *Geophysical Research Letters* 34: L06405. doi: 10.1029/ 2006GL029152.

Baas, A.C.W. and Nield, J.M. (2010). Ecogeomorphic state variables and phase–space construction for quantifying the evolution of vegetated aeolian landscapes. *Earth Surface Processes and Landforms* 35 (6): 717–731. doi: 10.1002/esp.1990.

Baddock, M.C., Livingstone, I., and Wiggs, G.F.S. (2007). The geomorphological significance of airflow patterns in transverse dune interdunes. *Geomorphology* 87 (4): 322–336. doi: 10.1016/j.geomorph.2006. 10.006.

Baddock, M.C., Wiggs, G.F.S., and Livingstone, I. (2011). A field study of mean and turbulent flow characteristics upwind, over, and downwind of barchan dunes. *Earth Surface Processes and Landforms* 36 (11): 1435–1448. doi: 10.1002/esp.2161.

Bagnold, R.A. (1941). *The Physics of Blown Sand and Desert Dunes*. London: Methuen.

Bagnold, R.A. (1953). The surface movement of blown sand in relation to meteorology. In: Desert Research, Proceedings of the International Symposium, Jerusalem, 7–14 May 1952, Research Council of Israel, Special Publication 2: 89–93.

Breed, C.S. and Grow, T. (1979). Morphology and distribution of dunes in sand seas observed by remote sensing. In: *A Study of Global Sand Seas*, vol. 1052 (ed. E.D.

McKee), 253–302. US Geological Survey Professional Paper.

Bristow, C.S., Lancaster, N., and Duller, G.A.T. (2005). Combining ground penetrating radar surveys and optical dating to determine dune migration in Namibia. *Journal of the Geological Society of London* 161: 315–361. doi: 10.1144/0016-764903-120.

Bullard, J.E., Thomas, D.S.G., Livingstone, I., and Wiggs, G.F.S. (1995). Analysis of linear sand dune morphological variability, southwestern Kalahari Desert. *Geomorphology* 11 (3): 189–203. doi: 10.1016/0169-555X(94)00061-U.

Butterfield, G.R. (1999). Near-bed mass flux profiles in aeolian sand transport: high resolution measurements in a wind tunnel. *Earth Surface Processes and Landforms* 24 (5): 393–412. doi: 10.1002/(SICI)1096-9837(199905)24:5<393::AID-ESP996> 3.0.CO;2-G.

Claudin, P., Wiggs, G.F.S., and Andreotti, B. (2013). Field evidence for the upwind velocity shift at the crest of low dunes. *Boundary-Layer Meteorology* 148 (1): 195–206. doi: 10.1007/s10546-013-9804-3.

Dong, Z., Wang, T., and Wang, X. (2004a). Geomorphology of the megadunes in the Badain Jaran Desert. *Geomorphology* 99 (1–2): 186–204. doi: 10.1016/j.earscirev. 2013.02.003.

Dong, Z., Guangting, C., He, X. et al. (2004b). Controlling blown sand along the highway crossing the Taklimakan Desert. *Journal of Arid Environments* 57 (3): 329–344. doi: 10.1016/j.jaridenv.2002.02.001.

Eckardt, F.D., Livingstone, I., Seely, M., and Von Holdt, J. (2013). The surface geology and geomorphology around Gobabeb, Namib desert, Namibia. *Geografiska Annaler Series A: Physical Geography* 95 (4): 271–284. doi: org/10.1111/geoa.12028.

Elbelrhiti, H., Claudin, P., and Andreotti, B. (2005). Field evidence for surface-wave-induced instability of sand dunes. *Nature* 437 (7059): 720–723. doi: 10.1038/nature04058.

Fenton, L.K., Michaels, T.I., and Beyer, R.A. (2014). Inverse maximum gross bedform-normal transport 1: how to determine a dune-constructing wind regime using only imagery. In: Proceedings: 3rd International Planetary Dunes Workshop, Lowell Observatory, Flagstaff, AZ, June 12-15, 2012. *Icarus* 230 (Special Issue): 5–14. doi: 10.1016/j.icarus.2013.04.001.

Frank, A. and Kocurek, G. (1996). Airflow up the stoss slope of sand dunes: limitations of current understanding. *Geomorphology* 17 (1–3): 47–54. doi: 10.1016/0169-555X(95)00094-L.

Fryberger, S.G. and Dean, G. (1979). Dune forms and wind regime. In: *A Study of Global Sand Seas*, vol. 1052 (ed. E.D. McKee), 137–170. United States Geological Survey Professional Paper.

Gay, S.P. (1999). Observations regarding the movement of barchan sand dunes in the Nazca to Tanaca area of Southern Peru. *Geomorphology* 27 (3–4): 279–293. doi: 10.1016/S0169-555X(98)00084-1.

Grove, A.T. (1958). The ancient ergs of Hausaland, and similar formations on the south side of the Sahara. *Geographical Journal* 124 (4): 528–533. doi: 10.2307/1790942.

Hersen, P., Andersen, K.H., Elbelrhiti, H. et al. (2004). Corridors of barchan dunes: stability and size selection. *Physical Review E* 69 (011304): 1–12. doi: 10.1103/PhysRevE. 69.011304.

Hersen, P. and Douady, S. (2005). Collision of barchan dunes as a mechanism of size regulation. *Geophysical Research Letters* 32 (21): 1–5. doi: 10.1029/2005GL024179.

Hesp, P.A. and Hastings, K. (1998). Width, height and slope relationships and aerodynamic maintenance of barchans. *Geomorphology* 22 (2): 193–204. doi: 10.1016/S0169-555X(97)00070-6.

Hesse, P.P. and Simpson, R.L. (2006). Variable vegetation cover and episodic sand movement on longitudinal desert sand dunes. *Geomorphology* 81 (3–4): 276–291. doi: 10.1016/j.geomorph.2006.04.012.

Hunter, R.E., Richmond, B.M., and Alpha, T.R. (1983). Storm-controlled oblique dunes of the Oregon coast. *Geological Society of America Bulletin* 94 (12): 1450–1465.

doi: 10.1130/0016-7606(1983)94<1450:SODOTO>2.0.CO;2.

Jerolmack, D.J., Ewing, R.C., Falcini, F. et al. (2012). Internal boundary layer model for the evolution of desert dune fields. *Nature Geoscience* 5 (3): 206–209. doi: 10.1038/ngeo1381.

Katsuki, A., Kikuchi, M., Nishimori, H. et al. (2011). Cellular model for sand dunes with saltation, avalanche and strong erosion: collisional simulation of barchans. *Earth Surface Processes and Landforms* 36 (3): 372–382. doi: 10.1002/esp.2049.

Kocurek, G. and Ewing, R.C. (2005). Aeolian dune field self-organization – implications for the formation of simple versus complex dune-field patterns. *Geomorphology* 72 (1–4): 94–105. doi: 10.1016/j.geomorph.2005.05.005.

Lancaster, N. (1989). *The Namib Sand Sea: Dune Forms, Processes and Sediments*. Rotterdam: A. A. Balkema.

Lancaster, N. (1991). The orientation of dunes with respect to sand-transporting winds: a test of Rubin and Hunter's gross bedform-normal rule. In: *NATO Advanced Research Workshop on Sand, Dust and Soil in Their Relation to Aeolian and Littoral Processes*, 47–49. Sandbjerg, Denmark: University of Aarhus.

Lancaster, N., Kocurek, G., Singhvi, A. et al. (2002). Late Pleistocene and Holocene dune activity and wind regimes in the western Sahara Desert of Mauritania. *Geology* 30 (11): 991–994. doi: 10.1130/0091-7613-31.1.e18.

Lancaster, N., Nickling, W.G., McKenna Neuman, C., and Wyatt, V.E. (1996). Sediment flux and airflow on the stoss slope of a barchan dune. *Geomorphology* 17 (1–3): 55–62. doi: 10.1016/0169-555X(95)00095-M.

Leenders, J.K., van Boxel, J.H., and Sterk, G. (2005). Wind forces and related saltation transport. *Geomorphology* 71 (3–4): 357–372. doi: 10.1016/j.geomorph.2005.04.008.

Leighton, C.L., Bailey, R.M., and Thomas, D.S.G. (2013). The utility of desert sand dunes as quaternary chronostratigraphic archives: evidence from the northeast Rub' al Khali. *Quaternary Science Reviews* 78: 303–318. doi: 10.1016/j.quascirev.2013.04.016.

Livingstone, I. (1988). New models for the formation of linear dunes. *Geography* 73: 105–115.

Livingstone, I. (1989). Monitoring surface change on a Namib linear dune. *Earth Surface Processes and Landforms* 14 (4): 317–332. doi: 10.1002/esp.3290140407.

Livingstone, I. (1993). A decade of surface change on a Namib linear dune. *Earth Surface Processes and Landforms* 18 (7): 661–665. doi: 10.1002/esp.3290180708.

Livingstone, I. (2003). A twenty-one year record of surface change on a Namib linear dune. *Earth Surface Processes and Landforms* 28 (9): 1025–1031. doi: 10.1002/esp.1000.

Livingstone, I., Bristow, C., Bryant, R.G. et al. (2010). The Namib Sand Sea digital database of aeolian dunes and key forcing variables. *Aeolian Research* 2 (2–3): 93–104. doi: 10.1016/j.aeolia.2010.08.001.

Livingstone, I. and Warren, A. (1996). *Aeolian Geomorphology: An Introduction*. London: Longman.

Livingstone, I., Wiggs, G.F.S., and Weaver, C.M. (2007). Geomorphology of desert sand dunes: a review of recent progress. *Earth Science Reviews* 80 (3–4): 239–257. doi: 10.1016/j.earscirev.2006.09.004.

Mainguet, M. and Chemin, M.C. (1984). Les dunes pyramidales du Grand Erg Oriental. *Travaux de l'Institut de Géographie de Reims* 59–60: 49–60.

McKee, E.D. (ed.) (1979). *A Study of Global Sand Seas*, vol. 1052. United States Department of the Interior. U.S. Geological Survey, Professional Paper.

McKenna-Neuman, C., Lancaster, N., and Nickling, W.G. (1997). Relations between dune morphology, air flow, and sediment flux on reversing dunes, Silver Peak, Nevada. *Sedimentology* 44 (6): 1103–1113. doi: 10.1046/j.1365-3091.1997.d01-61.x.

Momiji, H., Carretero-Gonzalez, R., Bishop, S.R., and Warren, A. (2002). Simulation of the effect of wind speed up in the formation of transverse dunefields. *Earth Surface Processes and Landforms* 25 (8): 905–918.

doi: 10.1002/1096-9837(200008)25:8<905::AID-ESP112>3.0.CO;2-Z.

Mulligan, K.R. (1988). Velocity profiles measured on the windward slope of a transverse dune. *Earth Surface Processes and Landform* 13 (7): 573–582. doi: 10.1002/esp.3290130703.

Nield, J.M. and Baas, A.C.W. (2008). Investigating parabolic and nebkha dune formation using a cellular automaton modelling approach. *Earth Surface Processes and Landforms* 33 (5): 724–740. doi: 10.1002/esp.1571.

Nield, J.M., Wiggs, G.F.S., Baddock, M.C., and Hipondoka, M.H.T. (2017). Coupling leeside grainfall to avalanche characteristics in aeolian dune dynamics. *Geology* 45 (3): 271–274.

Parsons, D.R., Walker, I.J., and Wiggs, G.F.S. (2004a). Numerical modelling of flow structures over idealised transverse aeolian dunes of varying geometry. *Geomorphology* 59 (1–4): 149–164. doi: 10.1016/j.geomorph.2003.09.012.

Parsons, D.R., Wiggs, G.F.S., Walker, I.J. et al. (2004b). Numerical modelling of airflow over an idealised transverse dune. *Environmental Modelling and Software* 19 (2): 153–162. doi: 10.1016/S1364-8152(03)00117-8.

Ping, L., Narteau, C., Dong, Z. et al. (2014). Emergence of oblique dunes in a landscape-scale experiment. *Nature Geoscience* 7 (2): 99–103. doi: 10.1038/ngeo2047.

Rubin, D.M. and Hunter, R.E. (1987). Bedform alignment in directionally varying flows. *Science* 237 (4812): 276–278. doi: 10.1126/science.237.4812.276.

Schwämmle, V. and Herrmann, H.J. (2003). Solitary wave behaviour of sand dunes. *Nature* 426 (6967): 619–620. doi: 10.1038/426619a.

Sterk, G., Jacobs, A.F.G., and van Boxel, J.H. (1998). The effect of turbulent flow structures on saltation sand transport in the atmospheric boundary layer. *Earth Surface Processes and Landforms* 23 (10): 877–887. doi: 10.1002/(SICI)1096-9837(199810)23:10<877:AID-ESP905>3.0.CO;2-R.

Stone, A.E.C. and Thomas, D.S.G. (2008). Linear dune accumulation chronologies from the southwest Kalahari, Namibia: challenges of reconstructing late Quaternary palaeoenvironments from aeolian landforms. *Quaternary Science Reviews* 27 (17–18): 1667–1681. doi: 10.1016/j.quascirev.2008.06.008.

Thomas, D.S.G., Knight, M., and Wiggs, G.F.S. (2005). Remobilization of southern African desert dune systems by twenty-first-century global warming. *Nature* 435 (7046): 1218–1221. doi: 10.1038/nature 03717.

Thomas, D.S.G. and Wiggs, G.F.S. (2008). Aeolian system response to global change: challenges of scale, process and temporal integration. *Earth Surface Processes and Landforms* 33 (9): 1396–1418. doi: 10.1002/esp.1719.

Tsoar, H. (1983). Dynamic processes acting on a longitudinal (seif) dune. *Sedimentology* 30 (4): 567–578. doi: 10.1111/j.1365-3091. 1983.tb00694.x.

Tsoar, H. (1989). Linear dunes – forms and formation. *Progress in Physical Geography* 13 (4): 507–528. doi: 10.1177/030913338901300402.

Tsoar, H. (1990). New models for the formation of linear sand dunes: a discussion. *Geography* 75: 144–147.

Tsoar, H., Blumberg, D.G., and Stoler, Y. (2004). Elongation and migration of sand dunes. *Geomorphology* 57 (3–4): 293–302. doi: 10.1016/S0169-555X(03)00161-2.

van Boxel, J.H., Sterk, G., and Arens, S.M. (2004). Sonic anemometers in aeolian sediment transport research. *Geomorphology* 59 (1–-4): 131–147. doi: 10.1016/j.geomorph.2003.09.011.

Walker, I.J. (2005). Physical and logistical considerations of using ultrasonic anemometers in aeolian sediment transport research. *Geomorphology* 68 (1–2): 57–76. doi: 10.1016/j.geomorph. 2004.09.031.

Walker, I.J. and Nickling, W.G. (2002). Dynamics of secondary airflow and sediment transport over and in the lee of transverse dunes. *Progress in Physical Geography* 26 (1): 47–75. doi: 0.1191/0309133302pp3.

Warren, A. and Kay, S.A.W. (1987). The dynamics of dune networks. In: *Desert Sediments: Ancient and Modern* (ed. L.E. Frostick and I. Reid), 205–212. Special Publication, 35. Oxford: Geological Society of London, Blackwell.

Wasson, R.J. and Hyde, R. (1983). Factors determining desert dune type. *Nature* 304 (5924): 337–339. doi: 10.1038/304337a0.

Weaver, C.M. and Wiggs, G.F.S. (2011). Field measurements of mean and turbulent airflow over a barchan sand dune. *Geomorphology* 128 (1–2): 32–41. doi: 10.1016/j.geomorph.2010.12.020.

Werner, B.T. (1995). Eolian dunes: computer simulations and attractor interpretation. *Geology* 23 (12): 1107–1110. doi: 10.1130/0091-7613(1995)023<1107:EDCSAA>2.3.CO;2.

Werner, B.T. and Kocurek, G. (1997). Bed-form dynamics: does the tail wag the dog? *Geology* 25 (9): 771–774. doi: 10.1130/0091-7613(1999)027<0727:BSFDD>2.3.CO;2.

Wiggs, G.F.S., Livingstone, I., Thomas, D.S.G., and Bullard, J.E. (1994). Effect of vegetation removal on airflow patterns and dune dynamics in the southwest Kalahari Desert. *Land Degradation and Rehabilitation* 5 (1): 13–24. doi: 10.1002/ldr.3400050103.

Wiggs, G.F.S., Livingstone, I., Thomas, D.S.G., and Bullard, J.E. (1995). Dune mobility and vegetation cover in the southwest Kalahari Desert. *Earth Surface Processes and Landforms* 20 (6): 515–529. doi: 10.1002/esp.3290200604.

Wiggs, G.F.S., Livingstone, I., and Warren, A. (1996). The role of streamline curvature in sand dune dynamics: evidence from field and wind tunnel measurements. *Geomorphology* 17 (1–3): 29–46. doi: 10.1016/0169-555X(95)00093-K.

Wiggs, G.F.S. and Weaver, C.M. (2012). Turbulent flow structures and aeolian sediment transport over a barchan dune. *Geophysical Research Letters* 39 (5): L05404. doi: 10.1029/2012GL050847.

7

Anchored Dunes

Patrick A. Hesp[1] and Thomas A.G. Smyth[2]

[1]*Flinders University, Adelaide, Australia*
[2]*Liverpool Hope University, Liverpool, UK*

7.1 Introduction

This chapter examines dunes that are 'anchored' to an obstacle, whose geometric characteristics then control the shape of the dunes. There are many kinds and sizes of obstacle: plants, organic debris of various kinds, pebbles, boulders, or isolated hills. The types of dune that are *not* formed in this way are covered in Chapter 6.

7.2 Nebkhas and Nebkha Fields

7.2.1 Generalities

Nebkhas (otherwise known as coppice dunes, shrub-coppice dunes or mounds, pimple mounds, and hummocks) are discrete dunes formed by the trapping of sand, silt, clay, and coarse particles up to the size of small gravel, and/or snow by a plant. The accumulation occurs upwind of, within, and in the lee of a plant. The material trapped by the plant may include organic products (e.g. seaweed, wrack, wood, and plant litter), animal faeces, body parts, or pumice. Nebkhas form in all aeolian environments, coast to desert, arid to tropical, and plain to mountain (Figures 7.1 and 7.2). They may be very small, ranging from a few centimetres in diameter (as in the initiation phase) to very large (up to 10 m + in height (Warren 1988).

The morphology of nebkhas is highly dependent on the morphology of the parent plant, including the density of its stems and leaves and the overall structural form (multistemmed or single stem, size, tall and erect vs low and compact, etc.). Dong Zhibao and colleagues (2008) showed that the flow structure strongly depended on the density of the foliage, because of the increasing presence and operation of 'bleed' flow through the plant as density decreases.

Nebkhas can be very important to the functioning of a dunefield or desert ecosystem, because they may be one of the few places in a dry environment that provide stable shelter, habitat, more moisture, lower temperature, and food (e.g. litter, roots, and leaves). They act as secure islands within a sea of sand. The degree of shelter, microclimate, and food that nebkhas may provide depends on the parent plant's species, form, and growth habit. Hesp and McLachlan (2000) found that in a coastal interdune plain and in general the

Aeolian Geomorphology: A New Introduction, First Edition. Edited by Ian Livingstone and Andrew Warren.
© 2019 John Wiley & Sons Ltd. Published 2019 by John Wiley & Sons Ltd.

Figure 7.1 Large nebkhas and shadow dunes seawards of parabolic-shaped sand streaks in Peru at 7°03 S, 44°W. The scale bar is 50 m. *Source:* Google Earth.

(a)

(b)

Figure 7.2 (a) Nebkhas and attendant shadow dunes; (b) Erosional nebkhas forming within and around live and dead plants. An echo dune has formed on the windward and lateral margins in (b). Scour around the front and sides of this dune may eventually lead to its destruction. Similar erosion patterns occur around solid objects. *Source:* Photos: P. Hesp.

numbers of amphipods and nematodes increased as the size of the parent plant and its nebkha increased, for *Gazania rigens* and *Arctotheca populifolia* nebkhas, but, above a certain size, decreased in *Arctotheca* nebkhas. The *Arctotheca* plant tended to become more open in growth habit over time and with increasing size, until it actually provided less shade, and thus, a less desirable habitat for animals.

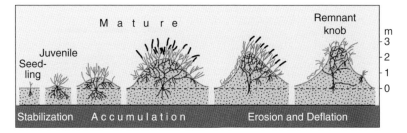

Figure 7.3 One possible evolutionary pathway for a nebkha, from initiation to eroding remnant knob, and then, not shown, possible stabilisation as vegetation manages to recolonise the knob. The evolutionary steps may stop at any point from juvenile to remnant knob, if the field is colonised by a plant species that is capable of stabilising the dune. *Source:* Based on Piotrewska (1991).

7.2.2 Evolutionary Trends

Figure 7.3 shows one possible evolutionary path that may be taken by nebkhas on coasts and some deserts: from incipient formation, to remnant knob, and in some cases to full erosion. Other nebkhas may become completely stable mounds, either in isolation, or as multiple forms in a nebkha field. In coastal environments, nebkhas commonly form on deflation plains, as the plains are colonised and stabilised by vegetation (see Hesp and Thom 1990, their Figure 7), or following small- to large-scale washover events on barrier islands (Figure 14 in Hesp 2011). These nebkhas may evolve over time into very large vegetated dunes (for example, 'tree islands' on in the west coast of the USA, in Cooper's (1958) terminology; or 'bush pockets' in South Africa).

The presence of nebkhas in some coastal and desert environments may indicate that intense grazing by animals is taking place. Parts of Newborough Warren in the UK were dominated by the presence of nebkhas in the 1950s (Ranwell 1959, 1960), but they disappeared following the almost total elimination of rabbits (Rhind et al. 2013).

7.3 Shadow Dunes

Shadow dunes are pyramidal through triangular to tear-drop-shaped dunes. They are formed downwind, or in the shelter of nebkhas, plants, or other obstacles (Bagnold 1941). Some shadow dunes may be termed 'lee dunes' because lee dunes tend to form in similar places. The highest portion of a shadow dune is situated immediately at the rear of the parent object, and usually tapers downwind (see Figure 7.2b). Many are ephemeral and some shift position to align with the prevailing wind. Shadow dunes are formed within a flow-separation envelope (or wake region), which is characterised by a mean flow that, while highly turbulent on a second-to-second scale, displays paired, symmetrically-opposed vortices at a larger scale (Hesp 1981; Dong et al. 2008). Figure 7.4 shows streamlines 0.25 m above the bed behind a model nebkha, and the mean flow structure for an incident wind of $5\,\mathrm{m\,s^{-1}}$ at the bed (Hesp and Smyth, 2017).

The length of a shadow is related to the width of the parent plant (or other obstacle). The width of the parent plant or obstacle controls the height of the shadow dune, because the height (h) is determined by the maximum angle of repose of the dune side slopes: h = w/2 tan θ, where θ is the angle of repose (Hesp 1981).

In some circumstances, shadow dunes may continue to grow downwind and become large and stable features (Xiao et al. 2015). Figure 7.1 shows long shadow dunes which may be relatively stable because of the dominance of a uni-directional wind that has allowed them to grow. Others form downwind of hills and mountains, some through mountain gaps (examples from the

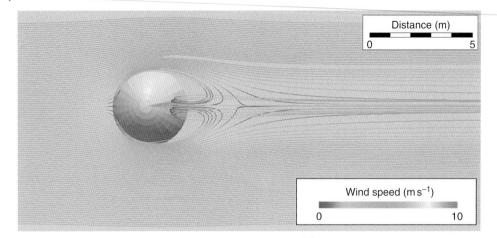

Figure 7.4 Streamlines at 0.25 m above the bed around a model nebkha, produced by a Reynolds-Averaged Navier–Stokes (RANS) computational fluid dynamics simulation. Mean flow comes from the left. The figure shows two symmetrically opposed reversing vortices forming behind the nebkha and two zones of low velocity in which the shadow dune would form. A small reversing flow separation zone is also produced upwind of the nebkha. *Source:* Hesp and Smyth, 2017. See insert for colour representation of this figure.

Moenkopi Plateau and Peru; Figures 5.12 and 5.13 in Greeley and Iversen 1985). These large lee dunes may be very stable. Some may extend downwind over time in the same way as in linear or longitudinal dune extension mechanisms (Rubin and Hesp 2009; Chapter 6).

Most plants are more porous than solid objects and do not experience the same degree of acceleration around their sides nor the degree of flow separation, and therefore do not produce the same degree of erosion on the flanks of the obstacle (Luo et al. 2012; see also Figures 8.6 and 8.7 in Bauer et al. 2013). The orientation of a shadow or lee dune to a particular incident wind also changes the degree of shelter or 'shadowing' and the resultant shadow dune morphology (Luo et al. 2012). Shadow dunes may occur as two teardrop or winged forms (Bagnold 1941) usually because the plant or obstacle is low, but wide, or because the shadow dune has not fully evolved. 'Sand streaks' are thin, elongated traces of sand whose characteristics are sometimes similar to those of shadow dunes (Greeley and Iversen 1985).

7.4 Foredunes

Foredunes have also been termed 'frontal dunes', 'retention ridges', 'dunas frontais', 'duna frontal', 'beach ridges', 'parallel dunes' and even 'transverse dunes' (see Hesp 1989 for review). The term 'foredune' has been used by some authors to refer to all dune types that occupy a foremost position on the shore or in front of a plant or obstacle (coastal or inland location). Most coastal geomorphologists restrict the term to mean coastal dunes formed by aeolian sand deposition within pioneer plant communities on the backshore of beaches (of any type – lake, lagoon, ocean, estuary, river etc). Foredunes are found in very many locations (from the Arctic to the Tropics) (Hesp 2002, 2012; Ruz and Hesp 2014). The plants on a foredune may be very small, very low-growing herbs (e.g. *Blutaparon* spp. in southern Brazil), or they may be quite large (like *Atriplex* spp. in Western Australia, or willow trees on the Canadian Great Lakes). They may form monospecific stands or be created by many species growing together. Actively forming foredunes occupy the foremost position, but not all the foremost dunes are foredunes (Hesp 2002). Other dune types may occupy the foremost position on eroding coasts (e.g. the Grand Dune du Pyla in France; parabolic dunefields in Madagascar – see below), or coasts where foredunes are unable to form (e.g. where transverse dunes are migrating alongshore) (Hesp 1999).

Two principal types are common: incipient (also called 'embryo' dunes) and established foredunes. Incipient foredunes are newly formed dunes dominated by pioneer plant species. Established foredunes are generally larger, more complex forms with greater plant diversity, height, width, and/or age (Hesp 2002). The ecogeomorphic state of established foredunes can vary considerably depending on the coastal barrier state (retrograding, prograding, or stationary), the sediment supply, the climatic regime, latitudinal position and beach-surfzone type (Hesp 1988, 2012; Davidson-Arnott 2010; Figure 7.5).

Foredunes form as sediment is transported by the wind from the exposed beach into a plant canopy. The increased drag and lower velocity imposed by the plants cause the sand to be deposited (Furieri et al. 2014). Over time, as the plants grow upwards and seawards, this process continues to build wider and higher ridges. The morphology of the resulting dune is dependent on plant density and distribution, species type, wind velocity, and sediment supply (Hesp 2002). As foredunes grow, the wind flow is accelerated over the ridge (Arens et al. 1995). Winds oblique to the coast are deflected towards a more perpendicular flow over larger dunes, this leads to sedimentation across a greater proportion of the dune especially in high winds. Occasional scarping by waves, scarp filling, and subsequent deposition on the crest of the dune lead to the development of higher foredunes. They may translate landwards on eroding coasts (Davidson-Arnott 2005; Hesp et al. 2015). This occurs because following scarping, and once the scarp fill is completed, the scarp fill provides a partially-vegetated or unvegetated aerodynamic ramp on which aeolian sand is easily transported up slope to the crest and beyond.

Swales are generally created as a foredune grows seaward. As plants grow out onto the backshore, the zone of maximum aeolian sand shifts seawards, leaving a zone of limited deposition in the lee. Swales thus develop as zones of low to

Figure 7.5 Established foredune formed in *Ammophila brevigulata* on Prince Edward Island, Canada. *Ammophila* spp. are discussed in Chapter 12. See insert for colour representation of this figure.

limited aeolian deposition and become deeper as seaward incipient foredunes become higher (Olson 1958; Hesp 2013a). Rapid shoreline progradation typically results in low foredune ridges and wide swales.

7.5 Lunette or Clay Dunes

Lunettes are typically crescentic or arcuate-shaped dunes formed on pan, playa, and lake margins, usually mimicking the plan shape of the adjacent palaeo or present shoreline, with the horns pointing upwind (Laity 2008; Figure 7.6). They are asymmetric with the steepest face on the windward side (Hills 1940; Holliday 1997). Some may have been slightly or significantly modified by wave erosion. Lunettes have also been termed 'clay dunes' (Price 1963; Bowler 1973) but not all lunettes are formed from clay.

As pans and playas vary from quite small to very extensive (Bowler 1973; Holliday 1997; De Deckker et al. 2011), so too do the lunettes formed downwind of them. Lunettes may be quite small (less than a metre in height) to

quite large ranging up to 20 to 40 m in some cases (Goudie and Wells 1995). While clay, sand, sandy clay, and loamy sand may be the most common sediments in lunettes, along with gravels in some circumstances, significant amounts of other minerals may be present including gypsum and calcium carbonate (Bowler 1973; Shaw and Thomas 1989; Holliday 1997). Most pans and playas have developed in palaeo-lacustrine basins, palaeo-drainages, interdunes, and on coastal plains (Goudie and Wells 1995).

Many lunettes have developed due to aeolian deflation of sediments adjacent to, and from within the playas or pans (Goudie and Wells 1995; Sabin and Holliday 1995). Bowler (1973) considered lunettes to have been formed by both beach deflation during higher lake levels and lake floor deflation during low stands. Since clays, and clay pellets, or clay aggregates are essentially sticky sediments, they may be deposited immediately downwind of the basins, pans, or playas where there is a slight change in surface morphology and wind speed. Once formed, flow deceleration or separation across the dune crest would lead to further deposition on the lunette. However,

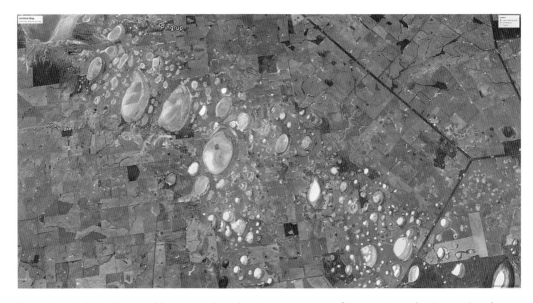

Figure 7.6 Multiple playas and lunettes in the Lake Pingarnup region of Western Australia. *Source:* Google Earth. See insert for colour representation of this figure.

on more sandy (or other non-clay sediment type) lunettes, it is difficult to imagine that without some trapping mechanism, a dune would form. A sheet deposit (or 'lake shadow', per Pillans 1987) would be more likely to develop as shown for some modern lunettes (Pillans 1987) and as indicated by Holliday (1997). Bowler et al. (2012) indicate that the Lake Mungo lunette was initially created by quartz sand deposition and the formation of a transverse dune. This was then overlaid by pelletal clay and sands.

Observations of some modern clay and sandy clay lunettes forming in Western and Southern Australia indicate that initiation begins with deposition of aeolian sediments (either clays or sands and/or mixtures of these) in salt-tolerant plants growing around the margins of the playas. This initiation mechanism is more common than perhaps indicated in some of the literature, as both Price (1963) and Bowler (1973) state that the lunettes in Texas and in SE Australia are initiated by aeolian sediment deposition in plants. Examination of the internal stratification of some lunettes indicating the initial structures are horizontal or low-angle dipping strata (e.g. Holliday 1997) would conform with this initiation mechanism. Once an initial dune form is created, positive feedback between form and flow would lead to the preferential deposition of sediments on the dune just as it does with the evolution of

protodunes to transverse dunes (Kocurek et al. 1992). Some lunettes found within parabolic dune deflation plains have formed in the same fashion as foredunes; that is, by aeolian deposition in plants according to Lees and Cook (1991). Thus, it is most likely there are multiple mechanisms by which lunette dunes can form.

7.6 Blowouts

Blowouts are erosional hollows typically found on vegetated to semi-vegetated sediment deposits. They are found in a range of biomes including deserts (Barchyn and Hugenholtz 2013a), semiarid regions (Guitérrez-Elorza et al. 2005), continental dunefields (Smith 1960), formerly glaciated regions and outwash plains (Koster 1988; Seppala 2004), and temperate grasslands (Hugenholtz and Wolfe 2006). They are also common in coastal dunes, on both eroding and accreting coastlines (Carter et al. 1990; Hesp 2002), from the tropics (Barbosa and Landim-Dominguez 2004) to the Arctic Circle (Black 1951; Hesp 2004; Ruz and Hesp 2014).

Descriptions of blowout forms include: cigar-shaped, v-shaped, scooped hollow, cauldron, and corridor (Ritchie 1972), pits, elongated notches, troughs, and broad basins (Smith 1960), and saucer, trough, and bowl (Hesp 2002). Cooper's (1958, 1967) primary

Figure 7.7 Bowl blowout at Cape Cod, MA, USA. The principal depositional lobe is on the right but due to the multi-directional wind regime in this environment, sand can be deposited all around the blowout rim. *Source:* P. Hesp.

classification of saucer, bowl, and trough, however, remains the most ubiquitously employed. Saucer blowouts are shallow depressions with an erosional lip, trough blowouts are deep, relatively narrow, longer slots, and bowls are deeper semi-circular depressions (Figure 7.6). All display connected depositional lobes. (Carter et al. 1990; Hesp 2002).

Blowouts form where vegetation cover is either naturally low or absent, or reduced, permitting moderate to high-velocity near surface winds to entrain and erode sediment to form a topographic depression. The steep walls of the depression are referred to as 'erosional walls' and the relatively flat erosional surface as the 'deflation basin'. The deposition of deflated sediment is also considered part of the landform and is referred to as a depositional lobe (Hesp 2002) or rarely, an apron (Barchyn and Hugenholtz 2013a). The reduction in vegetation cover may be caused by a range of factors including dune scarping (Hesp 2002), animal grazing and disturbance (Blanco et al. 2008), fire, trampling (Mir-Gual et al. 2012), vehicles (Catto et al. 2002), storm winds, and climate change (Hesp 2002) (Figure 7.7).

Once formed, a blowout characteristically expands downwind parallel to the prevailing wind direction (van der Meulen and Jungerius 1989). However, because of the multifaceted nature of the interactions between secondary wind flow, dune topography and vegetation, the shape of a blowout can be variable and complex (Gares and Nordstrom 1995; Hesp and Walker 2013; Smyth et al. 2012). Greater morphological complexity is often found in multi-directional wind regimes. Blowouts quite commonly begin as a cut though a foredune at the coast, or within a dunefield or dune types in coastal and desert and semi-to-arid terrain, but then as they grow, the direction of growth changes orientation towards that of the dominant wind (see the upside-down j-shaped blowouts in the foredune in Figure 7.8).

The enlargement and evolution of blowouts are controlled by the erosivity of winds within the deflation basin to entrain and remove

Figure 7.8 A trangressive and parabolic dunefield complex on the southern coast of Madagascar. It is likely that many of the parabolic dunes were formed following stabilisation of transgressive dunefields, as seen on the downwind and seaward margins of the active transgressive dunefield. *Source:* Google Earth.

sediment (Barchyn and Hugenholtz 2013b). This deepens the blowout, particularly through steepening the walls by avalanching. This in turn leads to lateral expansion (Carter et al. 1990). Within trough blowouts, winds are accelerated between the steep walls, resulting in the erosion of the deflation basin and upwind face of the depositional lobe (Hesp and Hyde 1996; Hesp and Pringle 2001; Hansen et al. 2009; Smyth et al. 2014). In saucer and bowl blowouts, a turbulent separation zone develops in the lee of the upwind erosional wall where wind flow enters the blowout. Towards the centre of the deflation basin, flow may reattach to the base, becoming realigned to the incident wind direction, before accelerating over the blowout rim (Gares and Nordstrom 1995; Wang et al. 2007; Hugenholtz and Wolfe, 2006; Hesp and Walker 2011; Smyth et al. 2011, 2012, 2013).

Surface changes in blowouts have been associated with wind speeds as low as $1.25\,\mathrm{m\,s^{-1}}$ (Jungerius et al. 1981) and sediment flux has been found to be greatest where wind flow is least variable, e.g. at the crest of the erosional wall/depositional lobe where wind flow is compressed and accelerated by the upwind slope (Hesp 2002; Smyth et al. 2014). Blowouts can grow in length or diameter up to hundreds of metres (Figure 6 in Hesp 2002), while others may cease to expand when erosion reaches an impenetrable stratum such as a cemented layer, a former pebble beach, an armoured surface or the water table (Ritchie 1972; Carter 1976; Hesp and Thom 1990; Gares 1992). Alternatively, expansion may be constrained by the reduction in the velocity of wind in the blowout as it becomes sufficiently broad, or, in some cases as the climate becomes more favourable to the growth of vegetation (Jungerius et al. 1991; Abhar et al. 2015).

7.7 Parabolic Dunes

Parabolic dunes may evolve from blowouts as blowouts extend downwind and form trailing ridges, or may form where active dunes, dune sheets, or dune fields migrate into a vegetated environment, or where vegetation begins to colonise parts of a dune landscape (Cooper 1967; Tsoar and Blumberg 2002; Baas and Nield 2007, 2010). Claims by some authors (e.g. Kelletat and Scheffers 2003) that some coastal examples are formed by tsunami waves have no basis in fact (Bourgeois and Weiss 2009).

Parabolic dunes are typically 'U' or 'V' shaped in planform and develop a form which is, wholly, or in part, controlled by the stabilisation of vegetation (Hack 1941; Melton 1940; Landsberg 1956; Jennings 1957). They have also been term edipsiloidal dunes, or hairpin dunes. They are characterised by trailing ridges of various length which typically terminate downwind in U- or V-shaped depositional lobes. In places the termination point has been blown out. Parabolic dunes with no depositional lobe have occasionally mistakenly been classified as linear dunes. The depositional lobes may be simple, relatively featureless sand sheets, or textured with a variety of dune forms (e.g. transverse dunes, barchanoidal dunes, etc.). Small to extensive deflation basins and plains occur between the trailing ridges and upwind of the depositional lobes. Parabolic dunes differ from blowouts in that they display trailing ridges (Hesp 2011). Parabolic dunes are found all over the world (see Yan and Baas 2015, for a review of locations).

Two principal subtypes of parabolic dune are common: long-walled types, and squat, elliptical types. Long-walled are as the name implies; they have trailing ridges which are relatively long. Elliptical types display much shorter trailing ridges. The multiple development of these leads to there being two principal subtypes of parabolic dunefields; long-walled types (e.g. Pye 1982; Tinley 1985) and imbricate types (Tinley 1985). Imbricate refers to the short-walled overlapping nature of the parabolic dunefield which resembles roof tiles. As with blowouts, there are significant numbers of other morphologies, sometimes very large (megadunes) and very complex types (see McKee 1979;

Davies 1980; David 1981; Pye and Tsoar 1990; their Figure 6.49; Trenhaile 1997; Wolfe and David 1997; Hugenholtz et al., 2009).

The form of parabolic dunes is primarily a function of wind regime, the speed of plant colonisation, and vegetation roughness. Long-walled parabolic dunes tend to develop in uni-modal or near uni-modal wind regimes, and where the vegetation cover is short. Higher, squat, or imbricate forms tend to develop where the vegetation cover is tall, as migration rates decrease as the vegetation height increases. Dunes may evolve from small, simple forms to very large, nested, digitate, comb, and complex types depending on sand supply, climate changes driving vegetation changes and particularly water table changes, wind regimes, the nature and number of re-activation phases or cycles, and vegetation cover and structure (e.g. Pye 1993; David et al. 1999; Hugenholtz and Wolfe 2005; Baas and Nield 2010; Hesp and Walker 2013). Migration rates vary considerably from a few cm yr^{-1} (Story 1982) to 13 m yr^{-1} (Anthonsen et al. 1996) and more depending on these factors and other local factors such as degree of snow cover, and rainfall amount and annual distribution.

7.8 Coastal Transgressive Sand Sheets and Dunefields

Transgressive dune sheets (i.e with no surficial dunes) and transgressive are aeolian deposits formed by the downwind (across or alongshore) movement of sand over vegetated to semi-vegetated terrain (Hesp and Thom, 1990). Transgressive dunefields are similar to many desert dunefields, contain a variety of vegetated and non-vegetated landforms, including barchan dunes, barchanoids, transverse dunes, rare star dunes, sand sheets, blowouts, parabolic dunes, trailing ridges, deflation ponds, basins, flats, and plains (or 'slacks'), gegenwalle ridges, dune track ridges (similar to gegenwalle ridges), nebkhas, shadow dunes, remnant knobs (erosional nebkhas and eroded remnants of other dune types; see Figure 7.9) and precipitation ridges (ridges formed on the margins of dunefields where sand 'precipitates' and avalanches into vegetation) (Cooper 1967; Hesp and Walker 2013). The presence and extent of many of these landforms depend at least on the degree of vegetation cover, climate regime and stage of evolutionary development (Hesp 2013b).

Transgressive dunefields are commonly colonised and stabilised naturally by vegetation over time (except perhaps in the most arid environments). Figure 7.10 shows this process taking place in a subtropical transgressive dunefield in Mexico. The rate of vegetation colonisation may be the most important factor in controlling the subsequent development of dune types and landform units. If stabilisation occurs at a moderate rate, the dunes can continue to migrate as portions of the dunes are colonised at different rates. Polygonal, digitate (digits, or finger-like projections) and crenate (like rounded teeth) ridge types develop (Figure 7.11a; Hesp 2013b). If the process is rapid, transverse dunes may be stabilised in place and with their form largely intact (Figure 7.11b). If the process is relatively slow, parabolic types may tend to develop as the transgressive dunefield stabilises (Hesp 2001; Figure 7.11c).

7.8.1 Semi-fixed (Stabilising) and Fixed (Stabilised) Desert Dunefields

The processes described above, whereby coastal dunes gradually become fixed or stabilised over time presumably apply to many of the world's semi-arid and arid inland and/or continental dunefields. Pioneer species colonise sand sheets, interdune plains, and the lower slopes of dunes first, and gradually spread to the highest driest portions (typically upper slopes and crests) over time. Barchans and barchanoids relatively easily convert to parabolic dunes where the vegetation first colonises the outside margins of the dunes (e.g. as at White Sands, a dunefield in New Mexico). The change to vegetated dunes typically takes place in concert with climate

Figure 7.9 Transgressive coastal dunefield system near Arcata, northern California, comprising: a foredune-blowout complex (i), blowout and parabolic dunes (ii), deflation plain (iii), active transgressive dunefield (iv) with transverse dunes, remnant knobs, and nebkhas (v), marginal precipitation ridge (vi), and relict (vegetated) transgressive dunefield (vii). *Source:* Photograph courtesy of Dave Kenworthy. See insert for colour representation of this figure.

Figure 7.10 Transverse and barchanoid-transverse dunes being colonised by vegetation. The dune form changes to a polygonal, complex, sometimes chaotic form as colonisation proceeds, and blowouts sometimes develop in association with the vegetation colonisation process. *Source:* P. Hesp.

(a) (b) (c)

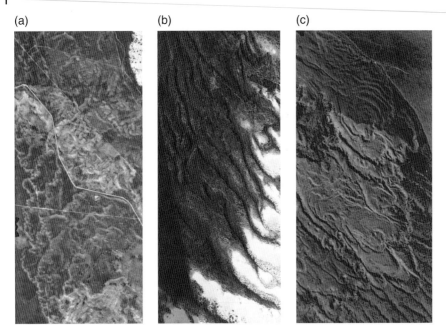

Figure 7.11 Three types of coastal transgressive dunefield morphologies which have developed following stabilisation by vegetation. (a) Digitate and crenate ridge types; (b) Transverse dune ridges; (c) Parabolic and sub-parabolic ridges. Examples from South and West Australian transgressive coastal dunefields. *Source:* Google Earth.

change, and a trend to higher rainfall, higher water tables, lower wind speeds, etc. (e.g. Thomas and Tsoar 1990; Swezey 2001; Thomas et al. 2005; Tsoar 2005; Singhvi et al. 2010). The original dune form may be preserved largely intact as, for example, it appears to have been in some of the Australian linear dunes systems, or in the example in Figure 7.11b where the rapidity of vegetation colonisation is such to almost completely capture the transverse dune in place. As the rate of vegetation colonisation slows, the original dune form is increasingly modified from the original form (e.g. the change from barchanoidal to polygonal form in Figure 7.10). The very extensive comb-like or rake parabolic dunes in the Thar desert (Warren 2013) may well have evolved from the gradual and spatially differential growth of vegetation up and across the downwind margins of originally mobile sand sheets and dunefields dominated by transverse dunes and barchanoids.

7.9 Echo Dunes

Echo dunes are dunes which 'echo' (partially to fully) the shape of a cliff, steep slope, or scarp that they form upwind of. A reversing vortex typically forms in the flow-separation zone on the upwind side of a scarp or cliff (Greeley and Iversen 1985; Thomas 1989; Figure 7.12), or even sometimes upwind of dense vegetation, and the echo dune is formed within that zone. Echo dunes are said to form when the slope/cliff/scarp has a gradient greater than 50–55° (Laity 2008). The vortex keeps a zone immediately adjacent to the cliff largely clear of sand. Sand is deposited a short distance upwind where the vortex edge occurs and the wind-speed drops.

Echo dunes echo the shape of their parent cliffline or steep topographic feature, be it straight, or curved. If the obstacle is narrow across the wind, the echo dune may form a semi-circular arc around the obstacle, merging into flanking dunes that trail away

Figure 7.12 A sinuous echo dune formed in front of a concave and notched cliff by winds approaching from the top left corner of the photograph. A pronounced corkscrew reversing flow vortex was produced at the cliff face resulting in a sand-free corridor and along-dune transport (note ripple orientations). The dune was extending by downwind accretion on the down-wind tip or nose. *Source:* P. Hesp.

downwind. These features have been termed wrap-around dunes by some authors (Thomas 1989; Thom et al. 1992). They may then merge into lee dunes or shadow dunes.

Figure 7.13 shows the flow structure and wind speeds over a vertical cliff/scarp, and for two cases where small or large echo dunes have formed upwind of the cliff. In the case of the vertical cliff, a separation envelope forms in front of the scarp, containing a reversing vortex. A high-speed jet flow occurs over the crest of the obstacle and a flow separation region, also with reversing flow, occurs downwind of the crest (cf. Bowen and Lindley 1977; Tsoar 1983). Wind-tunnel modelling suggests that an echo dune forms within an upwind zone of d/h = 0.3–2, where d is the distance upwind from the scarp/cliff, and h is the height of the scarp/cliff (Tsoar 1983). Once an echo dune has formed upwind of the cliff (Figure 7.13b), the flow is modified by topographically-forced acceleration up the echo dune's own upwind slope, and a more confined but pronounced flow separation zone between the dune and the cliff is formed. The

pronounced reversing flow tends to maintain a swale (topographic low) between the cliff face and the dune, and, thus, echo dunes may maintain their position and shape for considerable time. Oblique winds may produce corkscrew vortices in the swale and thus modify the shape of the dune, but the echo dune is still maintained. The flow-separation region on the top of the cliff is now smaller.

As the echo dune grows larger (Figure 7.13c), the zone of lee flow separation also becomes larger and acts to maintain the echo dune at a distance from the scarp or cliff. The incident flow streams across the dune and the zone of flow separation on the scarp/cliff crest is then markedly reduced. Echo dunes may also form downwind of topographic barriers due to the operation of oscillating lee waves, according to Thomas (1989).

In coastal situations most echo dunes are ephemeral. The sediment in the cliff or scarp will dry and slump to the angle of repose. In addition, slump blocks (either vegetated or unvegetated) may fall off the scarp, reducing its gradient and introducing greater

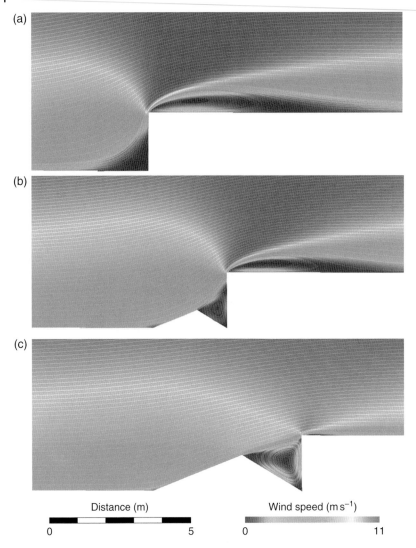

(a)

(b)

(c)

Distance (m)   Wind speed (m s^{-1})

0   5   0   11

Figure 7.13 Computational fluid dynamics modelled flow over a scarp/cliff (a) small echo dune and scarp/cliff (b), and large echo dune and scarp/cliff (c). The velocity of the wind entering on the right is 5 m s^{-1} at 25 cm above the surface and a log profile was present at the outset. Flow separation zones with reverse flow are present at the foot of the scarp, immediately downwind of the scarp crest, and in the swale between the echo dune and the scarp/cliff. *Source:* Hesp and Smyth (2017). See insert for colour representation of this figure.

roughness and topographic variability to the base of the dune scarp (Carter et al. 1990). Once this occurs, the degree of flow separation at the base of the scarp decreases, and the swale formed in the lee of the echo dune may gradually fill with sand. An aerodynamic ramp is then created which allows sediment to be transported to the dune crest and beyond (Hesp et al. 2009).

7.10 Climbing, Clifftop, Falling, and Lee Dunes

At lower angles on the windward slopes of solid obstacles, a separation zone and reversing flow do not develop (Bowen and Lindley 1977; Tsoar 1983), although there is often a small zone of reduced flow at the base of the windward slope, as seen in Figure 7.13c.

On these gentler slopes, climbing dunes may form. Sand may also be carried up windward valleys, slots, and canyons in the scarp or cliff and, if it reaches the plateau, may form climbing and clifftop dunes (Evans 1962).

Sand ramps may be distinguished from climbing dunes by their lower gradient slopes (3–11°), greater size, and composite nature. However, an alternative view is that a range of climbing dune types develops, of various sizes and slopes, depending on age, grain size, and underlying topography, and that therefore sand ramps are merely a certain class of climbing dune (Laity 2008).

If a climbing dune has reached the crest of the topographic feature it is ascending, clifftop dunes commonly form. Figure 7.13a shows that as the wind flow crosses the crest of the scarp, a flow separation region or zone, with reverse flow near the ground, may develop. Sand transported up the climbing dune rapidly falls out within this zone to form cliff-top dunes. Cliff-top dunes are very common along cliffed coasts where they were formed during the post-glacial marine transgression (Jennings 1967). They are also common in some continental deserts (e.g. Mojave Desert, SE California, Arizona, Idaho, and Argentina).

If the lee slope of the underlying or bedrock topography declines downwind, falling dunes may also form. If the lee slope is cliffed or very steep, a strong flow reversal may also occur there, and dunes will form within this zone (Luo et al. 2012). Decelerated flow and a separation zone develop downwind of obstacles such as hills, and the falling dune system initially forms within this zone (Figure 7.14; Koscielniak 1973). As with other dunes formed around obstacles, the form of a falling dune is strongly controlled by the morphology of the obstacle topography (Al-Enezi et al. 2008). Continued supply of wind-blown sand may increase the height and length of the falling dunes beyond the original flow deceleration or separation zone (see Figure 5.23 in Greeley and Iversen 1985).

Lee dunes also develop in the lee, or downwind, of gaps between topographic obstacles, such as hills or mountains. The morphology

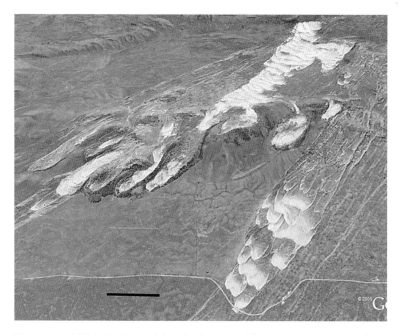

Figure 7.14 NW St. Anthony, Idaho. Climbing and falling parabolic dunes and dunefields have been steered around and over the hilly terrain. Note how the dunefield dominated by transverse dunes is converted to parabolic dunes as vegetation colonises the dunescape. Scale bar is 0.5 km long. *Source:* Google Earth.

of a lee dune is typically related to the morphology of the obstacle, or the width of the gap between obstacles (Sutton and McKenna Neumann 2008; Luo et al. 2012, 2014). Lee dunes may be single, double, or multiple dunes depending on the nature of the topography (discrete or elongate, nature of the gaps and number of them, local or regional sand supply, regional, and local wind field). It has been argued that lee dunes do not advance or elongate once they have achieved a steady-state form (Laity 2008), but some, at least, appear to continue to extend downwind over time, and if sand is still being supplied, may continue to grow.

7.11 The Influence of Topography on Wind Flow and Dune Orientation and Migration

Topography variably influences dune orientation and migration rate. Often it acts to steer wind flow around an object or through a gap, thus, altering the dune orientation and path from the general regional dune orientations and migration paths. Figure 7.14 illustrates how the higher peaks of St Anthony, Idaho, steer the parabolic dunes and active dunefields around the topography compared to the parabolic dunes unaffected by the hills to the right.

7.12 Conclusion

Dome dunes, barchans, barchanoids, transverse dunes, star dunes, parabolic dunes, sand sheets, small- to large-scale mobile (active) dunefields, nebkhas and nebkha fields, and blowouts are common in deserts, semi-arid regions, and on coasts. Climbing and falling dunes, cliff top dunes, lee dunes, echo dunes, and shadow dunes also occur in all dunescapes. Climate changes likely drive similar morphological changes in the active dunescapes regardless of whether coastal or continental desert, When they are partially to finally stabilised by vegetation, they may commonly evolve towards a finite set of morphologies regardless of being 'coastal' or 'desert'.

Acknowledgements

Thanks to Ian and Andrew for their fine editorial assistance. Thanks also to the School of the Environment, Flinders University, for support.

Further Reading

Cooke et al. (1993) cover anchored dunes in desert environments. Hesp (2000) provides a beginner's guide to understanding coastal dunes. Nordstrom et al. (1990) have an excellent, wide-ranging presentation of coastal dune geomorphology and processes around the world. In a general earth-science text, Bridge and Demicco (2008) have a perfect blend of dynamics, bedform evolution, and explanations of how landforms evolve and change.

References

Abhar, K.C., Walker, I.J., Hesp, P.A., and Gares, P.A. (2015). Spatial-temporal evolution of aeolian blowout dunes at Cape Cod. *Geomorphology* 236: 148–162. doi: 10.1016/j.geomorph.2015.02.015.

Al-Enezi, A., Pye, K., Misak, R., and Al-Hajraf, S. (2008). Morphologic characteristics and development of falling dunes, Northeast Kuwait. *Journal of Arid Environments* 72 (4): 423–439. doi: 10.1016/j.jaridenv.2007.06.017.

Anthonsen, K.L., Clemmensen, L.B., and Jensen, J.H. (1996). Evolution of a dune from crescentic to parabolic form in response to short-term climatic changes: Rabjerg Mile, Skagen Odde, Denmark. *Geomorphology* 17 (1–3): 63–78. doi: 10.1016/0169-555X(95)00091-I.

Arens, S.M., van Kaam-Peters, H.M.E., and van Boxel, J.-H. (1995). Air flow over fore dunes and implications for sand transport. *Earth Surface Processes and Landforms* 20 (4): 315–332. doi: 10.1002/esp.3290200403.1.

Baas, A.C.W. and Nield, J.M. (2007). Modelling vegetated dune landscapes. *Geophysical Research Letters* 34: L06405. doi: 10.1029/2006GL029152.

Baas, A.C.W. and Nield, J.M. (2010). Ecogeomorphic state variables and phase-space construction for quantifying the evolution of vegetated aeolian landscapes. *Earth Surface Process and Landforms* 35 (6): 717–731. doi: 10.1002/esp.1990.

Bagnold, R.A. (1941). *The Physics of Blown Sand and Desert Dunes*. London: Chapman and Hall.

Barbosa, L.M. and Landim-Dominguez, J.M. (2004). Coastal dune fields at the São Francisco River strandplain, northeastern Brazil: morphology and environmental controls. *Earth Surface Processes and Landforms* 29 (4): 43–456. doi: 10.1002/esp.1040.

Barchyn, T.E. and Hugenholtz, C.H. (2013a). Dune field reactivation from blowouts: Sevier Desert, UT, USA. *Aeolian Research* 11: 75–84. doi: 10.1016/j.aeolia.2013.08.003.

Barchyn, T.E. and Hugenholtz, C.H. (2013b). Reactivation of supply-limited dune fields from blowouts: a conceptual framework for state characterization. *Geomorphology* 201: 172–182. doi: 10.1016/j.geomorph.2013.06.019.

Bauer, B.O., Walker, I.J., Baas, A.C.W. et al. (2013). Critical reflections on the coherent flow structures paradigm in aeolian geomorphology. In: *Coherent Flow Structures at the Earth's Surface* (ed. J.G. Venditti, J.L. Best, M. Church and R.J. Hardy), 111–134. Chichester: Wiley-Blackwell.

Black, R.F. (1951). Eolian deposits of Alaska. *Arctic* 4 (2): 89–111. doi: 10.14430/arctic3938.

Blanco, P.D., Rostagno, C.M., del Valle, H.F. et al. (2008). Grazing impacts in vegetated dune fields: predictions from spatial pattern analysis. *Rangeland Ecology and Management* 61 (2): 194–203. doi: 10.2111/06-063.1.

Bourgeois, J. and Weiss, R. (2009). 'Chevrons' are not mega-tsunami deposits: a sedimentologic assessment. *Geology* 37 (5): 403–406.

Bowen, A.J. and Lindley, D. (1977). A wind-tunnel investigation of the wind speed and turbulence characteristics close to the ground over various escarpment shapes. *Boundary-Layer Meteorology* 12 (3): 259–271. doi: 10.1007/BF00121466.

Bowler, J.M. (1973). Clay dunes: their occurrence, formation and environmental significance. *Earth-Science Reviews* 9 (4): 315–338. doi: 10.1016/0012-8252(73)90001-9.

Bowler, J.M., Gillespie, R., Johnston, H., and Boljkovac, K. (2012). Wind v water: glacial maximum records from the Willandra Lakes. In: *Peopled Landscapes: Archaeological and Biogeographic Approaches to Landscapes*, vol. 34 (ed. S. Haberle and B. David), 271–296. Terra Australis: ANU E Press.

Bridge, J.S. and Demicco, R.V. (2008). *Earth Surface Processes, Landforms and Sediment Deposits*. Cambridge: Cambridge University Press.

Carter, R.W.G. (1976). Formation, maintenance and geomorphological significance of an aeolian shell pavement. *Journal of Sedimentary Petrology* 46 (2): 418–429. doi: 10.1306/212F6F8C-2B24-11D7-8648000102C1865D.

Carter, R.W.G., Hesp, P.A., and Nordstrom, K.F. (1990). Erosional landforms in coastal dunes. In: *Coastal Dunes: Form and Process* (ed. K.F. Nordstrom, N.P. Psuty and R.W.G. Carter), 217–249. Chichester: Wiley.

Catto, N., MacQuarrie, K., and Hermann, M. (2002). Geomorphic response to late Holocene climate variation and anthropogenic pressure, northeastern Prince Edward Island, Canada. *Quaternary International* 87 (1): 101–117. doi: 10.1016/S1040-6182(01)00065-9 Available at: http://linkinghub.elsevier.com/retrieve/pii/S1040618201000659.

Cooke, R.U., Warren, A., and Goudie, A.S. (1993). *Desert Geomorphology*. New York: CRC Press.

Cooper, W.S. (1958). Coastal sand dunes of Oregon and Washington Geological Society of America. *Memoir* 72.

Cooper, W.S. (1967). Coastal dunes of California. *Geological Society of America, Memoir* 104.

David, P.P. (1981). Stabilized dune ridges in northern Saskatchewan. *Canadian Journal of Earth Science* 18 (2): 286–311. doi: 10.1139/e81-022.

David, P.P., Wolfe, S.A., Huntley, D.J., and Lemmen, D.S. (1999). Activity cycle of parabolic dunes based on morphology and chronology from Seward sand hills, Saskatchewan. In: *Holocene Climate and Environmental Change in the Palliser Triangle: A Geoscientific Context for Evaluating the Impacts of Climate Change on the Southern Canadian Prairies*, vol. 534, 223–238. Ottawa: Geological Survey of Canada Bulletin.

Davidson-Arnott, R.G.D. (2005). Conceptual model of the effects of sea level on sandy coasts. *Journal of Coastal Research* 21 (6): 1166–1172. doi: 10.2112/03-0051.1.

Davidson-Arnott, R.G.D. (2010). *Introduction to Coastal Processes and Geomorphology*. Cambridge: Cambridge University Press.

Davies, J.L. (1980). *Geographical Variation in Coastal Development*, 2e. London: Longman.

De Deckker, P., Magee, J.W., and Shelley, J.M.G. (2011). Late Quaternary palaeohydrological changes in the large playa Lake Frome in Central Australia, recorded from the mg/ca and Sr/ca in ostracod valves and biotic remains. *Journal of Arid Environments* 75 (1): 38–50. doi: 10.1016/j.jaridenv.2010.08.004.

Dong, Z., Luo, W., Qian, G., and Lv, P. (2008). Wind tunnel simulations of the three-dimensional airflow patterns around shrubs. *Journal of Geophysical Research* 113: F202016. doi: 10.1029/2007JF000880.

Evans, J.R. (1962). Falling and climbing sand dunes in the Cronese ('Cat') mountains, San Bernardino County, California. *The Journal of Geology* 70 (1): 107–113. doi: 10.1007/s40333-014-0074-9.

Furieri, B., Harion, J.L., Milliez, M. et al. (2014). Numerical modelling of aeolian erosion over a surface with non-uniformly distributed roughness elements. *Earth Surfaces Processes and landforms* 39 (2): 156–166. doi: 10.1002/esp.3435.

Gares, P.A. (1992). Topographic changes associated with coastal dune blowouts at Island Beach State Park, NJ. *Earth Surface Processes and Landforms* 17 (6): 589–604. doi: 10.1002/esp.3290170605.

Gares, P.A. and Nordstrom, K.F. (1995). A cyclic model of foredune blowout evolution for a leeward coast: Island Beach, NJ. *Annals of the Association of American Geographers* 85 (1): 1–20. doi: 10.1111/j.1467-8306.1995.tb01792.x.

Goudie, A.S. and Wells, G.L. (1995). The nature, distribution and formation of pans in arid zones. *Earth Science Reviews* 38 (1): 1–69. doi: 10.1016/0012-8252(94)00066-6.

Greeley, R. and Iversen, J.D. (1985). *Wind as a Geological Process on Earth, Mars, Venus and Titan*. Cambridge: Cambridge University Press.

Gutiérrez-Elorza, M., Desir, G., Gutiérrez-Santolalla, F., and Marín, C. (2005). Origin and evolution of playas and blowouts in the semiarid zone of Tierra de Pinares (Duero Basin, Spain). *Geomorphology* 72 (1–4): 177–192. doi: 10.1016/j.geomorph.2005.05.009.

Hack, J. (1941). Dunes of the western Navajo country. *Geographical Review* 31 (2): 240–264.

Hansen, E., Vries-Zimmerman, D., S., van Dijk, D., and Yurk, B. (2009). Patterns of wind flow and aeolian deposition on a parabolic dune on the southeastern shore of Lake Michigan. *Geomorphology* 105 (1–2): 147–157. doi: 10.1016/j.geomorph.2007.12.012.

Hesp, P.A. (1981). The formation of shadow dunes. *Journal of Sedimentary Research* 51: 101–111. doi:10.1306/212F7C1B-2B24-11D7-8648000102C1865D.

Hesp, P.A. (1988). Morphology, dynamics and internal stratification of some established foredunes in Southeast Australia. *Sedimentary Geology* 55 (1): 17–42. doi: 10.1016/0037-0738(88)90088-7.

Hesp, P.A. (1999). The beach backshore and beyond. In: *Handbook of Beach and Shoreface Morphodynamics* (ed. A.D. Short), 145–170. London: Wiley.

Hesp, P.A. (2000). Parabolic dunes. In: *South of the North: Manawatu and its Neighbours* (ed. B.G.R. Saunders), 40–41. Wellington, NZ: Massey University Press.

Hesp, P.A. (2001). The Manawatu Dunefield: environmental change and human impacts. *New Zealand Geographer* 57 (2): 33–40.

Hesp, P.A. (2002). Foredunes and blowouts: initiation, geomorphology and dynamics. *Geomorphology* 48 (3): 245–268. doi: 10.1016/S0169-555X(02)00184-.

Hesp, P.A. (2004). Coastal dunes in the tropics and temperate regions: location, formation, morphology and vegetation processes. In: *Coastal Dunes, Ecology and Conservation* (ed. M. Martinez and N. Psuty), 29–49. Berlin: Springer-Verlag.

Hesp, P.A. (2011). Dune coasts. In: *Treatise on Estuarine and Coastal Science 3, Estuarine and Coastal Geology and Geomorphology* (ed. E. Wolanski, D. McLusky, B.W. Flemming and J.D. Hansom), 193–221. London: Academic Press.

Hesp, P.A. (2012). Surfzone-beach-dune interactions. In: NCK-days 2012. Crossing borders in coastal research.jubilee conference proceedings 20th NCK-days (ed. W.M. Kranenburg, E.M. Horstman, and K.M. Wijnberg), 35–40. Netherlands Centre for Coastal Research (Nederlands Centrum voor Kustonderzoek – NCK, http://nck-web.org) 4th lustrum (20th anniversary).

Hesp, P.A. (2013a). A 34-year record of foredune evolution, Dark Point, NSW, Australia. Proceedings ICS Conference. *Journal of Coastal Research* (Special Issue) 65: 1295–1300.

Hesp, P.A. (2013b). Conceptual models of the evolution of transgressive dunefield systems. *Geomorphology* 199: 138–149. doi: 10.1016/j.geomorph.2013.05.014.

Hesp, P.A., Giannini, P.C.F., Martinho, C.T. et al. (2009). The Holocene Barrier Systems of the Santa Catarina Coast, Southern Brazil. In: *Geology and Geomorphology of Holocene Coastal Barriers of Brazil: Lecture Notes in Earth Sciences, 107: 93–133* (ed. S.R. Dillenburg and P.A. Hesp). Berlin: Springer.

Hesp, P.A. and Hyde, R. (1996). Flow dynamics and geomorphology of a trough blowout. *Sedimentology* 43 (3): 505–525. doi: 10.1046/j.1365-3091.1996.d01-22.x.

Hesp, P.A. and McLachlan, A. (2000). Morphology, dynamics, ecology and fauna of *Arctotheca populifolia* and *Gazania rigens* nabkha dunes. *Journal of Arid Environments* 44: 155–172. doi: 10.1006/jare.1999.0590.

Hesp, P.A. and Pringle, A. (2001). Wind flow and topographic steering within a trough blowout. *Journal of Coastal Research* 34 (SI): 597–601.

Hesp, P.A. and Smyth, T.A.G. (2017). Nebkha flow dynamics and shadow dune formation. *Geomorphology* 282: 27–38.

Hesp, P.A. and Thom, B.G. (1990). Morphology and evolution of transgressive dune fields. In: *Coastal Dunes: Processes and Morphology* (ed. K. Nordstrom, N. Psuty and R.W.G. Carter), 253–288. Chichester: Wiley.

Hesp, P.A. and Walker, I.J. (2012). Three-dimensional æolian dynamics within a bowl blowout during offshore winds: Greenwich dunes, Prince Edward Island, Canada. *Aeolian Research* 3: 389–399. doi: 10.1016/j.aeolia.2011.09.002.

Hesp, P.A. and Walker, I.J. (2013). Aeolian environments: coastal dunes. In: *Treatise on Geomorphology*, Aeolian Geomorphology, vol. 11 (ed. J. Shroder, N. Lancaster, D.J. Sherman and A.C.W. Baas), 109–133. San Diego, CA: Academic Press.

Hesp, P.A. Ruz, M-H, Hequette, A., Marin, D., and Miot da Silva, G. (2015). Geomorphology and dynamics of a traveling cuspate foreland, Authie estuary, France. *Geomorphology* 254: 104–120.

Hills, E.S. (1940). The lunette, a new landform of aeolian origin. *Australian Geographer* 3: 15–21.

Holliday, V.T. (1997). Origin and evolution of lunettes on the High Plains of Texas and New Mexico. *Quaternary Research* 47 (1): 54–69. doi: 10.1006/qres.1996.1872.

Hugenholtz, C. and Wolfe, S. (2005). Biogeomorphic model of dunefield activation and stabilization on the northern Great Plains. *Geomorphology* 70 (1–2): 53–70. doi: 10.1016/j.geomorph.2005.03.011.

Hugenholtz, C.H. and Wolfe, S.A. (2006). Morphodynamics and climate controls of two aeolian blowouts on the northern Great Plains, Canada. *Earth Surface Processes and Landforms* 31 (12): 1540–1557. doi: 10.1002/esp.1367.

Hugenholtz, C.H., Wolfe, S.A., Walker, I.J., and Moorman, B.J. (2009). Spatial and temporal patterns of aeolian sediment transport on an inland parabolic dune, Bigstick Sand Hills, Saskatchewan, Canada. *Geomorphology* 105 (1–2): 158–170. doi: 10.1016/j.geomorph.2007.12.017.

Jennings, J.N. (1957). On the orientation of parabolic or U-dunes. *Geographical Journal* 123: 474–480.

Jennings, J.N. (1967). Cliff-top dunes. *Australian Geographical Studies* 5 (1): 40–49.

Jungerius, P.D., Verheggen, A.J., and Wiggers, A.J. (1981). The development of blowouts in 'De Blink', a coastal dune area near Noordwijkerhout, the Netherlands. *Earth Surface Processes and Landforms* 6 (3–4): 375–396. doi: 10.1002/esp.3290060316.

Jungerius, P.D., Witter, J.V., and van Boxel, J.H. (1991). The effects of changing wind regimes on the development of blowouts in the coastal dunes of the Netherlands Landscape. *Ecology* 6 (1): 241–248. doi: 10.1016/j.geomorph.2005.05.009.

Kelletat, D. and Scheffers, A. (2003). Chevron-shaped accumulations along the coastlines of Australia as potential tsunami evidences? *Science of Tsunami Hazards* 21 (3): 174–188.

Kocurek, G., Townsley, M., Yeh, E. et al. (1992). Dune and dune-field development on Padre Island, Texas, with implications for Interdune deposition and water-table-controlled accumulation. *Journal of Sedimentary Petrology* 62 (4): 622–635. doi: 10.1306/D4267974-2B26-11D7-8648000102C1865D.

Koscielniak, D.E. (1973). Unusual aeolian deposits on a volcanic terrain near St. Anthony, Idaho. Masters' thesis, State University of New York at Buffalo.

Koster, E.A. (1988). Ancient and modern cold-climate aeolian sand deposition: a review. *Journal of Quaternary Science* 3 (1): 69–83. doi: 10.1002/jqs.339003010.

Laity, J. (2008). *Deserts and Desert Environments*. Chichester: Wiley-Blackwell.

Landsberg, S.Y. (1956). The orientation of dunes in Britain and Denmark in relation to wind. *Geographical Journal* 122: 176–189.

Lees, B.G. and Cook, P.G. (1991). A conceptual model of lake barrier and compound lunette formation. *Palaeogeography, Palaeoclimatology, Palaeoecology* 84 (1–4): 271–284. doi: 10.1016/0031-0182(91)90048-V.

Luo, W., Dong, Z., Qian, G., and Lu, J. (2012). Wind tunnel simulation of the three-dimensional airflow patterns behind cuboid obstacles at different angles of wind incidence, and their significance for the formation of sand shadows. *Geomorphology* 139–140: 258–270. doi: 10.1016/j.geomorph.2011.10.027.

Luo, W., Dong, Z., Qian, G., and Lu, J. (2014). Near-wake flow patterns in the lee of adjacent obstacles and their implications for the formation of sand drifts: a wind tunnel simulation of the effects of gap spacing. *Geomorphology* 213: 190–200. doi: 10.1016/j.geomorph.2014.01.008.

McKee, E.D. (1979). A Study of Global Sand Seas. In: *Geological Survey Prof Paper 1052*. Washington, DC: US Govt Printing Office.

Melton, F.A. (1940). A tentative classification of sand dunes: its application to dune history of the High Plains. *Journal of Geology* 48: 113–174.

Mir-Gual, M., Pons, G.X., Martin-Prieto, J.A. et al. (2012). Geomorphological and ecological features of blowouts in a western Mediterranean coastal dune complex: a case study of the Es Comú de Muro beach-dune system on the island of Mallorca, Spain. *Geo-Marine Letters* 33 (2–3): 29–141. doi: 10.1007/s00367-012-0289-7.

Nordstrom, K., Psuty, N., and Carter, R.W.G. (eds.) (1990). *Coastal Dunes: Processes and Morphology*. Chichester: Wiley.

Olson, J.S. (1958). Lake Michigan dune development: 2 Plants as agents and tools in geomorphology. *Journal of Geology* 66: 345–351.

Pillans, B. (1987). Lake shadows – Aeolian clay sheets associated with ephemeral lakes in basalt terrain, southern New South Wales. *Search* 18: 313–315.

Piotrewska, H. (1991). The development of the vegetation in the active deflation hollows of the Leba Bar (N. Poland). *Fragmenta Floristica, et Geobotanica* 35: 172–215.

Price, W.A. (1963). Physicochemical and environmental factors in clay-dune genesis. *Journal of Sedimentary Petrology* 33 (3): 766–778. doi: 10.1306/74D70F24-2B21-11D7-8648000102C1865D.

Pye, K. (1982). Morphological development of coastal dunes in a humid tropical environment, Cape Bedford and Cape Flattery, North Queensland. *Geografiska Annaler A* 64 (3–4): 213–227. doi: 10.2307/520647.

Pye, K. (1993). Late quaternary development of coastal parabolic megadune complexes in northeastern Australia. In: *International Association of Sedimentology*, vol. 16 (ed. K. Pye and N. Lancaster), 23–45. Special Publication.

Pye, K. and Tsoar, H. (1990). *Aeolian Sand and Sand Dunes*. London: Unwin Hyman.

Ranwell, D.S. (1959). Newborough Warren, Anglesey: I. The dune system and dune slack habitat. *The Journal of Ecology* 47 (3): 571–601. doi: 10.2307/2257291.

Ranwell, D.S. (1960). Newborough Warren, Anglesey: III. Changes in the vegetation on parts of the dune system after the loss of rabbits by myxomatosis. *Journal of Ecology* 48 (1): 117–141. 385–395.

Rhind, P., Jones, R., and Jones, L. (2013). The impact of dune stabilization on the conservation status of sand dune systems in Wales. In: *Restoration of Coastal Dunes* (ed. M.L. Martinez, J.B. Gallego-Fernandez and P.A. Hesp), 125–143. Berlin: Springer.

Ritchie, W. (1972). The evolution of coastal sand dunes. *Scottish Geographical Magazine* 88 (1): 19–35. doi: 10.1080/00369227208736205.

Rubin, D.M. and Hesp, P.A. (2009). Multiple origins of linear dunes and implications for dunes on Titan. *Nature Geoscience* 2 (9): 653–658. doi: 10.1038/ngeo610.

Ruz, M.-H. and Hesp, P.A. (2014). Geomorphology of high-latitude coastal dunes. In: *Sedimentary Coastal Zones from High to Low Latitudes: Similarities and Differences*, vol. 388 (ed. I.P. Martini and H.R. Wanless), 199–212. London: Geological Society of London. doi: 10.1144/SP388.17.

Sabin, T.J. and Holliday, V.T. (1995). Playas and lunettes on the Southern High Plains: morphometric and spatial relationships. *Annals of the Association of American Geographers* 85 (2): 286–305. doi: 10.1111/j.1467-8306.1995.tb01794.x-i1.

Seppälä, M. (2004). *Wind as a Geomorphic Agent in Cold Climates*. London: Cambridge University Press.

Shaw, P.A. and Thomas, D.S.G. (1989). Playas, pans and salt lakes. In: *Arid Zone Geomorphology* (ed. D.S.G. Thomas), 184–205. London: Belhaven.

Singhvi, A.K., Williams, M.A.J., Rajaguru, S.N. et al. (2010). A ~200 ka record of climatic change and dune activity in the Thar Desert, India. *Quaternary Science Reviews* 29 (23–24): 3095–3105. doi: 10.1016/j.quascirev.2010.08.003.

Smith, H.T.U. (1960). Physiography and photo interpretation of coastal sand dunes. Final Report Contract NONR – 2242(00), Office of Naval Research, Geographical Branch.

Smyth, T.A.G., Jackson, D.W.T., and Cooper, J.A.G. (2011). Computational fluid dynamic modelling of three-dimensional airflow over dune blowouts. *Journal of Coastal Research* 64: 314–318. doi: 10.1016/j.geomorph.2012.07.014.

Smyth, T.A.G., Jackson, D.W.T., and Cooper, J.A.G. (2012). High resolution measured and modelled three-dimensional airflow over a coastal bowl blowout. *Geomorphology* 177–178: 62–73. doi: 10.1016/j.geomorph.2012.07.014.

Smyth, T.A.G., Jackson, D.W.T., and Cooper, J.A.G. (2013). Three-dimensional airflow patterns within a coastal blowout during fresh breeze to hurricane force winds. *Aeolian Research* 9: 111–123. doi: 10.1016/j.geomorph.2012.07.014.

Smyth, T.A.G., Jackson, D.W.T., and Cooper, J.A.G. (2014). Airflow and aeolian sediment transport patterns within a coastal trough blowout during lateral wind conditions. *Earth Surface Processes and Landforms* 39 (14): 1847–1854. doi: 10.1002/esp.3572.

Story, R. (1982). Notes on parabolic dunes, winds and vegetation in northern Australia. CSIRO Australian Division of Water and Land Resources, Technical Paper 43: 1–34.

Sutton, S.L.F. and McKenna Neuman, C. (2008). Sediment entrainment to the lee of roughness elements: effects of vortical structures. *Journal of Geophysical Research* 113: F02S09. doi: 10.1029/2007JF000783.

Swezey, C. (2001). Eolian sediment responses to late quaternary climate changes: temporal and spatial patterns in the Sahara. *Palaeogeography, Palaeoclimatology, Palaeoecology* 167: 119–155.

Thom, B.G., Shepherd, M., Ly, C., et al. (1992). Coastal geomorphology and quaternary geology of the Port Stephens – Myall Lakes Area. A.N.U, Department of Biogeography and Geomorphology, Monograph No. 6, 300.

Thomas, D.S.G. (1989). Aeolian sand deposits. In: *Arid Zone Geomorphology* (ed. D.S.G. Thomas), 232–261. London: Belhaven Press.

Thomas, D.S.G., Knight, M., and Wiggs, G.F.S. (2005). Remobilization of southern African desert dune systems by twenty-first century global warming. *Nature* 435 (7046): 1218–1221. doi: 10.1038/nature03717.

Thomas, D.S.G. and Tsoar, H. (1990). The geomorphological role of vegetation in desert dune systems. In: *Vegetation and Erosion* (ed. J.B. Thornes), 471–489. Chichester: Wiley.

Tinley, K.L. (1985). Coastal dunes of South Africa. South African National Science Program Report No. 109, CSIR.

Trenhaile, A.S. (1997). *Coastal Dynamics and Landforms*. Oxford: Clarendon Press.

Tsoar, H. (1983). Wind tunnel modelling of echo and climbing dunes. In: *Eolian Sediments and Processes, Developments in Sedimentology* (ed. M.E. Brookfield and T.S. Ahlbrandt), 247–260. Amsterdam: Elsevier.

Tsoar, H. (2005). Sand dunes mobility and stability in relation to climate. *Physica A* 357 (1): 50–56. doi: 10.1016/j.physa.2005.05.067.

Tsoar, H. and Blumberg, D.G. (2002). Formation of parabolic dunes from barchan and transverse dunes along Israel's Mediterranean coast. *Earth Surface Processes and Landforms* 27 (11): 1147–1161. doi: 10.1002/esp.417.

van der Meulen, F. and Jungerius, P.D. (1989). Landscape development in Dutch coastal dunes: the breakdown and restoration of geomorphological and geohydrological processes. In: Gimingham, C.H., Ritchie, W., Willetts, B.B. and Willis, A.J. (eds), Symposium: Coastal Sand Dunes, Proceedings of the Royal Society of Edinburgh B 96:219–229.

Wang, S., Hasi, E., Zhang, J., and Zhang, P. (2007). Geomorphological significance of air flow over saucer blowout of the Hulun Buir sandy grassland. *Zhongguo sha mo (Journal of Desert Research)* 27 (5): 745–749.

Warren, A. (1988). A note on vegetation and sand movement in the Wahiba Sands. *Journal of Oman Studies* 3 (Special Report): 251–255.

Warren, A. (2013). *Dunes*. Wiley-Blackwell: Chichester.

Wolfe, S.A. and David, P.P. (1997). Parabolic dunes: examples from the Great Sand Hills, southwestern Saskatchewan. *The Canadian Geographer* 41 (2): 207–214. doi: 10.1111/j.1541-0064.1997.tb01160.x.

Xiao, J.H., Qu, J., Yao, Z.Y. et al. (2015). Morphology and formation mechanism of sand shadow dunes on the Qinghai-Tibet plateau. *Journal of Arid Land* 7 (1): 10–26. doi: 10.1007/s40333-014-0074-9.

Yan, N. and Baas, A.C.W. (2015). Parabolic dunes and their transformations under environmental and climatic changes: towards a conceptual framework for understanding and prediction. *Global and Planetary Change* 124: 123–148. doi: 10.1016/j.gloplacha.2014.11.010.

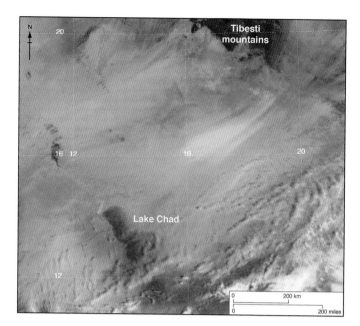

Figure 1.1 The Bodélé low-level jet in operation, raising a cloud of dust from the now-dry surface of a Holocene Lake. (MODIS true colour composite (channels 3, 4, and 1) image for 12:45UTC, 4 March 2005).

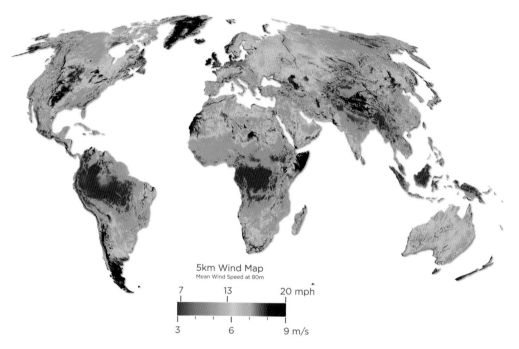

Figure 1.8 Global distribution of wind speed. The map shows average annual wind speed at 80 m above-ground at a 5 km spatial resolution. Wind-speed declines towards the ground, at a rate determined by, among other things, surface roughness (a much fuller explanation is given in Chapter 2). Thus, the map shows only the global pattern of relative differences in wind speed. The map is available at: http://www.3tier.com/static/ttcms/us/images/support/maps/3tier_5km_global_wind_speed.pdf. The average wind speeds are based on over 10 years of hourly data developed with computer model simulations that create realistic wind fields. The wind speed dataset compares well with observations from over 4000 NCEP-ADP network stations. Source: The second section of this caption is a shortened version of text provided by Vaisala Inc., 2001.

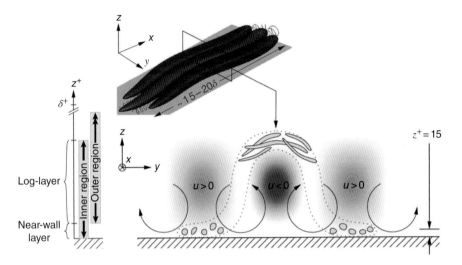

Figure 2.2 Very large-scale motions or 'super-structures' in the inner boundary layer. Grey patches indicate vortex filaments, both very close to the surface (cross-sections) and at the top of the inner region (sideways). *Source:* Modified from Marusic et al. (2010).

Figure 2.10 Streamers moving and meandering over a beach (image with enhanced colour contrasts).

Figure 3.2 Wind-eroded features in Antarctica. (a) Polished and pitted boulder surfaces, Victoria Valley. (b) A wind-abraded boulder, Antarctic coast. *Source:* Photos: C. Bristow.

Figure 3.5 Examples of ventifact microfeatures from north-west Ireland: (a) facets and keel, (b) pits developed on facets, (c, d) grooves. *Source:* Photos: J. Knight.

Figure 3.7 Examples of yardangs. (a) A field of yardangs from the Bodélé Depression, Chad, (b) a close-up of a single yardang from that field. *Source:* Photos: C. Bristow. (c) An individual 15 m-high yardang, from the Dunhuang Yardang National Geopark, China, (d) a group of parallel yardangs from the same field. *Source:* Photos: J. Bullard.

Figure 3.8 Stone pavement (reg) surfaces from (a) Jordan. *Source:* Photo: J. Bullard, and (b) Simpson Desert, Australia. *Source:* Photo: C. Bristow.

Figure 3.9 Photos of (a) a deflation basin developed on a wind-eroded diatomite surface, Bodélé Depression, Chad. *Source:* Photo: C. Bristow, and (b) a pan in the linear dunefield between William Creek and Coober Pedy, South Australia. Note the linear dune in the background. *Source:* Photo: D. Nash.

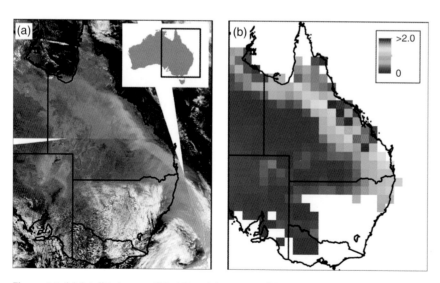

Figure 4.1 (a) Satellite image of 'Red Dawn' dust event; (b) MODIS Aerosol Optical Depth (AOD) of 'Red Dawn' dust event.

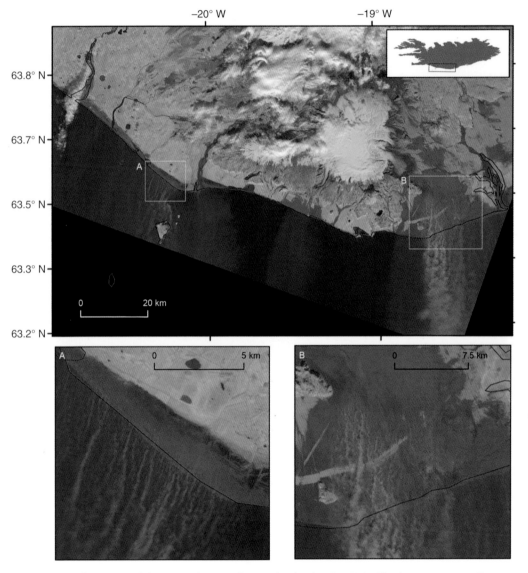

Figure 4.2 Satellite image of dust storm from southern Iceland, 5 October 2004. The dust sources are all proglacial floodplains; the plumes extend more than 500 km south over the Atlantic Ocean.

Figure 4.6 Dust storm approaching the town of Bedourie in western Queensland, Australia, on 4 December 2014.
Source: Photograph reproduced courtesy of Maggie Den Ronden.

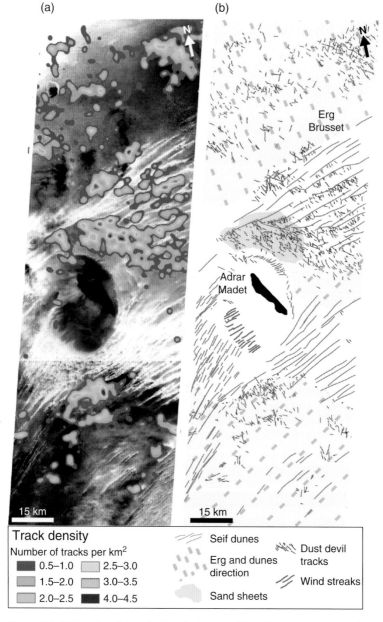

(a)

(b)

N

N

Erg
Brusset

Adrar
Madet

15 km

15 km

Track density

Number of tracks per km²

- ■ 0.5–1.0
- ■ 1.5–2.0
- ■ 2.0–2.5
- □ 2.5–3.0
- ■ 3.0–3.5
- ■ 4.0–4.5

Seif dunes

Erg and dunes
direction

Sand sheets

Dust devil
tracks

Wind streaks

Figure 4.9 (a) Density of dust devil tracks in part of the Ténéré Desert, Niger, for 26 May 2001 shown over the ASTER green band mosaic. (b) Geomorphological map of 26 May 2001 ASTER coverage indicating dust devil tracks. *Source:* Redrawn from Rossi and Marinangeli (2004).

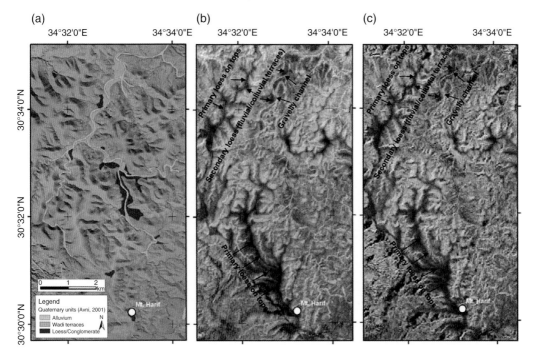

Figure 5.1 A map of loess as identified from remote sensing images in the western Negev highlands in Israel (Crouvi et al. 2009, Figure 5). The images show: (a) Quaternary units (alluvium, wadi terraces, loess/conglomerate) mapped by Avni (2001) at a scale of 1 : 50 000, projected on a rectified air photo; (b) an Advanced Spaceborne Thermal Emission and Reflection Radiometer (ASTER) image identifying primary loess by the dark red to dark purple shades; and (c) a Landsat Thematic Mapper (TM) image identifying primary loess by the brown, purple, red, and blue shades. Fluvial/colluvial loess terraces are in shades of green in ASTER (b) or blue in Landsat TM (c), depending on the clast- and vegetation-coverage, while channels with high gravel cover, or incised in bedrock, are shown in light purple in ASTER (b) or yellow to orange colours in Landsat TM (c). In both remote sensing images (b and c), north-facing slopes are greenish, probably because of the higher vegetation coverage compared to the south-facing slopes which are pink. *Source:* Crouvi et al. (2009).

Figure 5.5 Luochuan in the central Chinese Loess Plateau.

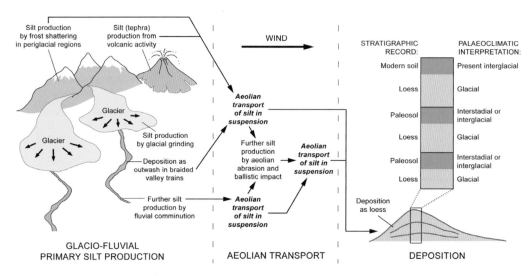

Figure 5.7 'Glacial' model of loess formation involving the production of silt-sized grains primarily by glacial grinding, prior to further transport and deposition. *Source:* Synthesised and modified from Muhs and Bettis Jr. (2003), Muhs (2007), and Muhs (2013).

Figure 5.8 Dust in suspension on the Delta River floodplain, near Delta Junction, Alaska (July 2004). *Source:* Photograph courtesy of E.A. Bettis III.

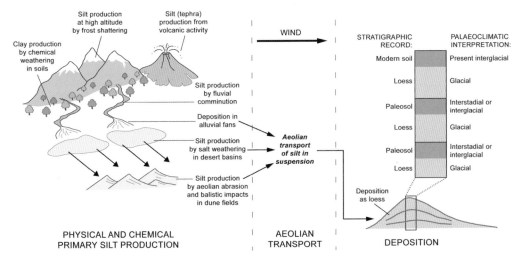

Figure 5.9 'Desert' model of loess formation involving the production of silt-sized grains primarily by various non-glacigenic processes, prior to further transport and deposition. *Source:* Synthesised and modified from Muhs and Bettis Jr. (2003), Muhs (2007), and Muhs (2013).

Figure 6.12 Transverse dunes on the flanks of a large linear dune in the northern Namib sand sea near Sossusvlei. *Source:* Giles Wiggs.

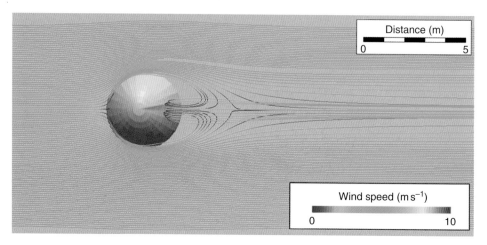

Distance (m)

0   5

Wind speed (m s^{-1})

0   10

Figure 7.4 Streamlines at 0.25 m above the bed around a model nebkha, produced by a Reynolds-Averaged Navier–Stokes (RANS) computational fluid dynamics simulation. Mean flow comes from the left. The figure shows two symmetrically opposed reversing vortices forming behind the nebkha and two zones of low velocity in which the shadow dune would form. A small reversing flow separation zone is also produced upwind of the nebkha. *Source:* Hesp and Smyth, 2017.

Figure 7.5 Established foredune formed in *Ammophila brevigulata* on Prince Edward Island, Canada. *Ammophila* spp. are discussed in Chapter 12.

Figure 7.6 Multiple playas and lunettes in the Lake Pingarnup region of Western Australia. *Source:* Google Earth.

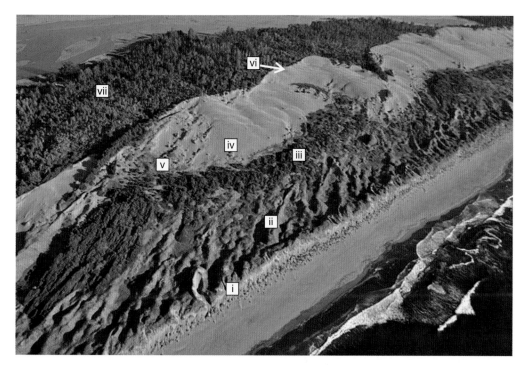

Figure 7.9 Transgressive coastal dunefield system near Arcata, northern California, comprising: a foredune-blowout complex (i), blowout and parabolic dunes (ii), deflation plain (iii), active transgressive dunefield (iv) with transverse dunes, remnant knobs, and nebkhas (v), marginal precipitation ridge (vi), and relict (vegetated) transgressive dunefield (vii). *Source:* Photograph courtesy of Dave Kenworthy.

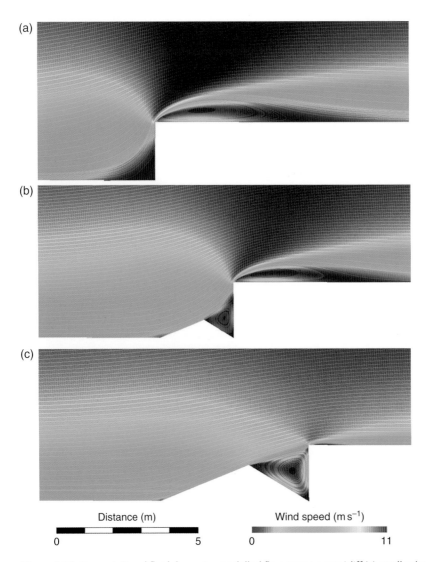

Distance (m)

0   5

Wind speed (m s^{-1})

0   11

Figure 7.13 Computational fluid dynamics modelled flow over a scarp/cliff (a) small echo dune and scarp/cliff (b), and large echo dune and scarp/cliff (c). The velocity of the wind entering on the right is 5 m s^{-1} at 25 cm above the surface and a log profile was present at the outset. Flow separation zones with reverse flow are present at the foot of the scarp, immediately downwind of the scarp crest, and in the swale between the echo dune and the scarp/cliff. *Source:* Hesp and Smyth in (2017).

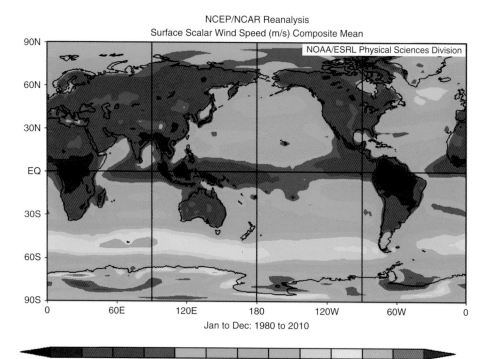

NCEP/NCAR Reanalysis
Surface Scalar Wind Speed (m/s) Composite Mean

NOAA/ESRL Physical Sciences Division

Jan to Dec: 1980 to 2010

2   4   6   8   10   12   14

Figure 8.4 Mean surface wind speed (NCEP/NCAR reanalysis data: http://www.esrl.noaa.gov/psd/data/ reanalysis/reanalysis.shtml). No threshold wind speed is used and there is no weighting according to velocity, unlike sand drift potential values.

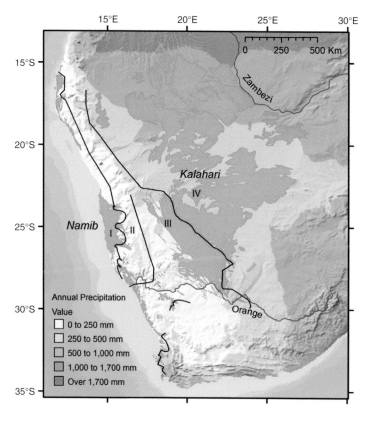

Figure 8.5 Southern African dune expression and climate. Dunefields are shown in orange superimposed over the distribution of annual average rainfall. Simplified boundaries between four morphological classes of dunes are shown:
(I) continuous sharp crests with slip faces and with little or no perennial vegetation;
(II) discontinuous segments of sharp crests with slip faces but sparse cover of perennial vegetation elsewhere;
(III) distinct crests but only rare patches of bare crest or slip faces and perennial vegetation present on crests, flanks, and interdunes; and (IV) indistinct crests with very rare or no bare areas and continuous cover of perennial shrubs and trees.

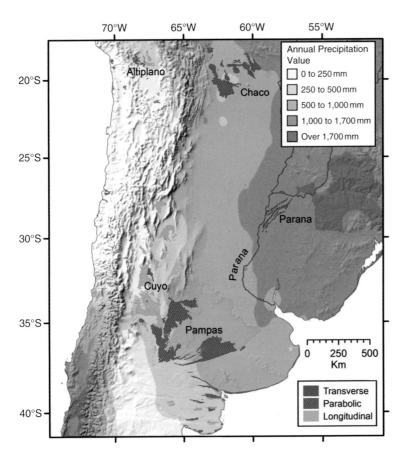

Figure 8.6 South American sand seas and large dunefields. Dunefields are shown in orange superimposed over the distribution of annual average rainfall (from WorldClim data). Dominant dune types within each dunefield are indicated (but often more than a single morphology is present).

(a)

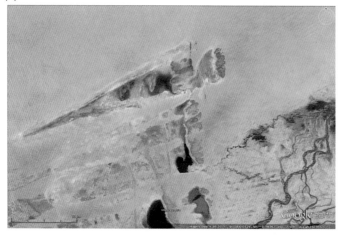

(b)

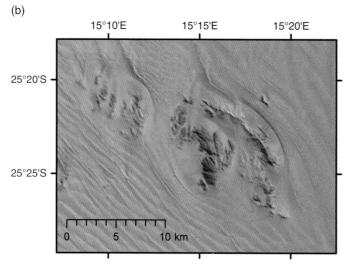

Figure 8.8 Examples of topographic effects on dune patterns; (a) blocking of dunes (from right to left) in the El Mreye sand sea, Mali, creating dune-free basins behind bedrock ridges, (b) deflection of dunes around inselbergs in the Namib sand sea.

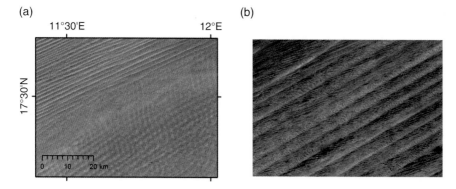

Figure 8.12 Complex dune patterns, Tenere Desert, Niger. (a) shaded digital elevation model (DEM) of longitudinal dunes (left) and superimposed longitudinal and transverse dunes (right). (b) Superimposed broad and narrow longitudinal dunes and intervening transverse features.

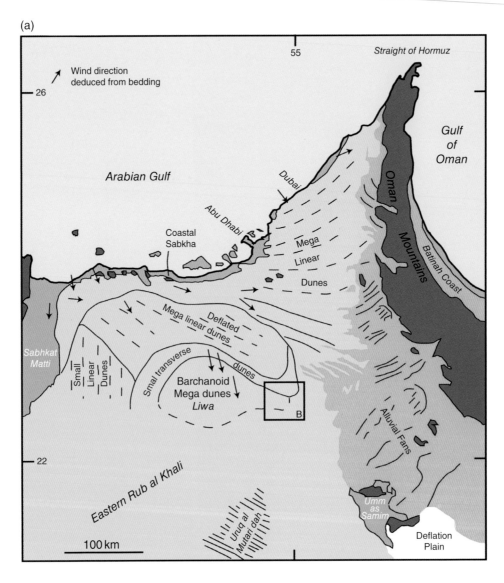

(a)

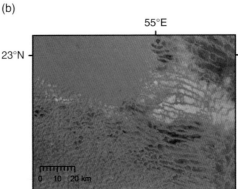

(b)

Figure 8.17 (a) Patterns of different dune types in the northern Rub' al Khali (modified from Glennie and Singhvi 2002), (b) patterns of dunes transitioning from closely spaces (upper left) to widely spaced linear dunes (right) and crescentic/transverse dunes (lower left).

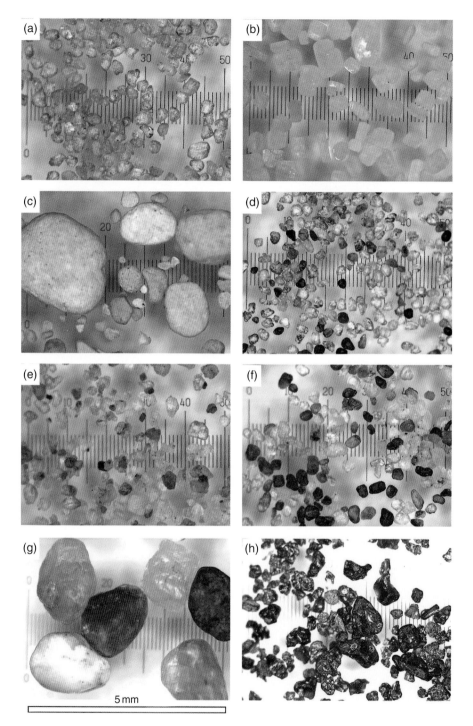

Figure 9.3 Colour photomicrographs of grains of dune sand from locations around the world demonstrating the wide range of sand colours and the variety of grains; the field of view in each photograph is 5 mm. Sample A, from the Grand Erg Occidental in Algeria, is a typical Sahara Desert dune sand with a red colour (H7.5yr 7/6 on a Munsell Soil Colour Chart); Sample B from White Sands National Monument is an off-white (H10yr 8/1), because it is composed of white and transparent gypsum crystals; Sample C from the Bodélé Depression in Chad (H2.5yr 6/2) is composed of grains of pale grey diatomite with a minor component of quartz; Sample D, from the crest of a linear dune in Namibia (H7.5yr 5/4), contains a mix of clear and red-coloured quartz grains and almost black pyroxene; Sample E from the Strzelecki desert in Australia (H10yr 7/4) is primarily composed of quartz; Sample F is from the Packard dunes in Victoria Valley, Antarctica, and contains a mixture of quartz, feldspar, pyroxene, and rock fragments, resulting in grey colours H10yr 4/1; Sample G is coarser-grained and contains more pyroxene, giving it a darker colour (H10yr 5/1), although that is difficult to see in the small number of grains illustrated; Sample H from Iceland is a very dark grey (H5yr 2.5/1) because it is primarily composed of basaltic rock fragments and volcanic glass (H).

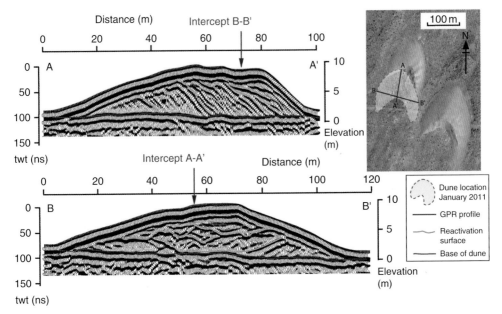

Box Figure 9.1 The internal structure of a barchan dune, near Tarfaya in southern Morocco, imaged with groundpenetrating radar (GPR). Section A-A′ is parallel to the direction of the migration of the dune and shows steeply-dipping foresets that record the former position of the lee side of the slipface. The blue lines pick out reactivation surfaces where reflections are truncated marking erosion and reshaping of the dune due to changes in the wind direction. Section B-B′ is roughly perpendicular to the direction of dune migration and shows a strike section with the preservation of the horns of the dune at the base, overlain by concave reflections from the trough-shaped dune slipface. The inset is a Google Earth™ image from May 2005 together with the position of the dune at the time of the survey in January 2011, showing that this 9 m high dune has moved 100 m downwind within six years. Source: From Bristow and Mountney (2012).

Figure 9.11 (a) Cross-stratification in eolianite Bermuda; The height of the face is around 5 m. At the base of thesection, partially covered by stone blocks, is a palaeosol horizon, across which the dune has migrated from left to right. (b) fossil mould of a palm tree preserved within eolianite, Bermuda.

Figure 10.4 A large sand ramp that has accumulated on the eastern margin of the Namib Desert, southern Africa. The ramp of aeolian sediment extends from left (east) to right (west) of the photograph and is over 2 km in length.

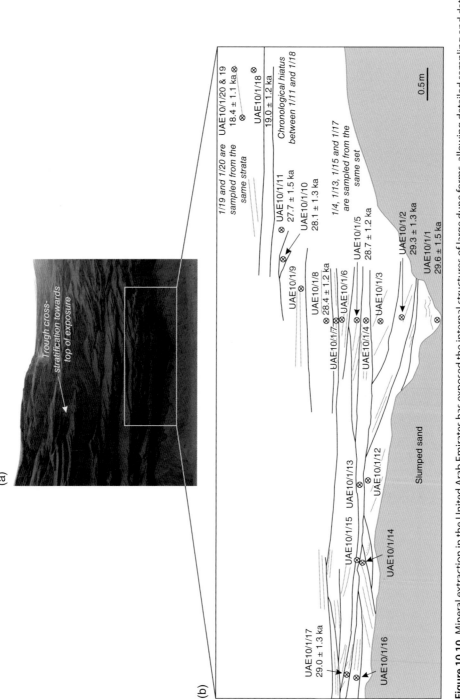

(a)

Trough cross-stratification towards top of exposure

(b)

UAE10/1/17
29.0 ± 1.3 ka

UAE10/1/16

UAE10/1/15   UAE10/1/13

UAE10/1/14

UAE10/1/12

Slumped sand

UAE10/1/11
27.7 ± 1.5 ka

UAE10/1/10
28.1 ± 1.3 ka

UAE10/1/9

UAE10/1/8
28.4 ± 1.2 ka

UAE10/1/6

UAE10/1/7

UAE10/1/5
28.7 ± 1.2 ka

UAE10/1/4

UAE10/1/3

UAE10/1/2
29.3 ± 1.3 ka

UAE10/1/1
29.6 ± 1.5 ka

1/19 and 1/20 are sampled from the same strata   UAE10/1/20 & 19
18.4 ± 1.1 ka

UAE10/1/18
19.0 ± 1.2 ka

Chronological hiatus between 1/11 and 1/18

1/4, 1/13, 1/15 and 1/17 are sampled from the same set

0.5m

Figure 10.10 Mineral extraction in the United Arab Emirates has exposed the internal structures of large dune forms, allowing detailed sampling and dating to take place for sediments deep within dune bodies. For Leighton et al. (2013a), a series of samples were collected for OSL dating, taken from the exposed sedimentary structures in several dunes. 'Bounding surfaces' are apparent stratigraphical breaks in a dune, separating sediments that have accumulated at different times. In this example, two bounding surfaces, shown as AB1 and AB2, separate sediments below that were dated as 27 000 years or older, from those above that were 19 000 years and younger. However, other surfaces, within the older lower sediments in the dune, do not have equivalent significance as features demarcating long breaks in deposition.

Figure 11.11 Yardangs on Mars. Section of HiRISE image ESP_023051_1865. North is towards top of image.

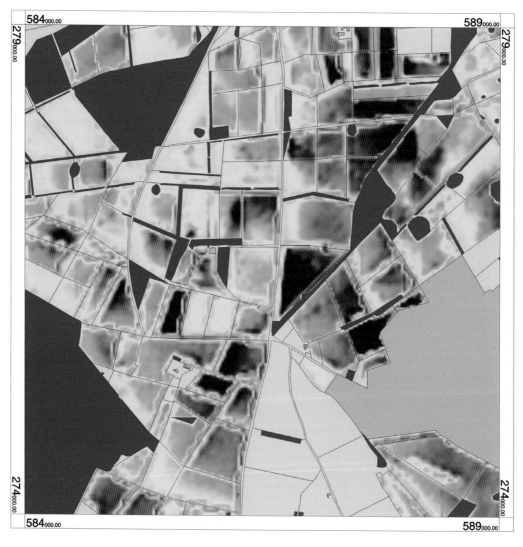

Figure 12.5 Wind erosion in part of Suffolk calculated for the period 1970–1998. *Source:* Böhner et al. (2003), Figure 4.

8

Sand Seas

Paul Hesse

Macquarie University, Sydney, Australia

8.1 Introduction

Sand seas are the fullest expression of aeolian landscapes, being so extensive that only the dunes themselves are visible from their interior. They challenge navigation, communication, exploitation, and development. Cooke et al. (1993) and Livingstone and Warren (1996) followed Wilson (1973) in defining sand seas as dunefields larger than $30\,000\,km^2$ in area, a cut-off point that Wilson had based on the global size-distribution of large sand bodies (as they were known at that time). Despite the vast improvements in access to high-resolution, global remote-sensing imagery since Wilson's research, and early Landsat images used in the landmark Global Sand Seas monograph (McKee 1979), there has been no global re-assessment that includes the areas not well captured in these early analyses, such as in the Southern Hemisphere or China.

The broad patterns found by Wilson are essentially duplicated on the areas of dunes mapped at the highest possible resolution from available satellite images (Esri imagery layer) in the south-west Kalahari and for all of South America (except coastal Peru) and North America (Figure 8.1). Whereas the number of distinct dune areas falls with increasing size (grouped into log-area classes), the total area of dunes increases greatly for very large dunefields. However, there does not seem to be any clear natural break that would warrant a numerical threshold. For very small dune areas ($<60\,km^2$, not mapped by Wilson in his analysis of Saharan Africa), there was an observed decrease in the number of patches. Apart from mapping and resolution artefacts, this probably reflects what Bagnold (1941) observed to be the tendency of sand to self-accumulate: the ultimate source of dunes and also of dunefields.

In this discussion, all very large dunefields will be considered, regardless of their apparent activity. While some bodies of dunes are quite obviously forming and re-forming over time-scales of years and decades (e.g. Besler et al. 2013; Lorenz et al. 2013, Chapter 10) and are indisputably active or mobile, there is a broad 'grey area' separating active dunes from completely fixed, stable or fossil dunes (Tsoar and Møller 1986; Wiggs et al. 1995; Hesse 2016; Chapter 10). Most sand seas contain a mixture of small active dunes and much bigger and probably older dunes (Livingstone and Thomas 1993). Dunes in many deserts show evidence of repeated episodes of rapid accumulation separated by long hiatuses measured in thousands or tens of thousands of years, and so any assessment of modern activity is also an artificial distinction along the range of behaviours of dunes and sand seas (e.g. Fujioka et al. 2009). The additional impact of modern land-use practice (Mainguet et al. 2008) has produced

Aeolian Geomorphology: A New Introduction, First Edition. Edited by Ian Livingstone and Andrew Warren.

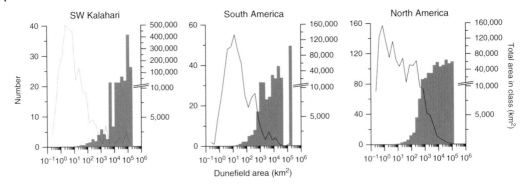

Figure 8.1 Size distribution of dune areas in the Kalahari (731 925 km²), South America (excluding the Pacific coast) (284 500 km²) mapped in geographic information system (GIS) from satellite imagery, and North America (Wolfe, S.A., Gillis, A., Robertson, L., 2009. Late Quaternary eolian deposits of northern North America: Age and extent. Open File, 6006. Geological Survey of Canada) (501 163 km²) collated from geological maps, show the same patterns first demonstrated by Wilson (1973) for Saharan dunefields but without a clear break between large dunefields and sand seas.

highly divergent patterns of activity over small distances that may not clearly reflect 'natural' or equilibrium conditions.

8.2 The Distribution of Sand Seas

There are dunes, dunefields, and possibly a sand sea (of ice dunes, Chapter 1) in the Arctic and the Antarctic (Wolfe 2007), but quartzose sand seas are more limited in their distribution. The deserts are associated with the descending arm of the mid-latitude Hadley cells, but most sand seas extend to higher latitudes (Figure 8.2 and Chapter 1). In South America and Australia, the Pampas and Mallee sand seas extend to 37.5°S and 36°S, respectively, mirrored by the Nebraska Sand Hills (43°N) and Erg Occidental (34.5°N) in North America and North Africa. However, the much more extreme continental climates (and ice-free history) of the interior of the Eurasian land mass have allowed development of sand seas up to 60°N in western Siberia and 48°N in north-eastern China.

Likewise, the equator-ward extent of the sand seas is relatively uniform in the southern hemisphere continents (17–18°S) but ranges from 31°N in North America to 24°N in Asia (India), 15°N in Arabia and 11°N in North Africa. The Llanos del Orinoco dunefield of Venezuela extends to 6°N in the northern hemisphere portion of South America. These variable extents demonstrate both the degree to which the subtropical deserts are enlarged by the enhanced continental climates of the large land masses and the ability of sand seas to form outside the subtropical belt, providing that they meet the necessary conditions, now or in the past (Chapter 1).

Three conditions are necessary for the development of a sand sea: (i) a *supply* of transportable sand; (ii) sufficiently *strong winds* to entrain and move sand particles; and (iii) the *availability* of sand particles for transport, such that they are not protected or bound together. Availability is most commonly governed by aridity, either permanent or periodic (Figure 8.3; Kocurek 1998). Neither all lowlands nor the most wind-prone regions are entirely covered in sand dunes. There are also sand dunes outside large basins and the windiest areas. Only those areas with a combination of all three necessary conditions are the sites of sand dune formation and this scales from the smallest isolated dune to the largest sand sea.

There are many examples of sand dunes and dunefields adjacent to localised sources of sand. These so-called source-bordering dunes represent the combination of sufficiently strong winds with a supply of easily erodible particles (see Figure 8.3) fed by the

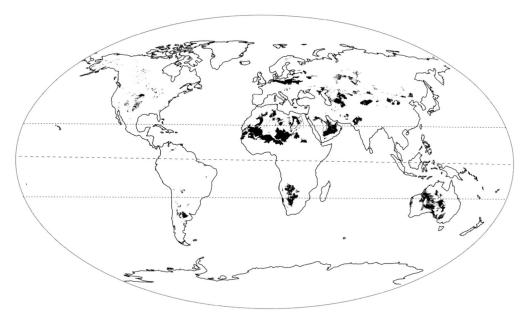

Figure 8.2 Map of global sand seas and large dunefields (shown in black) (Hesse et al. 2015). The North European sandbelt is a very large area in which sand dunes occur but are relatively restricted in area.

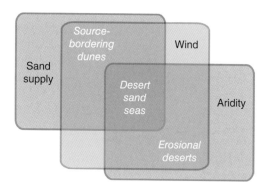

Figure 8.3 Conditions necessary for sand sea formation.

sorting and disturbance occurring in rivers (riparian dunes) (Chapter 1), lakes (shoreline dunes), and at the coast (Chapter 7). However, few areas of accumulation of aeolian sands in humid areas grow to the size of sand seas, largely because of the cover of vegetation, and therefore the limited availability of loose sand. Conversely, most desert areas are devoid of sand dunes because of the absence of transportable sand particles (see Figure 8.3). Even in the deserts, upland and lowland, there are large areas of bare surface, where winds are strong

enough to remove all the sand, and where there is a limited supply of sand either transported to the site or weathered *in situ*. There are several factors affecting the three conditions necessary for sand sea formation, some factors impacting on more than one of these conditions in complex and diverse ways. These conditions are summarised in Table 8.1.

8.3 Climatic Factors

8.3.1 Wind

As wind-blown bed forms, dunes necessarily require winds above some threshold, but the global distribution of dunes and sand seas is not closely linked to patterns of climatic wind strength (Figure 8.4). In Australia, as for other continents, sand drift potential (Fryberger and Dean 1979) is greatest at the coast and decreases inland (Kalma et al. 1988) and the largest sand seas in Australia occur in these low drift potential areas. In the windiest coastal areas around the world there are often narrow bands of coastal dunes (Chapter 7), but on arid coastlines with onshore winds,

Table 8.1 Factors affecting the conditions necessary for dune and sand sea formation.

	Supply	Availability	Wind
Climate (esp. precipitation)	• Production of particles through weathering, fluvio-glacial and mass wasting in basin margins • Transport and concentration of particles in basins	• Aridity suppresses soil moisture, vegetation cover, biotic crusts and pedogenic cements • Aridity reduces reworking of wind-blown sediments by water erosion	• Both precipitation and wind relatively low in continental interiors • Global circulation patterns (direction, seasonality, strength)
Topography	• Sand accumulation in basins • Efficient runoff on slopes	• Aridity related to topographic barriers and continentality • Evaporitic cements in basins	• Acceleration, blocking and deflection around topographic obstacles
Sorting process	• Fluvial, coastal and lacustrine environments sort sand from cohesive and unerodible particles	• Fluvial, coastal and lacustrine disturbance provide local sources of availability	
Lithology	• Presence of coarse grains, especially quartz, in substrates of the basin catchment areas		
Tectonic setting	• Rapid sediment supply from mountain belts • Creation of long-term sediment sinks (basins) • Long accumulation, weathering and sorting times in continental interiors		

sand supply and a receptive topography, there is a continuum from small areas of coastal dunes to large dunefields and sand seas (see Figure 8.2). The Peruvian and Skeleton Coast dunefields in Namibia are both in very windy and very arid areas, and have some large sand seas. Conversely, the Taklamakan Desert sand sea, one of the largest and most active in the world, has quite low total drift potential (Wang et al. 2002). Similar conditions occur in some other large Central Asian sand seas (Suslov 1961).

It is evident, then, that wind speed is rarely the limiting factor in the development of sand seas: if deposits of bare sand are available, then most areas of the world have sufficiently strong winds to move sand and create dunes. However, the fact that, at present, some sand seas are not very active, or degrading, may mean that winds in these areas were stronger in the past or the climate was more arid.

8.3.2 Precipitation

Could the occurrence of sand seas in the continental interiors have more to do with precipitation than sand drift potential? Aridity has several direct and indirect effects on sand supply and dune formation.

Wilson (1973) observed that active ergs (dunefields and sand seas) today fall within the present 150 mm mean annual precipitation isohyet. The much larger area of vegetated and partly vegetated sand seas outside this area is concentrated in areas that today

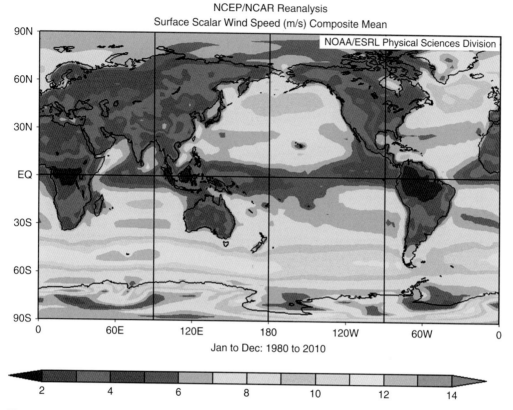

Figure 8.4 Mean surface wind speed (NCEP/NCAR reanalysis data: http://www.esrl.noaa.gov/psd/data/reanalysis/reanalysis.shtml). No threshold wind speed is used and there is no weighting according to velocity, unlike sand drift potential values. See insert for colour representation of this figure.

are arid or semi-arid, but extend into more humid zones as well (Figure 8.5). Whereas this may indicate that some sand seas experienced greater aridity in the past and are today inactive, it is also true that some modes of dune activity are possible under mildly arid conditions (Lancaster 1988; Wiggs et al. 1995).

The southern African sand seas, in particular, show a pattern of activity and form related to the pattern of rainfall distribution (Figure 8.5; the relationship between precipitation and dune movement with respect to southern Africa is discussed in detail in Chapter 10).

In the hyper-arid (mean annual rainfall less than 100 mm) Namib Desert in Namibia and Angola, the dunes have more or less continuous active sharp crests with slip faces, but little visible perennial (shrub or tree) vegetation (in high resolution Esri imagery layer viewed at a scale of 1:8000 or higher). The dunes range from barchan, crescentic, and linear to star (Livingstone 2013) and many are compound, with superimposed dune types and orientations. On the inland margins of the Namib Desert, where mean rainfall is slightly higher, perennial vegetation becomes more visible and the dunes show segments of sharp crests and slip faces interspersed with crest segments without clearly visible slip faces, but distinct crests with some shrubs. The two dune states are not clearly separated but interspersed in patches, sometimes related to dune type and apparent sand thickness. For example, patches of crescentic dune ridges are less vegetated than adjacent areas of linear dunes.

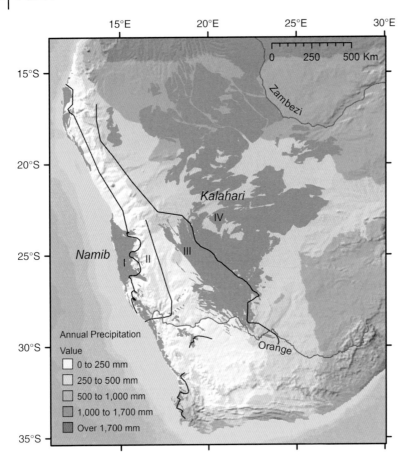

Figure 8.5 Southern African dune expression and climate. Dunefields are shown in dark grey (orange) superimposed over the distribution of annual average rainfall. Simplified boundaries between four morphological classes of dunes are shown: (I) continuous sharp crests with slip faces and with little or no perennial vegetation; (II) discontinuous segments of sharp crests with slip faces but sparse cover of perennial vegetation elsewhere; (III) distinct crests but only rare patches of bare crest or slip faces and perennial vegetation present on crests, flanks, and interdunes; and (IV) indistinct crests with very rare or no bare areas and continuous cover of perennial shrubs and trees. See insert for colour representation of this figure.

Beyond the Namib, in the south-western Kalahari and northern coastal strip of South Africa, where mean annual rainfall is up to 250 mm, dunes are well covered by sparse perennial vegetation and have only patches of bare crest and slip faces, but quite distinct narrow crests. Vegetation and sand movement in this zone respond to inter-annual climatic variability (Bullard et al. 1997), maintaining periodic vertical accumulation and along-axis sand transport (Telfer and Thomas 2007; Stone and Thomas 2008; Telfer 2011). Dunes in this zone are *overwhelmingly* simple linear types, with areas of

network and dendritic dunes (Bullard and Nash 1998), and crescentic ridges or barchans are rare or absent. In the areas of the northern and eastern Kalahari, and the southern coastal strip of South Africa, the dunes are almost continuously vegetated with perennial vegetation and have broad, indistinct crests (see Figure 8.5). The dunes are almost exclusively linear in form, but bare crests are absent and the higher rainfall (up to 1000 mm yr^{-1} or more) in the summer is very probably degrading the dunes. Patches of now well-vegetated parabolic dunes and blowouts are also found in these marginal areas. In the extreme north of the Kalahari the dunes have a degraded appearance and show the emergence of lacustrine/pan processes in the interdunes, reshaping the dune outlines, or streams eroding along the interdunes.

The broadly accepted model for dune types (Wasson and Hyde 1983) proposes that sand availability (equivalent sand thickness) and wind variability determine dune type (Chapter 6, Figure 6.3). However, in many sand seas in marginally humid climates (Tripaldi and Forman 2007; Hesse 2011; Halfen et al. 2016), there is a distinct transition from compound crescentic and linear structures towards linear dunes and then parabolic dunefields, reflecting the increasing limitation of vegetation with increasing rainfall. As a consequence, parabolic and blowout morphologies dominate the many isolated dunefields over presently more humid areas in South America, including large parts of the Pampas sand sea complex (Figure 8.6). Small areas of riparian dunes are forming today in Bolivia under very seasonal but high mean annual precipitation (around $1000\,\mathrm{mm\,yr^{-1}}$) (Latrubesse et al. 2012). These dunes, and the larger area of forested dunes around them, are predominantly parabolic in form. Simple linear dunes are predominant in drier areas with mean annual rainfall less than around $400\,\mathrm{mm\,yr^{-1}}$ and limited areas of compound crescentic ridges and linear dunes are found

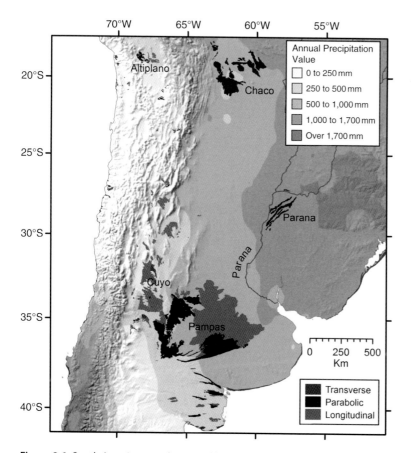

Figure 8.6 South American sand seas and large dunefields. Dunefields are shown in black and dark grey superimposed over the distribution of annual average rainfall (from WorldClim data). Dominant dune types within each dunefield are indicated (but often more than a single morphology is present). See insert for colour representation of this figure.

in areas with mean annual rainfall less than $250\,mm\,yr^{-1}$ and greater sand thickness.

Precipitation also plays a positive role in the construction of sand seas, because of the importance of fluvial systems in delivering sand to dunes (Chapter 1). In particular, many river basins are characterised by large distributive fluvial systems (DFS) (Weissman et al. 2010), which have clearly contributed sand to aeolian dunes (Davidson et al. 2013). Despite this potential, the actual contribution to dune development also depends on how sand is distributed over the DFS surface or its availability from the channels. Although there are examples of sand seas (Kara kum, Kyzyl kum) derived from DFS (Maman et al. 2011), there are probably more examples in Asia where there is apparently insufficient fluvial sorting of sand to supply dunes, despite the apparently high rates of sediment delivery. Hyper-arid areas are noted for their low contributions of sediment to streams, not only because of low sediment transfer, but also low weathering rates and particle supply.

The destructive role of precipitation is felt in the reworking of aeolian deposits. For example, the mounds of saltated sand formed by agricultural wind erosion rarely survive reworking by splash and wash processes for long. Sand dunes are therefore a manifestation not only of the occurrence of aeolian transport, but also the balance between constructive aeolian saltation and destructive runoff erosion. While runoff erosion is rarely of importance in hyper-arid deserts, it is clearly important in more humid sand sea margins where indistinct, low-angle, and discontinuous dunes are common (Hesse 2011; Roskin et al. 2011).

In the arid to semi-arid areas of Australia, pedogenesis also plays a role in stabilising sand dunes (Hesse 2011). In particular, calcium carbonate cements develop in the subsoils of dunes in the southern half of the continent (Fitzsimmons et al. 2009). The calcium is largely derived from marine sources (Quade et al. 1995), and the calcium load is highest in the south because of the strong, onshore westerly winds from the Southern Ocean (Figure 1.5). Respired carbon dioxide within the soil provides the carbonate ions in the formation of the calcium carbonate (Quade et al. 1995). The survival of calcium carbonate within the soil depends on the balance between leaching and biomass respiration along the climatic gradient. In the most arid areas, and adjacent to groundwater lake basins, gypsum (calcium sulphate) dominates over calcium carbonate. Other sand seas (as in southern Africa) apparently lack the same kinds of palaeosol (Thomas and Burrough 2016) possibly because of their latitude (beyond the reach of the westerlies), continental-interior positions or the low biomass that accompanies hyper-aridity (Figure 1.5).

8.4 Topography

Bedrock topography exerts a very strong influence on the distribution and formation of sand seas. Although dunes and small dunefields exist on steeper slopes and among mountains, for example, in the US south-west, the South American Cordillera (Figure 8.6), and central Asia, sand seas are largely found in basins and other topographic lowlands. Lowlands provide the accommodation space for sand contributed by the surrounding catchments and slopes by streams and, in general, basins have gentler winds than the surrounding uplands.

The most extensive active sand sea in the world is the Rub' al Khali in Arabia. The Rub' al Khali and other sand bodies connected to it occupy a topographic lowland between the Hijaz mountains and the Yemen/Oman uplands (Glennie and Singhvi 2002). Most of the sand in these sand seas has been contributed from weathering of Palaeozoic sandstones in these mountains (Figure 8.7). Other basement, carbonate, and basic igneous rocks of the Arabian Peninsula have contributed less sand to these dunes. However, in the coastal areas of the Persian Gulf, shelf sediments are the dominant source of sand, originating in the Zagros Mountains and

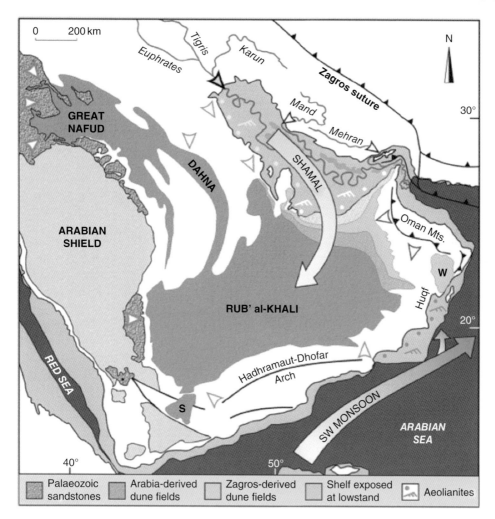

Figure 8.7 Provenance of Arabian sands (Garzanti et al. 2013, Figure 10). Short arrows indicate points of sediment input. Long arrows represent dominant seasonal winds.

Tigris-Euphrates River when it flowed across the Gulf when sea levels were lower (Garzanti et al. 2013) (Figure 8.7). The connection between plate tectonics, tectonic uplift, and fluvial erosion and transport in Egypt is discussed, briefly, in Chapter 1.

Runoff from slopes and by rivers redistributes weathering products from soils to the basin margins and floors. The efficiency of sediment delivery increases with the steepness of the slope, but decreases with decreasing precipitation, in turn linked to vegetation. Sediment derived from catchments on the eastern margin of the Namib

sand sea contributes locally to the sand dunes (Garzanti et al. 2012; Stone 2013), but the rivers are mostly too small to reach far into the dunefields before they are blocked by dunes. Instead the Namib sands can be traced to the mouth of the Orange River. The most arid desert area on Earth, the Atacama, has no internal runoff, and virtually no sediment delivery from slopes to valley floors. As a consequence, it is largely a dune-free desert.

Some deserts are created, or their aridity enhanced, by topographic barriers. For example, the rain-shadow created by the southern Andes enhances the aridity of Argentinean Patagonia (Garreaud 2009), although low sand supply by the Andean rivers apparently limits dune formation. In the tropics, the prevalence of descending dry air of the Hadley-cell circulation lessens the rain-shadow effect (Garreaud 2009). Conversely, ranges in the continental interiors can have elevated precipitation because of a mild orographic effect which contributes runoff and sediment delivery to the surrounding basins.

A more widely discussed effect of topography on dunes and sand seas is the effect on sand-moving surface winds by topographic acceleration, deflection, and blocking. Wilson (1973) described the possible impacts of topographic barriers on wind: acceleration and deflection, both of which would result in substantially sand-free uplands across which sand is transported rapidly. Most sand seas show clear examples of the deflection of sand-transporting winds, and therefore dunes, around topographic obstacles (Figure 8.8; Chapter 7),

(a)

(b)

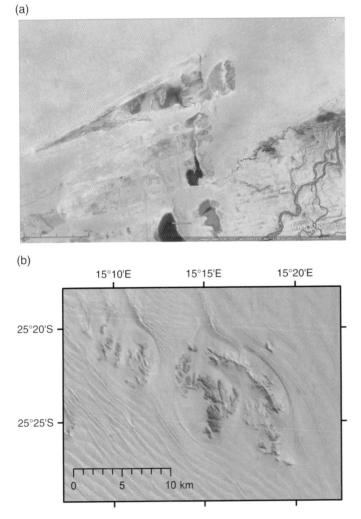

Figure 8.8 Examples of topographic effects on dune patterns; (a) blocking of dunes (from right to left) in the El Mreye sand sea, Mali, creating dune-free basins behind bedrock ridges, (b) deflection of dunes around inselbergs in the Namib sand sea. See insert for colour representation of this figure.

as around boulders in a river. Wilson (1971) envisaged long-distance sand flow over very large topographic barriers in the Sahara, a concept supported by El-Baz et al. (2000). However, a series of sand-provenance studies in Australia (Pell et al. 1999, 2000, 2001) has shown that even relatively low topographic barriers between sand seas appear to mark distinct boundaries in sand composition. It is conceivable that sand is more easily reworked by runoff on the steeper hills and returned to the upwind foot-slope, but in central Australia there is only very minor accumulation of sand upwind of blocking bedrock ridges, indicating little overall downwind transport (Hollands et al. 2006).

Topographic depressions in deserts are commonly the sites of accumulation of salts from groundwater or runoff evaporation. Gypsum is a common evaporitic mineral which can also form transportable particles and be blown into dunes on the margins of lakes or groundwater window depressions. The largest area of gypsum dunes is the White Sands of New Mexico (e.g. Baitis et al. 2014). Within the Australian sand seas, gypsum dunes form in topographic depressions among a broader field of quartz-rich dunes (Chen et al. 1995). The gypsum is frequently found to have been remobilised and re-crystallised into gypcrete, creating extraordinarily stable, non-erodible dunes. Their lower mobility also appears to give rise to distinctive network and mound morphologies among the surrounding oriented (mostly linear) quartz dunes (Hesse 2011).

8.5 Sorting Processes

While sand supply is critical to the formation of sand seas, it is also clear that the rate of sediment supply to basins is not closely linked to the distribution of sand seas and sand dunes. For example, the Gangetic foreland basin has only minor sand dune development despite extensive semi-arid areas over the rapidly accumulating fluvial plain.

Likewise, large-scale sediment delivery to the wind-blown glacial outwash plains of the arid Patagonian steppe desert has not been accompanied by widespread dune formation (Figure 8.2; Darvill et al. 2014). The large sand seas of Arabia, Africa, and Australia mostly occur in basins with low sediment-delivery rates today. Effective sorting processes, other than by the wind, seem to be necessary to turn unsorted sediment into sand supply suitable for the generation of sand seas.

In Patagonia, small dunefields occur only where there is an additional sorting mechanism, usually fluvial, which separates sand from large, unestertion clasts (Jerolmack and Brzinski 2010). In North America, dunefields related to the Laurentide ice sheet are small in comparison to fluvial source-bordering dunefields of the southern Great Plains and the south-western USA (Halfen et al. 2016). Globally, the largest area of glacially-linked dunes may be in northern Europe (Zeeberg 1998; Kasse 2002; Chapter 1), similar to the situation in Canada (Wolfe 2007).

Deflation of sand from rivers is a common contributing mechanism to dunefields and sand seas. The Colorado River of the south-western United States is the major source of sand to the Gran Desierto and Parker dunefields (Muhs et al. 2003). The Thar Desert of India and Pakistan has formed in the margins of the distal Indus plain from sand-bed (palaeo-) channels of rivers draining from the Himalayas and local (shield) sources (Singhvi and Kar 2004). A combination of the climatic gradient and sediment sorting in the distal fluvial environment appears to have provided conditions suitable for sand sea formation not found in the proximal DFS. Davidson et al. (2013) link dunes to intermediate and distal DFS surfaces where fluvial processes have already sorted sand-sized sediment. In drylands, the tendency for DFS rivers to terminate in dunefields is pronounced (Hartley et al. 2010).

Coastal areas are commonly associated with dunes because some experience both strong winds but also the very effective sorting action of waves. Several desert sand seas and large dunefields are at least partly coastal in origin, for example, the Namib (discussed earlier). Coastal sands were the largest source for the Eyre dunefield and supply parts of the Gascoyne and Mallee dunefields in Australia (Hesse 2010). Rivers have also contributed sand to the interior margins of the Gascoyne and Mallee dunefields. Regressive Late Miocene to Early Pleistocene shoreline dunes (McLaren et al. 2011), later reworked into parabolic and linear dunes, are the origin of most dunes in the Mallee dunefield. Eocene shoreline dunes are partly preserved and reworked into linear dunes in the Eyre – Great Victoria Desert sand seas.

Lake shorelines are another situation in which waves are effective in sorting sand and supplying it to sand dunes. Dunefields are found around large freshwater lakes in humid climates (Hansen et al. 2010) as well as bordering saline lakes in deserts (Baitis et al. 2014). Although no sand seas are derived solely from lake shoreline deflation, such dunes contribute to many sand seas, especially in endorheic basins.

Wind is a powerful sorting agent, but only to the degree that surfaces are initially erodible. Both large non-erodible and fine binding components can suppress entrainment by wind. Sorting by other agents, especially water, is an important pre-condition for wind erosion in many situations. It is also likely that pedogenesis can achieve the same outcome, for example, by illuviation, contributing to sand-rich topsoils susceptible to reworking by wind (Hesse 2011).

8.6 Geological Factors

8.6.1 Lithology

One important factor affecting sand supply is the lithology of the source rocks in the basin. While there are several mechanisms by which sands evolve (become texturally mature), either before or after formation into sand dunes (Muhs 2004; Jerolmack and Brzinski 2010), the initial formation of sand-sized particles from rock weathering is fundamental to the supply of sand for dune construction. The provenance of sand in the Namib Desert (Garzanti et al. 2012) is from a very geologically-diverse catchment and is sorted, concentrated, and transported by the Orange River to the coast where it is reworked into sand dunes, volumetrically overwhelming local sand sources.

There are many large dune-free areas within the generally low-relief, arid interior of Australia (Figure 8.9). Many of these areas are associated with lithologies which are inherently quartz- and sand-poor sources: limestones, shales and mudstones, alluvial clay plains, and basic igneous rocks. The most notable of these areas is the Nullarbor Plain, which forms the surface of the Miocene Eucla Basin, which is an extensive limestone plain. However, the neighbouring coarse-grained clastic sedimentary rocks of the Officer Basin, Canning Basin, and Musgrave Block have yielded abundant quartz sand to form the Great Victoria and Great Sandy sand seas. Most intriguing is the Yilgarn Craton, comprised largely of granite and minor greenstone (basic igneous) belts. Based on geochemical characteristics, the Yilgarn has not contributed significant sand to the surrounding sand seas (Pell et al. 1999). The extensive granite, low topography and long weathering history of the craton has yielded only small areas of dunes, scattered widely in the largely relict drainage system (Hesse 2010). The relative absence of dunes, despite the dominance of granite substrate, may be due to the formation of cemented lateritic profiles, restricting the release of sand particles.

8.6.2 Tectonic Setting

Tectonic setting is, in general, a prime driver of erosion rates and sediment delivery. While fluvial erosion rates in basins do show higher values for tectonically active areas than inactive areas (Portenga and Bierman 2011), this difference in sediment delivery

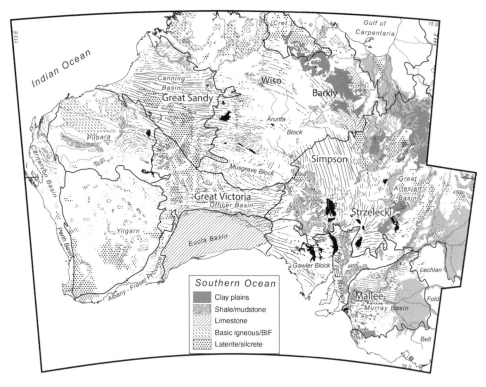

Figure 8.9 Outcrop lithology and sand seas in Australia. Non-sand bearing lithologies and the boundaries of major geological provinces are shown (italics, solid black lines). Sand seas are indicated by the orientation of longitudinal dunes (grey lines). *Source:* Modified from Hesse (2010).

rates is not mirrored by the global distribution of aeolian sand seas (Figure 8.2). The simple presumption of high sediment supply from tectonically active mountains is complicated in deserts because of the lower present rates of fluvial erosion and lower rates of weathering. The complex tectonic boundary extending from Tibet through northern China is host to several significant sand seas (Figure 8.10), which occupy basins and other lowlands between steep mountain belts. However, the tectonically quiescent Sahara is home to much more extensive sand seas (Figure 1.12).

Both the Asian collision zone and the South and North American cordillera are host to dunefields, with variable sediment thickness, set in small intramontane basins. Sand seas are found in some large epi-suture basins in northern China and Mongolia (Ortindag, Horqin, Hulun Buir) (Yang et al.

2012) with deeper and more extensive sediment sequences. Foreland basins associated with these tectonically active mountain belts are the setting for several large sand seas in Asia (Karakum, Kyzylkum, Muyunkum, Taukum, Taklamakan, Gurbantunggut, and Qaidam), Arabia (northern parts of the Rub' al-Khali), northern Africa (Grand Erg Oriental, Grand Erg Occidental, Iguidi, Check-Adrar, Raoui), India (Thar, Thal), and South America (Chaco, Cuyo).

Globally, cratonic areas of southern and northern Africa and Australia account for the majority of the world's sand seas. This pattern probably reflects long periods of sediment accumulation and sorting (both physical and chemical) and the large accommodation space in the lowlands of these continents. Some of these sand seas are located in intra-cratonic sag basins with extensive, but relatively thin, Neogene sediment cover,

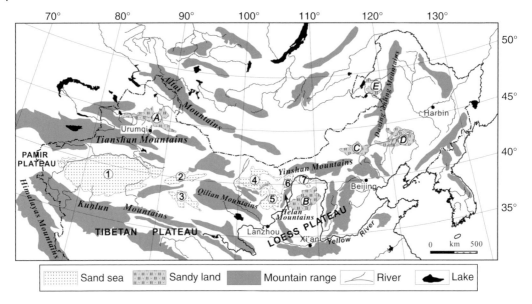

Figure 8.10 Tectonic setting of Chinese sand seas in the Asian collision zone with sand seas in foreland and episuture basins between mountain ranges (Yang et al. 2012). Mostly active sand seas (1) Taklamakan, (2) Kumtag, (3) Qaidam, (4) Badain Jaran, (5) Tengger, (6) Ulan Buh, (7) Hobq; mostly stable sand seas (A) Gurbantuggut, (B) Mu Us, (C) Hunshandake, (D) Horqin, (E) Hulun Buir.

such as the Simpson, Strzelecki, and Mallee dunefields of Australia (Hesse 2010), the Nefud of Arabia, or the Kalahari of Southern Africa, while others occupy topographic depressions without long-term Neogene sediment accumulation (e.g. Barkly and Wiso, Australia). Further along the spectrum, the Great Victoria and Great Sandy Desert sand seas cover broad areas of cratonic landscape with low hills and shallow valleys (Hesse 2010). The dunes appear to have been reworked from a range of alluvial and shoreline sources and now drape low piedmont hills.

The most extensive coastal, or coast-derived, dunefields are associated with passive tectonic margins. The Namib sand sea, of the west coast of southern Africa, and the Gascoyne and Mallee sand seas of Australia combine sediment delivered to the passive margin by rivers draining the hinterland, either directly or after reworking along the coastline. However, extensive coastal dune systems in the humid passive margins of the Atlantic coasts of South America, North America, and Europe, and Eastern Australia

do not transfer sand to large interior desert sand seas, but do rework coastal sands, supplied by rivers. The dunes of coastal Peru and northern Chile are a rare example of extensive dunefields developed in a fore-arc setting. The scarcity of similar examples may simply be because no other fore-arc coasts occur in the subtropical arid belts.

8.7 A Basin Model of Sand-Sea Formation

Terrestrial basins in a wide range of tectonic settings show several consistent morphological features of sediment input and transfer. Recently, these have been synthesised in a global analysis (Weissman et al. 2010) showing the dominance of DFS at a wide range of scales and in a wide range of settings, including those in deserts and with sand dunes present. DFS include alluvial fans, which may be relatively small and steep, and mega-fans (low-gradient and large) (Latrubesse et al. 2012). Streams in deserts commonly display avulsions, leading to breakdown and flood-outs (Tooth 1999; Ralph and Hesse 2010), resulting in the deposition and accumulation of sediment. Elaboration of the DFS model (Hartley et al. 2010; Davidson et al. 2013) has included consideration of the presence of dunes on DFS surfaces, partly limited by the presence of sandy sediment and partly by climate. Both the sharp transition from gravel to sand-bed channels (Jerolmack and Brzinski 2010) and from sand (bedload) to suspended load determine the zone on the DFS surface susceptible to dune formation.

Figure 8.11 shows the relationship of dunes to features of terrestrial basins that exhibit DFS. Basins range from those in which sand dunes are relatively rare and small (Figure 8.11a) to those in which dunes are extensive and may largely obscure any DFS (Figure 8.11c), but in all cases it is assumed that sand has been introduced in the past or at present by streams draining the surrounding mountains and piedmont, which has been reworked within the basin into sand dunes. The sequence of diagrams in Figure 8.11 may be viewed as either (i) a climatic sequence from less arid to more arid; (ii) a chronological sequence of aridification; or (iii) a sequence from lithologies yielding little sand to lithologies yielding much sand. (Chapter 1 explores some other aspects of the relationship between rivers and dunes.)

Many basins and DFSs contain few aeolian dunes (Figure 8.11a). In many places, relatively small areas of source-bordering dunes can be found either in riparian, shoreline (lacustrine) or coastal (marine) settings. These dunefields are usually narrow, and contain transverse ridges proximal to their sand source, but may also include secondary dunes, often parabolic, blown some distance downwind (Figure 8.11b). Many of the dunefields of the Great Plains in the western United States fit this pattern. In all these cases the sources of sand can be easily identified at the upwind limit of the dunefield, and the dunefields themselves are continuous from the source of their sand to their downwind limit. In better-developed aeolian systems, common in deserts, the dunefields blanket the DFS surfaces and even the axial (tributary) fluvial systems and axial distributary fluvial systems (Weissman et al. 2010) of the basin floor. Blanketing sand seas may be constrained by limitation of supply (such as the Cuyo sand seas in Argentina), so that a specific linear or point source of sand is not immediately evident. Both blanketing and source-bordering dunes can disconnect or redirect their parent streams. In deserts many streams have been blocked by dunes and develop ephemeral lake basins (Figure 8.11b). These lakes may develop their own shoreline dunes.

Basins subject to long-term aeolian processes, or extreme aridification, or abundant sand-accumulation may develop blanketing dunefields over almost the entire basin floor, making identification of the underlying original DFS difficult (Figure 8.11c). Many dunes in the Simpson Desert in Australia and Kalahari in southern Africa fit this description. In these areas, the fluvial network is commonly highly disconnected, such that the axial fluvial system may have become beheaded and desiccated.

Winds have moved many dunes far from the sources of their sand, disconnecting them from the sources of their sand, especially if the supply of sand is restricted. In the Badain Jaran Desert of China, great thicknesses of sand are piled at the downwind margins of the basin, but not in their lowest parts, thus moving them far from the sources of their sand (Dong et al. 2004, 2013). Downwind transport may also spread dunes onto neighbouring plains and piedmont slopes. This appears to have occurred in the Great Sandy Desert and Great Victoria Deserts in Australia (Hesse 2010).

Most of the now-stabilised inland dunes of North America and Europe were derived ultimately from glacial deposits, but these sands were then repeatedly redistributed by seasonal streams fed by glacial-melt-water and therefore also fit this pattern.

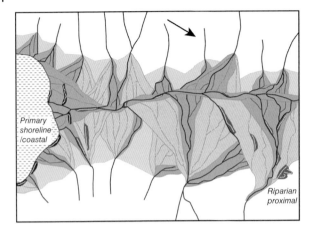

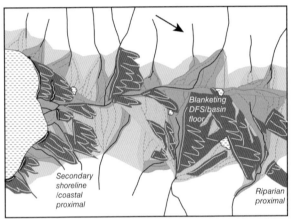

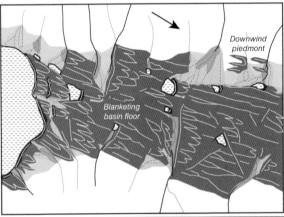

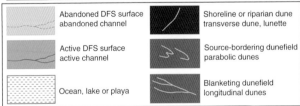

Figure 8.11 Development of dunes, dunefields, and sand seas in basins dominated by distributive fluvial systems (DFS). From top to bottom represents a transition from either less arid to more arid, or an evolutionary sequence of aridification, or from low sand supply to high sand supply. *Source:* Based on Weissman et al. (2010).

8.8   Dune Patterns and Their Origins

This section describes broad patterns of dunes in sand seas. Chapters 6 and 7 take a closer look at a basic set of dune patterns and their environmental associations.

Most sand seas extend across substantial gradients of rainfall, wind climate, and wind direction, and source their sand from diverse sources, themselves from a range of substrate types. As a result, sand seas show rich and complex patterns of dunes, of type, spacing, orientation, composition, and history.

8.8.1   Sediment Properties (Grain Size and Composition)

The Ténéré Desert of Niger exhibits several dune types within the same area (Warren 1972). Transverse linguoid zibars fill linear belts between linear (seif) dunes over a wide area. Warren (1972) showed that the zibars were composed of bimodal sand sizes, whereas the linear dunes were composed of complementary well-sorted unimodal sand. He proposed that the two dune forms had evolved simultaneously from the deflation of substrate alluvium. The resultant contrasting aeolian forms reflected the threshold of movement and mode of transport of the different sand fractions. Similar features have been noted in ripples of different sand sizes and may well explain many other complex dunefields.

The Ténéré Desert and neighbouring sand seas host spectacular examples of contrasting dune forms in close association, McKee's (1979) 'complex' dunes (Figure 8.12). Some of his 'compound' dunes (superimposed dunes of the same type) are shown in Figure 8.12b. These may also share the same origin in different sand fractions as in Warren's zibars, but they also have the appearance of being superimposed. It is not clear whether different dune types are formed simultaneously, as has been observed in some instances in flume experiments (Reffet et al. 2010), or sequentially (see following sections).

At a yet larger scale, the Mallee sand sea of south-eastern Australia exhibits large (broadly) parabolic dunefields set within a larger sea of short longitudinal dunes (Figure 8.13). Comprehensive optically-stimulated luminescence (OSL) single grain single-aliquot regeneration (SAR) dating of both sets of dunes shows that the parabolic dunes formed in several episodes simultaneously with growth of the longitudinal dunes (Lomax et al. 2011). The differences between the dune forms is most likely due to their

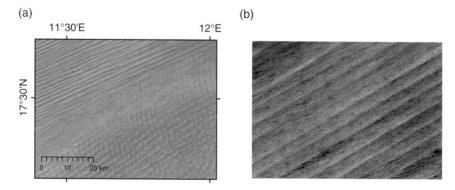

Figure 8.12   Complex dune patterns, Tenere Desert, Niger. (a) Shaded digital elevation model (DEM) of longitudinal dunes (upper left) and superimposed longitudinal and transverse dunes (upper right). (b) Superimposed broad and narrow longitudinal dunes and intervening transverse features. See insert for colour representation of this figure.

(a)

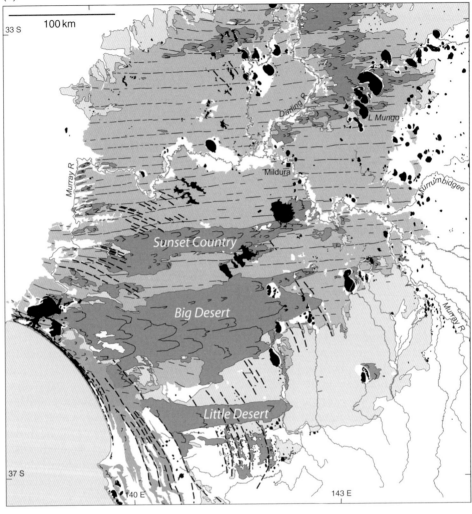

(b)

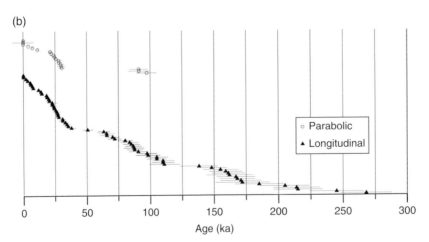

Figure 8.13 Mallee sand sea of Southeastern Australia, including parabolic (dark grey) and longitudinal (medium grey) dunes (Hesse 2010) (a) and ranked age plot of single grain OSL ages (data from Lomax et al. 2011) (b). Dashed black lines indicate Pliocene and Pleistocene strandlines. Black areas indicate lakes/playas.

different source materials: the parabolic dunes have low levels of cohesive impurities such as clay or carbonate and are derived from deflation of Miocene to Pleistocene strandlines, whereas the longitudinal dunes, where the sediments have relatively high levels of both clay and carbonate, are non-mobile constructive features created by deflation of a broader, probably fluvial, basin cover (Bowler and Magee 1978; Hesse 2011). The side-by-side occurrence and simultaneous formation of these dune types challenge the proposal that different environmental conditions accompanied their formation: strong unidirectional winds and the presence of vegetation to form parabolic dunes, and bi-directional or divergent winds and non-limiting vegetation for longitudinal dunes (following the classifications of Hack 1941 and Wasson and Hyde 1983) (Figure 6.2). Complex dunefields such as this demand more complex explanations involving a wider range of environmental conditions (Hesse 2011) or feedbacks between vegetation and dune growth (Hugenholtz and Wolfe 2006).

8.8.2 Climate Change and Changing Wind Regime

To distinguish between simultaneous and sequential development of dune formation requires sufficiently good chronological control on the sequence of formation, as is the case in the Mallee dunefield, but there are few such studies. Some of the most spectacular examples of complex dunefields, such as the Médanos Grandes in the northern Cuyo of central Argentina (Figure 8.14) appear to be contemporaneously active, but there is insufficient information to test this hypothesis (Tripaldi and Forman 2007). Others show a greater contrast between subdued and apparently active forms, suggesting successive development. The Gran Desierto dunefield in Mexico contains a broad range of overprinted dune types, including linear, crescentic, and star dunes (Beveridge et al. 2006) (Figure 8.15). In that sand sea, each dune type and area shows sequential OSL ages, suggesting formation under evolving wind climate (especially its directionality) and vegetation conditions. A similar history has led to several superimposed and differently aligned dunes in the western margin of the Sahara in Mauritania (Lancaster et al. 2002). Each of three sets of linear dunes was formed in short intervals of around 5000 years (Figure 8.16 and discussion in Chapter 1) followed by smaller and more distinct NNE–SSW linear dunes formed around 15–10 ka and then by still smaller north–south-trending linear dunes formed 5 ka ago. The sequential change in orientation of the dunes

Figure 8.14 Complex dunes comprising both barchanoid/crescentic ridges and smaller superimposed longitudinal dunes. Médanos Grandes, Argentina (Tripaldi and Forman 2007).

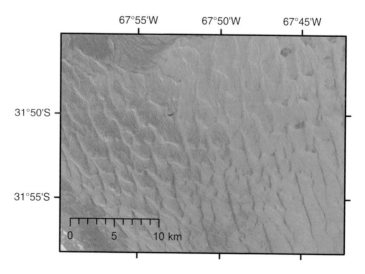

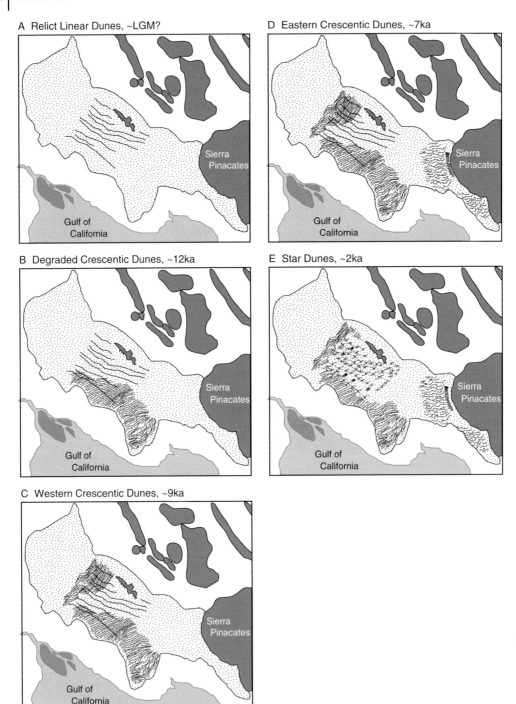

Figure 8.15 Development of the Gran Desierto dunefield (Beveridge et al. 2006) showing the sequential development of dune types with contrasting morphology and orientation in response to changing climate and sand availability.

(a)

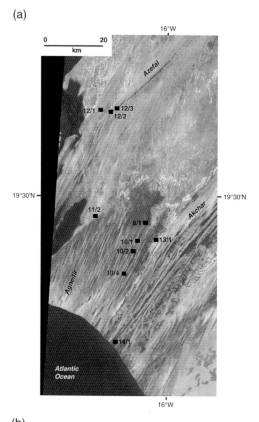

(b)

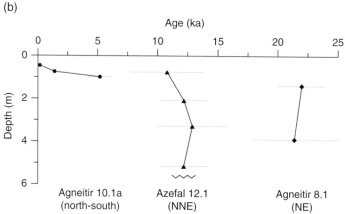

Figure 8.16 (a) Generations of superimposed linear dunes, Mauritanian Sahara (Lancaster et al. 2002), (b) age-depths profiles for selected dunes.

shows a trend from NE to SW resultant drift direction (RDD), reflecting the strongest winds, during the Last Glacial Maximum (LGM) towards N to S RDD in the Late Holocene. The decreasing dune size and slower accumulation rate of the youngest dunes may also reflect decreasing wind strength. Since each generation of dunes is supplied by erosion of older, underlying, dunes, the decrease in dune size may also represent a gradual reduction in sand supply.

8.8.3 Sand Supply and Flow Paths

Sand seas exhibit spatial variation in dune types, spacing, size, and orientation. One cause of spatial variation is the changing supply of sand (sand available for dune construction) along the sand transport pathway. Within the northern portion of the Rub' al Khali of Arabia there are distinct areas of contrasting dune types with relatively sharp boundaries between them (Figure 8.17). The abundant supply of sand from the north has led to the formation of closely spaced, small dunes in the north, which give way in the east and south to more widely spaced large dunes over a non-sand substrate to the east and south (Glennie and Singhvi 2002). As the sand supply decreases, the dominant dune type changes from transverse crescentic ridges to barchanoid ridges and barchans in the south and longitudinal dunes in the east (Figure 8.17), reflecting differences in the wind climate.

The reverse situation occurs in the central western Taklamakan Desert (Figure 8.18), where the bedrock Mazartag Ridge blocks the transport of sand from north to south, causing it to accumulate on the upwind side of the ridge (Sun et al. 2009). As sand supply increases, towards the ridge in this area, the dune morphology changes from longitudinal ridges to transverse crescentic ridges (Dong et al 2000), in agreement with other observations (Wasson and Hyde 1983). Downwind of the ridge, where sand is scarce, longitudinal dunes are re-established.

8.8.4 Climatic Gradients

Moisture gradients have the potential to affect dune processes and morphologies within sand seas (see also Section 1.2.1). In Argentina (see Figure 8.6), there is a transition from parabolic dunes and degraded linear dunes in the Pampas to areas dominated by sharper linear dunes, with fewer parabolic dunes, in the arid Tunuyan and Telteca dunefields of the Cuyo, to complex megadunes in the arid Médanos Grandes dunefield (see Figure 8.14) (Tripaldi and Forman 2007). Comparable patterns of dune preservation according to activity and type can also be seen in other sand seas at the margins of the arid zone of all continents.

At the spatial scale of sand seas, there are also often changes in the wind climate (strength, direction, seasonality). Within the Kumtagh Desert of China, dune orientation and morphology change dramatically over the relatively short transport path (Figure 8.19). In the upwind northern parts of the desert, winds are more variable in direction but stronger (Dong et al. 2012) and the dunefield is dominated by longitudinal dunes. Downwind, towards the southern margin, the wind direction becomes much less variable and the sand drift potential also decreases and dunes transition into transverse crescentic ridges.

Within the Australian dunefields certain dune types have distributions matching their position within the continental scale wind climate and whorl of dunes (Hesse 2010). Near the centre of the whorl, where there is greater variability of wind direction, mound and network types dominate. In the outer parts of the whorl, where there is still marked seasonality of wind directions, longitudinal dunes dominate. There are very few areas of transverse dunes, associated with unidirectional winds in Australia's continental deserts, but parabolic dunefields are present in the somewhat more uniform southern wind belts and tropical easterly belt which are also better vegetated because of higher rainfall.

8.9 Conclusion

Our knowledge of the world's sand seas is still limited by widely variable levels of scientific investigation. In recent decades scientific efforts in China, North America, and southern Africa, for example, have increased markedly while investigation in northern Africa and central Asia has lagged. Because of the variety of topographic, geological, and

(a)

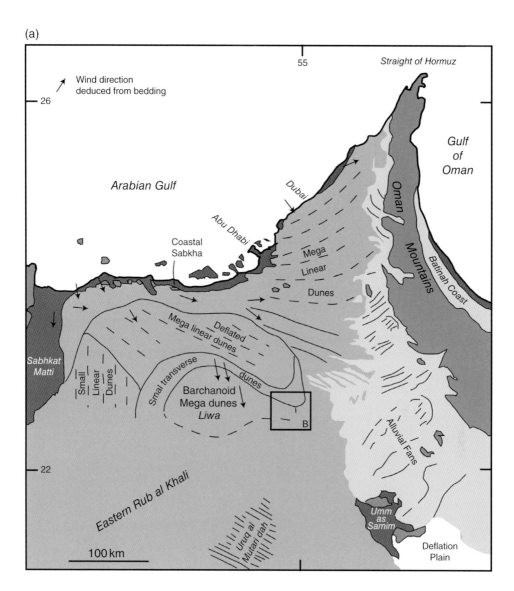

55　　　　　*Straight of Hormuz*

Wind direction
deduced from bedding

— 26

*Gulf
of
Oman*

Arabian Gulf

Dubai

Abu Dhabi

Coastal
Sabkha

Mega
Linear
Dunes

Oman

Batinah Coast

Mountains

*Sabhkat
Matti*

Mega linear dunes

Deflated

dunes

Small transverse

Small Linear Dunes

Barchanoid
Mega dunes
Liwa

B

— 22

Alluvial Fans

Eastern Rub al Khali

Uruq al
Mutaridah

*Umm
as
Samim*

Deflation
Plain

100 km

(b)

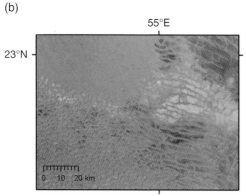

55°E

23°N

0　10　20 km

Figure 8.17 (a) Patterns of different dune types in the northern Rub' al Khali (modified from Glennie and Singhvi 2002), (b) patterns of dunes transitioning from closely spaces (upper left) to widely spaced linear dunes (right) and crescentic/transverse dunes (lower left). See insert for colour representation of this figure.

(a)

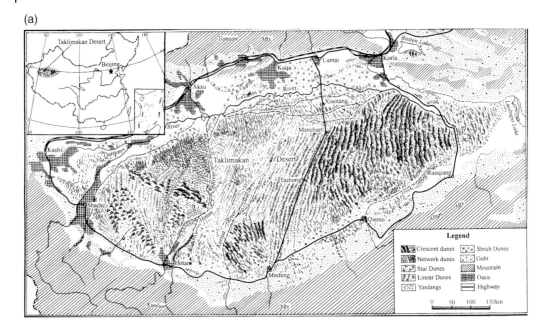

(b)

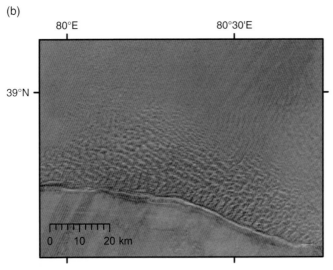

Figure 8.18 (a) Dune patterns in the Taklamakan Desert (Dong et al. 2000), (b) dune patterns upwind (top) and downwind (bottom) of a bedrock ridge change as sand thickness changes.

climatic settings of sand seas around the world, it is important to synthesise information from as wide a range of sand seas as possible to develop comprehensive models of their formation. For example, studies in the vegetated sand seas at the margins of the world's deserts have highlighted modes of dune behaviour and evolution that have considerably expanded the insights initially gained from the hyper-arid Sahara. Integrating dunefield development within the distributary fluvial system model for terrestrial basins, for example, increases both the power to generalise about the

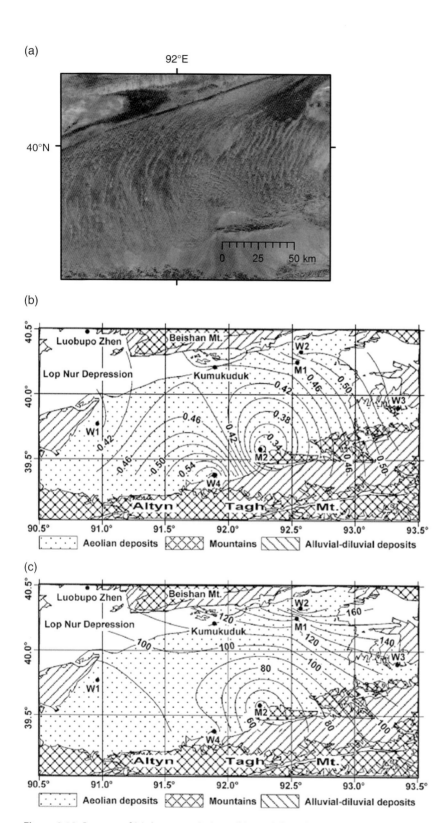

Figure 8.19 Patterns of (a) dune morphology, (b) variability of sand drift direction (RDP/DP), and (c) resultant drift potential in the Kumtagh sand sea of China (Dong et al. 2012).

development of sand seas, but also highlights those which do not appear to fit easily within that framework. The availability of high-resolution remote sensing images, digital elevation models, and global gridded climate data has considerably advanced our ability to understand the processes, rates, and history of formation of the world's sand seas.

Further Reading

The 2016 special issue of *Quaternary International* includes a suite of papers giving the latest regional/continental syntheses of sand sea evolution and late Quaternary chronology. These complement the recent provenance studies of dune sands in the Namib and Arabian sand seas by Garzanti and others. The classic works by McKee (1979), *A Study of Global Sand Seas*, and Wilson (1973), 'Ergs', retain relevance today.

References

Bagnold, R.A. (1941). *The Physics of Blown Sand and Desert Dunes*. London: Methuen.

Baitis, E., Kocurek, G., Smith, V. et al. (2014). Definition and origin of the dune-field pattern at White Sands, New Mexico. *Aeolian Research* 15: 269–287. doi: 10.1016/j.aeolia.2014.06.004.

Besler, H., Lancaster, N., Bristow, C.S. et al. (2013). Helga's dune: 40 years of dune dynamics in the Namib Desert. *Geografiska Annaler. Series A, Physical Geography* 95: 361–368. doi: 10.1111/geoa.12013.

Beveridge, C., Kocurek, G., Ewing, R.C. et al. (2006). Development of spatially diverse and complex dune-field patterns: Gran Desierto Dune Field, Sonora, Mexico. *Sedimentology* 53 (6): 1391–1409. doi: 10.1111/j.1365-3091.2006.00814.x.

Bowler, J.M. and Magee, J.W. (1978). Geomorphology of the Mallee region in semi-arid Northern Victoria and Western New South Wales. *Proceedings of the Royal Society of Victoria* 90 (1): 5–21.

Bullard, J.E. and Nash, D.J. (1998). Linear dune pattern variability in the vicinity of dry valleys in the Southwest Kalahari. *Geomorphology* 23 (1): 35–54. doi: 10.1016/S0169-555X(97)00090-1.

Bullard, J.E., Thomas, D.S.G., Livingstone, I., and Wiggs, G.F.S. (1997). Dunefield activity and interactions with climatic variability in the Southwest Kalahari Desert. *Earth Surface Processes and Landforms* 22 (2): 165–174. doi: 10.1002/(SICI)1096-9837(199702)22:2<165:AID-ESP687>3.0.CO;2-9.

Chen, X., Chappell, J., and Murray, A.S. (1995). High (ground) water levels and dune development in Central Australia: TL dates from gypsum and quartz dunes around Lake Lewis (Napperby), Northern Territory. *Geomorphology* 11 (4): 311–322. doi: 10.1016/0169-555X(94)00072-Y.

Cooke, R.U., Warren, A., and Goudie, A. (1993). *Desert Geomorphology*. London: UCL Press.

Darvill, C.M., Stokes, C.R., Bentley, M.J., and Lovell, H. (2014). A glacial geomorphological map of the southernmost ice lobes of Patagonia: the Bahia Inutil – San Sebastian, Magellan, Otway, Skyring and Rio Gallegos lobes. *Journal of Maps* 10: 500–520. doi: 10.1080/17445647.2014.89013.

Davidson, S.K., Hartley, A.J., Weissman, G.S. et al. (2013). Geomorphic elements on modern distributive fluvial systems. *Geomorphology* 180–181 (3): 82–95. doi: 10.1080/17445647.2015.1035346.

Dong, Z., Qian, G., Lu, P., and Hu, G. (2013). Investigation of the Sand Sea with the tallest dunes on Earth: China's BadainJaran Sand Sea. *Earth-Science Reviews* 120: 20–39. doi: 10.1016/j.earscirev.2013.02.003.

Dong, Z., Wang, T., and Wang, X. (2004). Geomorphology of the megadunes in the Badain Jaran Desert. *Geomorphology* 60 (1–2): 191–203. doi: 10.1016/j.earscirev.2013.02.003.

Dong, Z., Wang, X., and Chen, G. (2000). Monitoring sand dune advance in the Taklimakan Desert. *Geomorphology* 35 (3–4): 219–231. doi: 10.1016/S0169-555X(00)00039-8.

Dong, Z., Zhang, Z., Lv, P. et al. (2012). Analysis of the wind regime in context of dune geomorphology for the Kumtagh Desert, Northwest China. *Zeitschrift für Geomorphologie* 56 (4): 459–475. doi: org/10.1127/0372-8854/2012/0085.

El-Baz, F., Mainguet, M., and Robinson, C. (2000). Fluvio-aeolian dynamics in the north-eastern Sahara: the relationship between fluvial/aeolian systems and ground-water concentration. *Journal of Arid Environments* 44 (2): 173–183. doi: 10.1006/jare.1999.0581.

Fitzsimmons, K.E., Magee, J.W., and Amos, K.J. (2009). Characterisation of aeolian sediments from the Strzelecki and Tirari Deserts, Australia: implications for reconstructing palaeoenvironmental conditions. *Sedimentary Geology* 218 (1–4): 61–73. doi: 10.1006/jare.1999.0581.

Fryberger, S.G. and Dean, G. (1979). Dune forms and wind regime. In: *A Study of Global Sand Seas, Professional Paper 1052* (ed. E.D. McKee), 137–169. Washington, DC: USGS.

Fujioka, T., Chappell, J., Fifield, L.K., and Rhodes, E.J. (2009). Australian desert dune fields initiated with Pliocene-Pleistocene global climatic shift. *Geology* 37 (1): 51–54. doi: 10.1130/G25042A.1.

Garreaud, R.D. (2009). The Andes climate and weather. *Advances in Geosciences* 7: 1–9. doi: 10.5194/adgeo-22-3-2009.

Garzanti, E., Andò, S., Vezzoli, G. et al. (2012). Petrology of the Namib Sand Sea: long-distance transport and compositional variability in the wind-displaced Orange Delta. *Earth-Science Reviews* 112 (3–4): 173–189. doi: 10.1016/j.earscirev.2012.02.008.

Garzanti, E., Vermeesch, P., Ando, S. et al. (2013). Provenance and recycling of Arabian Desert Sand. *Earth-Science Reviews* 120: 1–19. doi: 10.1016/j.earscirev.2013.01.005.

Glennie, K.W. and Singhvi, A.K. (2002). Event stratigraphy, paleoenvironment and chronology of SE Arabian Deserts. *Quaternary Science Reviews* 21 (7): 853–869. doi: 10.1016/S0277-3791(01)00133-0.

Hack, J.T. (1941). Dunes of the Western Navajo country. *Geographical Review* 31 (2): 240–263.

Halfen, A.F., Lancaster, N., and Wolfe, S. (2016). Boundary conditions for aeolian activity in North American dune fields. *Quaternary International* 410: 75–95.

Hansen, E.C., Fisher, T.G., Arbogast, A.F., and Bateman, M.D. (2010). Geomorphic history of low-perched, transgressive dune complexes along the Southeastern shore of Lake Michigan. *Aeolian Research* 1 (3–4): 111–127. doi: 10.1016/j.aeolia.2009.08.001.

Hartley, A.J., Weissman, G.S., Nichols, G.J., and Warwick, G.L. (2010). Large distributive fluvial systems: characteristics, distribution, and controls on development. *Journal of Sedimentary Research* 80 (2): 167–183. doi: 10.2110/jsr.2010.01.

Hesse, P.P. (2010). The Australian desert dunefields: formation and evolution in an old, flat, dry continent. In: *Australian Landscapes*, vol. 346 (ed. P. Bishop and B. Pillans), 141–163. London: Geological Society, London, Special Publication. doi: 10.1144/SP346.9.

Hesse, P.P. (2011). Sticky dunes in a wet desert: formation, stabilisation and modification of the Australian desert dunefields. *Geomorphology* 134 (3–4): 309–325. doi: 10.1016/j.geomorph.2011.07.008.

Hesse, P.P. (2016). How do longitudinal dunes respond to climate forcing? Insights from 25 years of luminescence dating of the Australian desert dunefields. *Quaternary International* 410: 11–29. doi: 10.1016/j.quaint.2014.02.020.

Hesse, P., Lancaster, N. and Telfer, M.W. (2015). Digital mapping of the extent of global dune systems, EGU General

Assembly Conference Abstracts 17: 3638. Vienna: EGU.

Hollands, C.B., Nanson, G.C., Jones, B.G. et al. (2006). Aeolian-fluvial interaction: evidence for late quaternary channel change and wind-rift linear dune formation in the Northwestern Simpson Desert, Australia. *Quaternary Science Reviews* 25 (1–2): 142–162. doi: 10.1016/j.quascirev. 2005.02.007.

Hugenholtz, C.H. and Wolfe, S.A. (2006). Biogeomorphic model of dunefield activation and stabilization on the Northern Great Plains. *Geomorphology* 70: 53–70. doi: 10.1016/j.geomorph.2005.03.011.

Jerolmack, D.J. and Brzinski, T.A. (2010). Equivalence of abrupt grainsize transitions in alluvial rivers and Eolian Sand Seas: a hypothesis. *Geology* 38 (8): 719–722. doi: 10.1130/G30922.1.

Kalma, J.D., Speight, J.G., and Wasson, R.J. (1988). Potential wind erosion in Australia: a continental perspective. *Journal of Climatology* 8 (4): 411–428. doi: 10.1002/ joc.3370080408.

Kasse, C. (2002). Sandy aeolian deposits and environments and their relation to climate during the last glacial maximum and Late glacial in northwest and Central Europe. *Progress in Physical Geography* 26 (4): 507–532. doi: 10.1191/0309133302pp350ra.

Kocurek, G. (1998). Aeolian system response to external forcing factors, a sequence stratigraphic view of the Saharan region. In: *Quaternary Deserts and Climatic Change* (ed. A.S. Alsharhan, K.W. Glennie, L.G. Whittle and C.G.S.C. Kendall), 327–337. Rotterdam: Balkema.

Lancaster, N. (1988). The development of linear dunes in the Southwestern Kalahari, Southern Africa. *Journal of Arid Environments* 14 (3): 233–244. doi: 10.1016/0037-0738(88)90090-5.

Lancaster, N., Kocurek, G., Singhvi, A. et al. (2002). Late Pleistocene and Holocene dune activity and wind regimes in the Western Sahara Desert of Mauritania. *Geology* 30 (11): 991–994. doi: 10.1130/0091-7613-31.1.e18.

Latrubesse, E.M., Stevaux, J.C., Cremon, E.H. et al. (2012). Late Quaternary megafans, fans and fluvio-aeolian interactions in the Bolivian Chaco, tropical South America. *Palaeogeography, Palaeoclimatology, Palaeoecology* 356–357: 75–88. doi: 10.1016/j.palaeo. 2012.04.003.

Livingstone, I. (2013). Aeolian geomorphology of the Namib Sand Sea. *Journal of Arid Environments* 93: 30–39. doi: 10.1016/ j.jaridenv.2012.08.005.

Livingstone, I. and Thomas, D.S.G. (1993). Modes of linear dune activity and their palaeoenvironmental significance: an evaluation with reference to Southern African examples. In: *The Dynamics and Environmental Context of Aeolian Sedimentary Systems* (ed. K. Pye), 91–102. London: Geological Society. doi: 10.1144/ GSL.SP.1993.072.01.10.

Livingstone, I. and Warren, A. (1996). *Aeolian Geomorphology*. Harlow: Addison-Wesley Longman Ltd.

Lomax, J., Hilgers, A., and Radtke, U. (2011). Palaeoenvironmental change recorded in the palaeodunefields of the Western Murray Basin, South Australia – new data from single grain OSL-dating. *Quaternary Science Reviews* 30 (5–6): 723–736. doi: 10.1016/ j.quascirev.2010.12.015.

Lorenz, R.D., Gasmi, N., Radebaugh, J. et al. (2013). Dunes on planet Tatooine: observation of barchan migration at the *Star Wars* film set in Tunisia. *Geomorphology* 201: 264–271. doi: 10.1016/j. geomorph.2013.06.026.

Mainguet, M., Dumay, F., OuldElhacen, M.L., and Georges, J.-C. (2008). Changement de l'état de surface des ergs au nord de Nouakchott (1954–2000): conséquences sur la désertification et l'ensablement de la capitale. *Géomorphologie: Relief, Processus, Environnement* 14 (3): 143–152.

Maman, S., Blumberg, D.G., Tsoar, H. et al. (2011). The Central Asian ergs: a study by remote sensing and geographic information systems. *Aeolian Research* 3 (3): 353–366. doi: 10.1016/j. aeolia.2011.09.001.

McKee, E.D. (ed.), (1979). A Study of Global Sand Seas. Geological Survey Professional Paper, 1052. Washington, DC: USGS.

McLaren, S., Wallace, M.W., Gallagher, S.J. et al. (2011). Palaeogeographic, climatic and tectonic change in Southeastern Australia: the late Neogene evolution of the Murray Basin. *Quaternary Science Reviews* 30 (9–10): 1086–1111. doi: 10.1016/j.quascirev. 2010.12.016.

Muhs, D.R. (2004). Mineralogical maturity in dunefields of North America, Africa and Australia. *Geomorphology* 59 (1–4): 247–269. doi: 10.1016/j.geomorph.2003.07.020.

Muhs, D.R., Reynolds, R.L., Been, J., and Skipp, G. (2003). Eolian sand transport pathways in the Southwestern United States: importance of the Colorado River and local sources. *Quaternary International* 104: 3–18. doi: 10.1016/S1040-6182(02)00131-3.

Pell, S.D., Chivas, A.R., and Williams, I.S. (1999). Great Victoria Desert: development and sand provenance. *Australian Journal of Earth Sciences* 46 (2): 289–299. doi: 10.1046/j.1440-0952.1999.00699.x.

Pell, S.D., Chivas, A.R., and Williams, I.S. (2000). The Simpson, Strzelecki and Tirari Deserts: development and sand provenance. *Sedimentary Geology* 130 (1–2): 107–130. doi: 10.1016/S0037-0738(99)00108-6.

Pell, S.D., Chivas, A.R., and Williams, I.S. (2001). The Mallee Dunefield: development and sand provenance. *Journal of Arid Environments* 48 (2): 149–170. doi: 10.1006/ jare.2000.0751.

Portenga, E.W. and Bierman, P. (2011). Understanding Earth's eroding surface with [10]Be. *GSA Today* 21 (8): 4–10. doi: 10.1130/ G111A.1.

Quade, J., Chivas, A.R., and McCulloch, M.T. (1995). Strontium and carbon isotope tracers and the origins of soil carbonate in South Australia and Victoria. *Palaeogeography, Palaeoclimatology, Palaeoecology* 113 (1): 103–117. doi: 10.1016/0031-0182(95)00065-T.

Ralph, T.J. and Hesse, P.P. (2010). Downstream hydrogeomorphic changes along the Macquarie River, Southeastern Australia,

leading to channel breakdown and floodplain wetlands. *Geomorphology* 118 (1–2): 48–64. doi: 10.1016/j.geomorph. 2009.12.007.

Reffet, E., Courrech du Pont, S., Hersen, P., and Douady, S. (2010). Formation and stability of transverse and longitudinal sand dunes. *Geology* 38 (6): 491–494. doi: 10.1130/ G30894.1.

Roskin, J., Porat, N., Tsoar, H. et al. (2011). Age, origin and climatic controls on vegetated linear dunes in the Northwestern Negev Desert (Israel). *Quaternary Science Reviews* 30: 1649–1674.

Singhvi, A.K. and Kar, A. (2004). The aeolian sedimentation record of the Thar Desert. *Proceedings of the Indian Academy of Sciences (Earth and Planetary Sciences)* 113: 371–401. doi: 1007/BF02716733.

Stone, A.E.C. (2013). Age and dynamics of the Namib Sand Sea: a review of chronological evidence and possible landscape development models. *Journal of African Earth Sciences* 82: 70–87. doi: 10.1016/ j.jafrearsci.2013.02.003.

Stone, A.E.C. and Thomas, D.S.G. (2008). Linear dune accumulation chronologies from the Southwest Kalahari, Namibia: challenges of reconstructing late Quaternary palaeoenvironments from aeolian landforms. *Quaternary Science Reviews* 27 (17–18): 1667–1681. doi: 10.1016/ j.jaridenv.2012.01.009.

Sun, J., Zhang, Z., and Zhang, L. (2009). New evidence on the age of the Taklimakan Desert. *Geology* 37 (2): 159–162. doi: 10.1130/G25338A.1.

Suslov, S.P. (1961). *Physical Geography of Asiatic Russia*. San Francisco: WH Freeman and Company.

Telfer, M.W. (2011). Growth by extension, and reworking, of a South-western Kalahari linear dune. *Earth Surface Processes and Landforms* 36 (8): 1125–1135. doi: 10.1002/ esp.2140.

Telfer, M.W. and Thomas, D.S.G. (2007). Late Quaternary linear dune accumulation and chronostratigraphy of the Southwestern Kalahari: implications for aeolian palaeoclimatic reconstructions and predictions of future dynamics. *Quaternary Science Reviews* 26 (19–21): 2617–2630. doi: 0.1007/BF02716733.

Thomas, D.S.G. and Burrough, S.L. (2016). Luminescence-based dune chronologies in Southern Africa: analysis and interpretation of dune database records across the subcontinent. *Quaternary International* 410 (Part B): 30–45. doi: 10.1016/j.quaint. 2013.09.008.

Tooth, S. (1999). Downstream changes in floodplain character on the Northern Plains of arid Central Australia. In: *Fluvial Sedimentology VI. Special Publications of the International Association of Sedimentologists*, vol. 28 (ed. N.D. Smith and J. Rogers), 93–112. Oxford: Blackwell. doi: 10.1002/9781444304213.ch8.

Tripaldi, A. and Forman, S.L. (2007). Geomorphology and chronology of late Quaternary dune fields of Western Argentina. *Palaeogeography, Palaeoclimatology, Palaeoecology* 251 (2): 300–320. doi: 10.1016/j.palaeo.2007.04.007.

Tsoar, H. and Moller, J.T. (1986). The role of vegetation in the formation of linear sand dunes. In: *Aeolian Geomorphology* (ed. W.G. Nickling), 75–95. Boston: Allen and Unwin.

Wang, X., Dong, Z., Zhang, J., and Chen, G. (2002). Geomorphology of sand dunes in the Northeast Taklimakan Desert. *Geomorphology* 42 (3–4): 183–195. doi: 10.1016/S0169-555X(01)00085-X.

Warren, A. (1972). Observations on dunes and bi-modal sands in the Ténéré Desert. *Sedimentology* 19 (1): 37–44. doi: 10.1111/j.1365-3091.1972.tb00234.x.

Wasson, R.J. and Hyde, R. (1983). Factors determining desert dune type. *Nature* 304 (5924): 337–339. doi: 10.1038/304337a0.

Weissman, G.S., Hartley, A.J., Nichols, G.J. et al. (2010). Fluvial form in modern continental sedimentary basins: distributive fluvial systems. *Geology* 38: 39–42. doi: 10.1130/G30242.1.

Wiggs, G.F.S., Thomas, D.S.G., Bullard, J.E., and Livingstone, I. (1995). Dune mobility and vegetation cover in the Southwest Kalahari Desert. *Earth Surface Processes and Landforms* 20 (6): 515–529. doi: 10.1002/esp.3290200604.

Wilson, I.G. (1971). Desert sandflow basins and a model for the development of ergs. *The Geographical Journal* 137 (2): 180–199. doi: 10.2307/1796738.

Wilson, I.G. (1973). Ergs. *Sedimentary Geology* 10 (1): 77–106. doi: 10.1016/0037-0738(73)90001-8.

Wolfe, S.A. (2007). High latitude dune fields. In: *Encyclopedia of Quaternary Science*, vol. 2 (ed. S.A. Elias), 599–607. Amsterdam: Elsevier. doi: 10.1016/B978-0-12-409548-9.09433-1.

Yang, X., Li, H., and Conacher, A. (2012). Large-scale controls on the development of Sand Seas in Northern China. *Quaternary International* 250: 74–83. doi: 10.1016/j.quaint.2011.03.052.

Zeeberg, J. (1998). The European sand belt in Eastern Europe – and comparison of late glacial dune orientation with GCM simulation results. *Boreas* 27 (2): 127–139. doi: 10.1111/j.1502-3885.1998.tb00873.x.

9

Dune Sediments

Charles Bristow[1] and Ian Livingstone[2]

[1] *Birkbeck, University of London, London, UK*
[2] *University of Northampton, Northampton, UK*

9.1   Introduction

For fundamental physical reasons explained in Chapter 2, wind is very selective about the size of material that it is able to transport. As a fluid, air lacks the density to carry anything much larger than sand-sized quartz grains ((0.063-2 mm) in diameter) although it can transport larger particles of lower density. In addition, where the air becomes dense or moves at high speed, such as in the cold, windy conditions found in high mountains or near the poles, coarser particles can be moved. The finer material moved by wind – usually termed 'dust' by aeolian geomorphologists – is transported in suspension (see Chapter 4). Where it is deposited, it forms mantles of dust which may be sufficiently thick to become loess deposits (see Chapter 5). Sand-sized material, by contrast, is moved by mechanisms of saltation and creep which means that it stays close to the Earth's surface and creates bedforms – the ripples and dunes described in Chapters 6 and 7. This chapter is concerned with the sedimentary characteristics of those deposits.

Dune sediments hold very considerable amounts of information about dune processes: what we know about the geomorphological processes operating on contemporary dunes can help sedimentologists infer the environments in which the ancient wind-laid deposits found in the geological record were created. There are quite substantial accumulations of aeolian sandstones around the planet and these can only really be successfully used to infer palaeoenvironments if the formative processes are understood. However, there is still much to learn about the sedimentary characteristics of contemporary or recently active dunes, and most glaringly, although understandably, there is only a handful of studies of the internal structure of active dunes. This gap in our knowledge, though, is being redressed by the use of ground-penetrating radar which can image layers within dunes and is especially valuable when combined with luminescence dating of dune deposits.

Dune deposits reflect the huge diversity of dune forms found on Earth (see Chapters 6 and 7) and beyond (see Chapter 11). Nonetheless, many dune sands do share some common properties of mineralogical composition, shape, grain surface textures, colour, and size, as well as characteristics of the strata found in the dune stratigraphy. This chapter provides an overview of the characteristics of dune sediments.

Aeolian Geomorphology: A New Introduction, First Edition. Edited by Ian Livingstone and Andrew Warren.
© 2019 John Wiley & Sons Ltd. Published 2019 by John Wiley & Sons Ltd.

9.2 Dune Sands: Mineralogy, Shape, Surface Textures, and Colour

Dune sediments are primarily composed of sand-sized particles, in the range between 0.063 (63 μm) and 2 mm although the sediments may also include coarser 'granules' (2–4 mm) and finer silt and clay-sized particles. As Livingstone and Warren (1996) pointed out, the nineteenth-century view was that dune sands were golden yellow, very well rounded, 'millet seed' quartz grains, but subsequent work has shown that this is far from true in many cases.

9.2.1 Mineralogy

There was, however, some validity to the caricature: the vast majority of dune sands are indeed composed of quartz, as is a high proportion of coarse dust (including the material that makes up most loess). This is partly because quartz is abundant in the silicate rocks that weather to produce material that is moved by earth surface processes but also because quartz is physically hard and chemically stable. It is not soluble in the *Eh* and *pH* environment of most of the terrestrial surface, has a crystal lattice that is very difficult for hydrogen ions to penetrate, and is hard enough to resist rapid wearing down by physical processes.

The mineralogical composition of dune sands is primarily controlled by their source, and the sources of dune sands are many and varied. Sometimes those sources are very local to the dunes, but often the material has travelled very long distances. Either way, quartz can only dominate if quartz is what is supplied for the wind to transport. Other than quartz, dune-building materials include gypsum, diatoms, clays, calcium carbonate, and volcanic ash. Many of these non-quartzose sands behave much as quartz sand in transport, so that the dunes they form have much the same shapes and behave in much the same way, until they encounter

moisture. Gypsum sand is an example. The most extensive field of gypsum sand dunes is at White Sands in New Mexico (32°51′30N, 106°17′00W; Kocurek et al. 2007). Other gypsum dunes have been reported from the Great Salt Lake Desert in Utah (Jones 1953), south-central Tunisia (Drake 1997) (parabolic dunes composed of sand rich in gypsum are seen at 33°50′N, 8°54′E), northern Australia (Chen et al. 1995), Bolivia, and Saudi Arabia. Wetting dissolves some gypsum and can create cohesive bodies of sand (Kocurek et al. 2007). The degree of cementation varies with gypsum content and the frequency and degree of wetting; thorough wetting will immobilise gypsum dunes.

Diatomite (formed from diatoms, a type of phytoplankton found in water bodies) dominates the composition of sand-sized particles in some of the dunes in the central Bodélé Depression of northern Chad. The source is lacustrine, the lake having desiccated a few thousand years ago (Bristow et al. 2009). The low-density diatomite erodes into pellets and flakes up to a cm in diameter, which tumble about in high winds at up to half a metre above the surface. The fierce wind in the Bodélé breaks down the flakes into smaller and smaller particles and eventually to dust. Despite their large size, the diatomite particles have a very low density and behave much as quartz sand in big, 30 m high barchan dunes (Figure 9.1). Grey barchan dunes composed of diatomite pellets around 16°55′00N, 18°30′44E are thought to be the fastest moving dunes on Earth (Vermeesch and Drake 2008), due to the combination of strong winds and low-density particles.

It should be impossible for the wind to build dunes from clay. Clay particles are much more cohesive than sand so individual clay particles cannot be as easily entrained as sand (see Chapter 2). Yet many years ago Coffey (1909) correctly deduced that a mound of clay on the Gulf Coast of Texas was a dune, and that it had been built with sand-sized pellets of aggregated clay. The pellets had been entrained and moved by the wind as if they had been sand. Because these

Figure 9.1 Sand dunes in: (a) Iceland, (b) Chad, (c) Namibia, and (d) USA. While most sand dunes are composed of quartz sand, this is not always the case and other granular materials can be transported by the wind to form dunes including volcanic rock fragments in Iceland (a), pieces of diatomite eroded form a desiccated lake bed in Chad (b). The sand dunes in Namibia (c) are primarily composed of quartz sand. The dark patches are concentrations of heavy minerals including pyroxene and garnet, both of which have been shown to survive aeolian transport (Garzanti et al. 2015). The dunes at White Sands National Monument (d) are composed of gypsum.

pellets rapidly disaggregate in mechanical bombardment, they either form a dune close to their source, or are dispersed as dust. Percolating rainfall later disaggregates the pellets that have collected in the dune, which itself then becomes a coherent, static mass. Many more clay dunes have been identified since Coffey's insight.

Clay pellets are created in two situations: (i) from alluvium, especially shortly after flooding; and (ii) in ephemeral lakes, some coastal. Most clay dunes derive their pellets from shallow, ephemeral lakes, of which there are many in semi-arid environments. Some types of clay mineral, salts, plants, or algae bind the pellets, abetted by seasonal wetting and drying (Bowler 1973). Clay dunes, therefore, can only form where there is: (i) seasonal flooding (or as at Coffey's site on the Texas Gulf coast, tidal flooding), to bring the clay into the basin; (ii) a distinct dry season in which the pellets desiccate and aggregate; and (iii) a wind that is strong enough to entrain them, and constant enough to carry them to a dune. Infiltrating rainwater in the next wet season disintegrates the pellets in the dunes and welds it into a coherent mass (Dare-Edwards 1983). In Australia, infiltration of clay from dust as well as aggregates helps to stabilise dunes (e.g. Greene and Nettleton 1996, Hesse 2011).

Clay content has been observed to increase with depth (Churchward 1961) and dune age (Bristow et al. 2007a), suggesting that clays slowly infiltrate down through the dune (illuviation). The clay content of clay dunes may reach 77% (Bowler 1973). Some clay dunes, especially in coastal settings, as in Senegal, are very saline (Mohamedou et al. 1999). The cross-sectional shapes of clay dunes differ from those of sand dunes: they have much lower angle slopes, and steeper windward than lee faces. Most lake-marginal 'lunettes' are clay dunes. Many clay- and salt-cemented dunes, like those in the Mojave, are eroded into yardangs after cementation (Blackwelder 1934) or, because their salt content discourages vegetation, become deeply gullied by runoff.

Many clay lunettes were formed in more than one phase; many in the early Holocene or late Pleistocene (Goudie and Thomas 1985). Bowler (1973), building on this evidence, and believing that clay dune formation had a longer than seasonal rhythm, proposed that clay dunes were a response to a particular kind of climatic history, such as the transition from a wet to a dry climate, as has happened repeatedly in these environments.

Other rock-forming minerals are more susceptible to chemical weathering and only make up dune sands close to sources. For example, some coastal dunes, close to offshore deposits derived from shelly marine fauna, have high proportions of calcite and aragonite (calcium carbonate) (Brooke 2001). Occasionally deposits of volcanic sand act as further sources of aeolian sand (Edgett and Lancaster 1993).

9.2.2 Shape

The caricature suggests that grains moved by the wind will have a more or less spherical shape, but the reality turns out to be far more complex. Before even considering the reality, however, there is very considerable debate about how to characterise the shape (sometimes termed 'morphology' or 'form') of sedimentary particles. Blott and Pye (2008) provided a comprehensive review of the variety of descriptors and their dimensions, and their paper provided a schema for describing particle shape. Computer-based image analysis has greatly improved the facility to analyse shape data, although frequently those data still tend to be two-dimensional.

Our expectation might be that because sand grains transported by the wind to form dunes have presumably undergone considerable attrition as a result of collisions between grains, they will tend to be rather spherical; that is, their longest axis will not be much different in length from their shortest. Yet Stapor et al. (1983) found that low-sphericity particles had been selected for transport and Willetts et al. (1982) found that less spherical particles moved at slightly higher rates than more spherical ones when the shear stress was low.

We might also expect higher roundness in aeolian particles; that is, fewer sharp edges on the grain, sometime also expressed as 'angularity', because sharp edges and points get rounded off during transport. Yet Goudie and Watson's study of 21 600 particles from a range of sand dune types found plenty of low-sphericity, angular particles and this led them to conclude that the sphericity and roundness of aeolian sand grains had been overplayed (Goudie and Watson 1981). Garzanti et al. (2015) found that an increase in roundness was one feature that was associated with aeolian transport, indicating that abrasion is much more effective during sediment transport in air.

There are several possible explanations for why all aeolian sand grains are not well rounded. One is that, of course, the shape of the grain may be controlled by its initial form, particularly where the grain has not travelled very far from its source. If it started out elongate or angular, there may not have been enough collisions with other bodies to modify its form. Related to this, some dune types provide more opportunity for grain-on-grain bombardment than others. In transverse dunes (including barchans) which migrate, there is regular turnover of grains so that grains frequently find themselves in motion, whereas in dune types such as linear or star, only the grains at or close to the surface are in regular motion.

9.2.3 Surface Textures

Related to shape, observations of quartz grains using scanning electron microscopes (SEM) reveal microscopic surface features on sand grain surfaces. Since initial studies in the 1960s there have been many attempts to link the quartz grain surface features to the environments in which those grains have been transported and deposited (e.g. Krinsley and Donahue 1968; Krinsley and Doornkamp 1973). According to Krinsley and Trusty (1985), there are five types of surface features that are characteristic of aeolian desert sands, including: grain roundness; upturned plates

attributed to abrasion fatigue; elongate depressions; smooth surfaces: and arcuate, circular or polygonal fractures. Krinsley and Trusty (1985) noted differences between the abundance of these features between coastal aeolian sands, desert aeolian sands, and periglacial aeolian sands. It is also worth noting that surface textures can be masked by surface coatings and diagentic effects and, as a consequence, quartz grains are often subject to pre-treatment prior to examination (see Vos et al. 2014). In their review, Vos et al. (2014) included a case study based on sand samples from the Dune du Pyla in southern France where they compared surface textures on 25 quartz grains from a beach sand with those from the crest of the dune (Figure 9.2). The results showed a wide range of surface features common to both sample A from the dune crest and sample B from the beach. In their study there were only two features that were unique to the dune sand (1 and 10) which are angular outline and flat cleavage

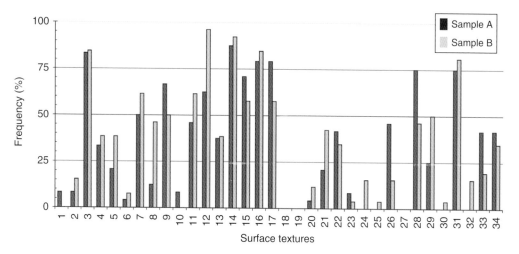

Figure 9.2 A comparison of quartz grain surface features from the crest of a coastal sand dune (Dune du Pyla in south-eastern France) and the adjacent beach taken from Vos et al. (2014). The dune-sand sample A (dark grey) is compared with the beach sand sample B (light grey). Surface features that are supposed to be associated with aeolian sands include: rounded outlines (3), graded arcs (11), upturned plates (14), crescentic percussion marks (15), bulbous edges (16), and elongated depressions (not included). Comparison of the frequency of surface textures 3, 11, 14, 15, and 16 shows that these microtextural features are common in both the dune sand sample A and the beach sand sample B. Of these, it appears that only the crescentic percussion marks are more abundant in the dune sand, while rounded outlines, graded arcs, bulbous edges, and upturned plates are slightly more frequent in the beach sand. The results suggest either extensive reworking of sand between the two environments, or that it is not possible to distinguish aeolian sand using these microtextural surface features.

surfaces, neither of which are listed in the common features of dune sands although they have a very low frequency (Figure 9.2). It is interesting to note that the features considered to be characteristic of desert aeolian sands by Krinsley and Trusty (1985) – rounded shape (3), upturned plates (14), elongate depressions (31), bulbous edges (16) and arcuate/circular/polygonal fractures (34) – were common in both samples (Figure 9.2), although we should note that Vos et al. (2014) did not include smooth surfaces in their analysis. The results suggested that grain surface textures should be used with caution: even if some surface features diagnostic of a given environment are present, there is often considerable reworking of sand as it moves through a series of environments which results in complex histories and inherited surface features.

9.2.4   Colour

Desert dune sands are often coloured pale yellow, buff, or shades of red/brown. This colour may, most straightforwardly, be attributed to the colour of the detrital grains, but is also frequently the result of surface coatings on the grains: most quartz sand grains themselves lack strong body colour and tend to be pale beige (Walker 1979). Sometimes, the intensity of the colour is inversely proportional to grain size, with fine sands being darker than coarse sands, potentially due to the surface area to volume ratio. The 'rusty' red/brown coating on desert sand grains is a variety of desert varnish that includes manganese, iron, and silica coatings. Studies of dune sands using a binocular microscope showed thin surface coatings of iron oxides with thicker concentrations within pits or indentations (Walker 1979). Various theories have been proposed for the source of the surface coatings, including clay minerals (Potter and Rossman 1977), or microbial oxidation of iron and manganese from dust (Dorn and Oberlander 1981).

Other changes in the colour of dune sands can be due to the presence of heavy minerals, in particular, pyroxene and garnet, which can become concentrated on the surface of dunes where lighter-coloured and less dense grains of quartz sand have been removed by deflation, leaving a lag of dense, darker-coloured grains. Garnet often has a dark reddish-purple colour and pyroxene appears black, giving a dark shading to dune surfaces where lighter-coloured minerals have been removed by the wind (Figure 9.1c). In contrast, the diatomite dunes of the Bodélé basin in Chad have pale-coloured shoulders due to concentrations of quartz sand after deflation of lower-density darker-coloured diatomite (Figure 9.1b). Samples of dune sands from deserts around the world vary in colour depending primarily on the detrital mineralogy and source rocks. Sample a (Figures 9.3 & 9.4) from the Grand Erg Occidental in Algeria is a mature quartz sand with a typical red colour attributed to iron oxide coatings (H7.5yr 7/6 on a Munsell Soil Colour Chart), sample b (Figures 9.3 & 9.4) from White Sands National Monument is off-white (H10yr 8/1) because it is composed of white and transparent gypsum

Figure 9.3 Colour photomicrographs of grains of dune sand from locations around the world demonstrating the wide range of sand colours and the variety of grains; the field of view in each photograph is 5 mm. Sample A, from the Grand Erg Occidental in Algeria, is a typical Sahara Desert dune sand with a red colour (H7.5yr 7/6 on a Munsell Soil Colour Chart); Sample B from White Sands National Monument is an off-white (H10yr 8/1), because it is composed of white and transparent gypsum crystals; Sample C from the Bodélé Depression in Chad (H2.5yr 6/2) is composed of grains of pale grey diatomite with a minor component of quartz; Sample D, from the crest of a linear dune in Namibia (H7.5yr 5/4), contains a mix of clear and red-coloured quartz grains and almost black pyroxene; Sample E from the Strzelecki desert in Australia (H10yr 7/4) is primarily composed of quartz; Sample F is from the Packard dunes in Victoria Valley, Antarctica, and contains a mixture of quartz, feldspar, pyroxene, and rock fragments, resulting in grey colours H10yr 4/1; Sample G is coarser-grained and contains more pyroxene, giving it a darker colour (H10yr 5/1), although that is difficult to see in the small number of grains illustrated; Sample H from Iceland is a very dark grey (H5yr 2.5/1) because it is primarily composed of basaltic rock fragments and volcanic glass (H). See insert for colour representation of this figure.

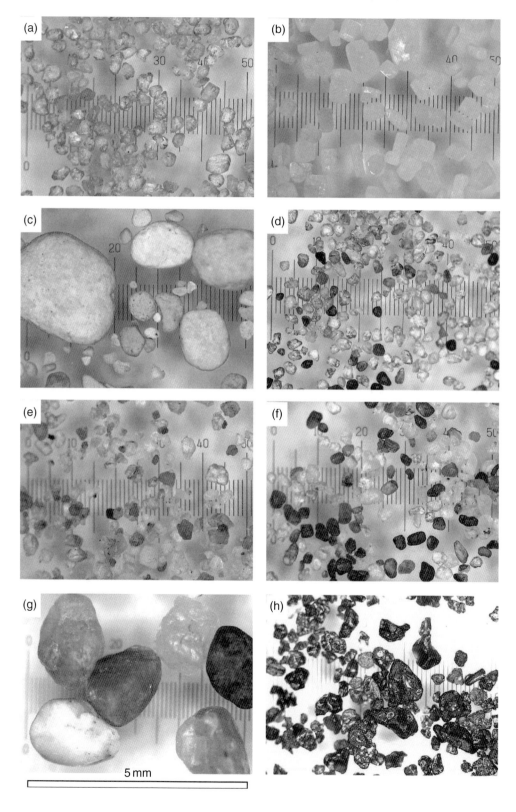

5 mm

crystals. Sample c (Figure 9.3) from the Bodélé Depression in Chad (H2.5yr 6/2) is composed of grains of pale grey diatomite with a minor component of quartz. The diatomite particles are very coarse grained sand (1-2 mm) and well rounded (Figures 9.3c & 9.4c). Sample d (Figures 9.3 & 9.4) from the crest of a linear dune in Namibia (H7.5yr 5/4) contains a mix of partially reddened quartz and almost black pyroxene, but is very well sorted despite the different minerals (Figures 9.3d & 9.4d). Sample e (Figures 9.3 & 9.4) from the Strzelecki desert in Australia (H10yr 7/4) is primarily composed of quartz but less mature than sample a, because the grains are more angular. Samples from the Dry Valleys of Antarctica (f and g) contain a mixture of quartz, feldspar, pyroxene and rock fragments resulting in grey colours H10yr 4/1 and H10yr 5/1 respectively. Sample h from Iceland is a very dark grey (H5yr 2.5/1) because it is primarily composed of basaltic rock fragments and volcanic glass (Figures 9.3 & 9.4h). This sand is less mature because it is not well sorted and some of the particles are subangular.

Dune colour changes across sand seas. This is often attributed to increases in surface coatings with the intensity of the colour being proportional to the time that the sand has remained within a desert environment. However, the patterns can also reflect climatic controls, particularly rainfall, the level of activity of the dune sands, the distance from the source of the sand, or some combination of these factors. For example, in the Namib sand sea, White et al. (2007) described a pattern of dune colour that intensified from the coast inland (west to east). Here the pattern has been attributed by various authors to increasing moisture from west to east, to the activity levels of different dune types and to the mixing of sands from different sources.

9.3 Particle-Size Characteristics

The sedimentary characteristic of dunes that has received greatest attention has been the size of the particles although, despite all this effort, it is possible to argue that little fundamental progress has been made. There is a range of controls on the size of aeolian sand particles. Often particles will have been weathered from rocks so that their size will be influenced by the size of crystals or grains in the parent rock as well as by the nature of the weathering and erosion. But far more fundamental is the capacity of the wind to transport the material provided. Material for dune-building is transported by the wind in two major modes: saltation where grains leap or make short jumps through the air before returning to the bed; and 'creep' where particles are rolled along the surface (these terms are explained in Chapter 2). This means that the sediments in aeolian dunes are generally restricted to a very small size range: too small and the material is carried off in suspension – and dunes provide a source of suspended material (see Chapter 4); too large and the wind is unable to entrain it. Occasionally larger particles with lower density, such as the diatomite grains in the Bodélé Depression (see Figures 9.3c and 9.4c), can be aerodynamically equivalent to smaller, denser grains of quartz sand. Generally, though, aeolian sand dune deposits are among the most size-selective sediments on Earth.

Methods for ascertaining particle size distributions have greatly improved as sieving and pipette analysis have been widely replaced by laboratory equipment using laser diffraction or electrostatic technology. These newer techniques offer the opportunity to provide particle size distributions at much higher resolution (typically hundreds of size classes rather than a handful) using much smaller samples but this change in method has created some difficulties of comparability between the techniques. While sieving will – at least in theory – allow particles to pass whose shortest axis is smaller than the sieve mesh size, the laser diffraction technique assumes that the sand grains are spherical and are composed of quartz. These differences in *a priori* assumptions mean that there is always some uncertainty in the results because some grains may not be composed of quartz and will not be spherical. In a comparison of sieve and laser granulometer

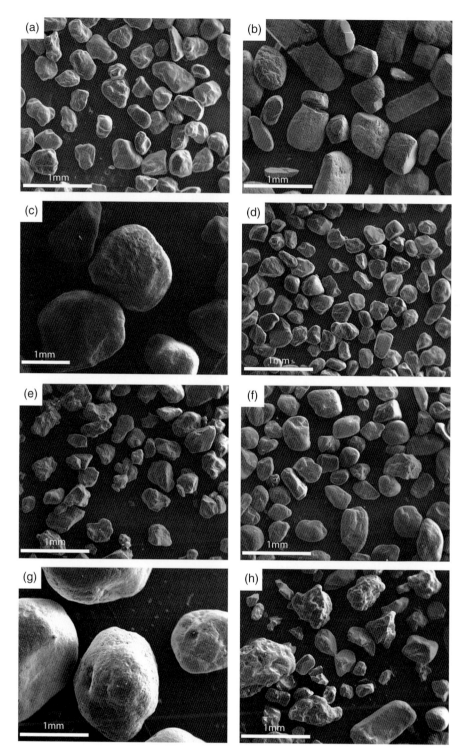

Figure 9.4 Scanning Electron Microscope (SEM) images of dunes sands with different grain size, shape and compositions, illustrating the variety that can be found within dune sands: (a) medium-grained quartz sand from a transverse dune, in the Grand Erg Occidental, Algeria; (b) coarse-grained gypsum sand from a transverse dune at White Sands National Monument, New Mexico, USA; (c) very coarse-grained diatomite, from a barchan dune in the Bodélé Depression, Chad; (d) medium- grained quartz sand from a linear dune in the Namib Sand Sea, Namibia, (e) less well-sorted fine-to-medium-grained quartz sand from a linear dune near Innamincka in the Strzelecki Desert, Australia; (f) medium-grained quartz and lithic sand from a reversing dune, Victoria Valley, Antarctica; (g) very coarse-grained sand, whaleback dune, Victoria Valley, Antarctica; (h) poorly sorted fine-to-coarse-grained volcanic lithic sand, from a dune near Kvensodull in Iceland. Note the variety of grainsizes from very coarse to fine-grained sand, demonstrating that the wind can transport particles of different dimensions. All the images are at the same scale, except for c.

results, Shi (1995) found that sieve results were consistently finer than those from the laser granulometer. Livingstone et al. (1999) concluded that the advantages of resolution and sample size outweighed any disadvantage from loss of comparability.

In early studies, Udden (1914) and Wentworth (1922) noted that the frequency distributions of particle sizes of natural sediments appeared to be log-normal, although it was not clear why this should be. It was also apparent that different geomorphological environments produced distinctive particle size distributions and that these distributions could be described using a suite of statistical measures (reviewed by Blott and Pye (2001)). The most commonly used of these statistical descriptors was developed by Folk and Ward (1957) who used Krumbein's (1938) log-transformation of the data (see Box 9.1). There has been considerable debate about whether log-normal transformations are the most appropriate and the use of log-hyperbolic plots and log-Laplace transformations has also been advocated, although log-normal transformations are still widely used (but see the discussion in Hartmann (2007)).

Box 9.1 Particle-Size Descriptors

The size definition of sand

For earth scientists, including geomorphologists, the term 'sand' denotes an exact particle size, regardless of mineralogy, which is usually taken to be between 0.063 and 2 mm. Sub-millimetre particle sizes are sometimes expressed as 1/1000s of mm, known as micrometres or microns (μm), so the lower size limit for sand is 63 μm. This size definition is drawn from a wider classification of particle sizes, based on pioneering work by Udden (1914) and Wentworth (1922). Particles smaller than sand (<63 μm) are termed silts and clays, often collectively called 'dust'. In aeolian geomorphology, this is generally the size fraction moved in suspension. Particles larger than sand (>2 mm) are termed granules; granules are only exceptionally moved by aeolian processes in terrestrial environments.

Describing particle size distributions

A variety of descriptors have been proposed for particle size distributions (reviewed by Blott and Pye 2001). Because of the wide recognition that many particle size distributions approximate a normal distribution when size is transformed logarithmically, many studies refer to particle size on a phi (ϕ) scale (Krumbein 1938) such that:

$$\text{particle size in phi } (\phi) = -\log_2 d$$

where d is the particle size in mm. On this scale, higher phi values represent finer particles. Sand-sized material lies between the limits 0ϕ (2 mm) and 4ϕ (63 μm).

Making use of the phi scale to logarithmically transform particle-size data, Folk and Ward (1957) proposed a group of descriptive statistical parameters for a particle-size distribution which is still widely used.

Graphic Mean $(M_Z) = (\phi16 + \phi50 + \phi84)/3$.

Inclusive Graphic Standard Deviation

$$(\sigma_I) = \frac{(\varphi84 - \varphi16)}{4} + \frac{(\varphi95 - \varphi5)}{6.6}.$$

Inclusive Graphic Skewness

$$(Sk_I) = \frac{\varphi16 + \varphi84 + 2\varphi50}{2(\varphi84 - \varphi16)} + \frac{\varphi5 + \varphi95 - 2\varphi50}{2(\varphi84 - \varphi16)}$$

Graphic Kurtosis $(K_G) = \dfrac{\varphi95 - \varphi5}{2.44(\varphi75 - \varphi25)}$

Blott and Pye (2001) provided a spreadsheet for calculating statistical descriptors from particle size distributions. However, there continues to be some debate (see main text) about the validity of various transformations of particle size data.

Particle size distributions of aeolian dune sands are generally finer, better sorted and less positively (coarse) skewed than fluvial, glacial or marine sediments. In a study of 506 aeolian sand samples (291 samples from coastal dunes, 175 from inland dunes, 40 from interdunes) analysed by Ahlbrandt (1979), the inland dune samples were characterised as moderately sorted to well-sorted fine to medium grained sands. Ahlbrandt noted an absence of very coarse (> 1.6 mm) and very fine particles (<0.1 mm) in the samples. This contrasts with the interdune samples that include both coarser granules (2–4 mm) and finer, clay sized particles, are often poorly sorted, and may have bimodal distributions. In a study of 1289 desert dune samples, Goudie et al. (1987) reported an average mean particle size of 200 µm while Ahlbrandt's samples had means in the range 175–225 µm. Attempts to discriminate aeolian sands from other sands such as beach sands and river sands on the basis of their grain size characteristics have met with mixed results (see Ahlbrandt 1979), which may in part be due to recycling of sands between one environment and another, as well as limitations imposed by the sediment available for transport by the wind. Recent results from cosmogenic dating suggest that sand can remain within desert environments for millions of years (Vermeesch et al. 2010), and most dune sand grains will have been entrained, transported, and deposited hundreds, thousands or even hundreds of thousands of times before they finally come to rest within an aeolian sediment.

However, because several processes or mechanisms are operating in combination in any geomorphological environment, most particle size distributions appear to represent a compound of several distributions related to those processes. In addition, sampling will often mix particles from different layers in laminated sands. Occasionally the constituent populations are sufficiently distinctive to allow recognition of more than one mode in the distribution. In the zibar – i.e. low-relief dunes without slipfaces – of the Ténéré Desert, it appears that larger grains protect smaller ones and intermediate sand sizes have been removed, giving a bimodal deposit with modes around 350 and 125 µm. Another example of a bimodal 'lag' deposit where an intermediate population has been removed was described from the deflationary corridors between linear dunes in Australia (Crocker 1946). For lots of sediments the overlap between modes may be much greater so that although two or more modes are present, it is difficult to distinguish the constituent populations. Leys et al. (2005) have shown that the MIX model of MacDonald and Green (1988) can distinguish component populations from aeolian sediments. It might be therefore possible in the future to distinguish not only lag populations but also different saltation populations on a single dune (Figure 9.5).

9.3.1 Particle Size Patterns

It appears that saltation of quartz grains tends preferentially to select particles around 125 µm (McTainsh et al. 2013). Because most source material is a mix of sands of this size with coarser material, dune sands are usually finer than surrounding interdune deposits. But on linear dunes, Livingstone and Warren (1996) were able to find examples of three

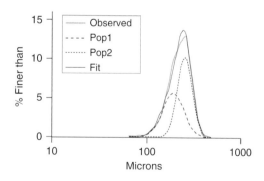

Figure 9.5 Particle size frequency distribution (PSD) of a dune surface sample from a linear dune in the Namib Desert. Two populations are resolved from the PSD by MIX analysis. *Source:* From McTainsh et al. (2013).

distinctive particle size patterns: where crest sands were *finer* than surrounding inter-dunes, where the crests were *coarser* and where there was *no difference*. By far the majority of linear dunes have finer crests, exemplified by the linear dunes of the Namib sand sea (Watson 1986; Livingstone 1987). On the other hand, in the Kalahari (Thomas 1988) and in Australia (Wasson 1983), crest samples are coarser than the surrounding interdunes. This probably reflects the rela-tively fine nature of the original material: saltation has moved the particles around 125 μm, leaving finer material behind in the interdunes. A *no difference* pattern was described by Sneh and Weissbrod (1983), by Livingstone et al. (1999) in the Kalahari, and by Wang et al. (2003) in the Taklimakan. In each of these cases, although mean size seemed similar, the sorting (represented by standard deviation) of the sands did demon-strate statistically significant variations between crests, dune flanks, and interdunes.

As well as varying across individual dunes, particle size also varies across sand seas. The principles driving the patterns are the same at this larger scale. The wind preferentially moves the fraction that is most readily trans-ported by saltation. So where the source material – often alluvial or beach deposits (see Chapter 8) – contains coarser material, the dune sands will become finer, better sorted, and less skewed as they move further into the dunefield. This is exemplified by the Namib Desert on the south-west coast of Africa. Here much of sand has been brought from the interior of southern Africa to the coast by the Orange River and has then been transported from the coast into the dunefield in a northerly direction (Garzanti et al. 2012). Lancaster (1989) was able to describe a broad pattern of decreasing particle size, better sorting, and less skewness from south to north in the Namib. He also described this pattern in the direction of transport in the south-west Kalahari (Lancaster 1986). Often, however, dunefields incorporate multiple episodes of dune building or the mixing of sands from more than one source. In these cases, there may be sufficient variability to

make it difficult to discern a clear pattern (e.g. Goudie et al. 1987; Buckley 1989).

9.4 Dune Structure

Sand dunes are piles of unconsolidated sand shaped by the wind. Dunes have a wide variety of forms (see Chapters 6 and 7), range in size from 1 to around 300 m in height and are mobile. As they migrate and accumulate, sand is eroded from the upwind (stoss) side of the dune, transported over the crest of the dune and deposited on the downwind (lee) side of the dune. The layers of sand that accumulate on the downwind (lee) side of a sand dune include grain-fall lamination, wind ripple lamination, and sand-flow strata (see below). Changes in wind strength and wind direction result in changes in dune morphology which may be represented by layers of sediment enclosed by erosion surfaces, termed bounding surfaces (see below). The packages of sediments, the small scale sedimentary structures they con-tain and their bounding surfaces record depo-sition and erosion on dunes and can be used to reconstruct dune morphodynamics.

9.4.1 Sedimentary Structures in Sand Dunes

Depositional units greater than 1 cm in thick-ness are known as beds; their boundaries are bedding planes. Layers that are less than 1 cm thick are termed laminae. Layers that are inclined at an angle to the bedding are called cross-strata: this includes cross-lamination (< 1 cm thickness) and cross-bedding (> 1 cm thickness); packages of cross-strata are called sets. Cross-stratification is very common in aeolian sediments and ubiquitous in dune deposits because it is formed by deposition on the downwind margin of migrating sand dunes and ripples. Layers of sand accumulat-ing on the top of the dune can be termed topsets, layers formed by sediment deposited on the leeside of the dune are termed fore-sets, and layers of sediment deposited at the base of the dune, beneath the slipface are termed toesets or bottomsets (Figure 9.6).

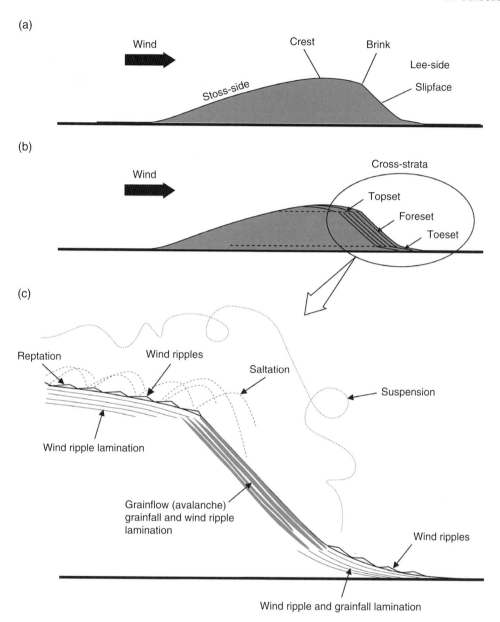

Figure 9.6 Sketch of a dune in cross-section, indicating its parts (a) upwind stoss-side, crest, brink and downwind lee-side with a slipface. (b) As a dune migrates, the surface morphology on the downwind side is preserved by layers of sediment (strata) arranged as topset, foreset, and toeset. (c) The processes that occur on a migrating sand dune, over which wind ripples migrate up the stoss-side, over the crest, and deliver sand to the slipface. In addition, saltating sand grains are projected beyond the brink and land on the upper part of the slipface as 'grainfall', while some are temporarily held in suspension and carried further downwind. Recirculating winds on the leeside of the dune form wind ripples at the toe of the foreset and sometimes on the slipface. Erosion on the stoss-side of the dune and repeated avalanches of sand down the slipface on the leeside cause the dune to migrate downwind.

Trenches through dunes reveal layers of sand that record successive positions of the dune surface preserved by deposition and usually picked out by changes in grainsize. Close examination reveals three common types of layering produced by different processes: sandflow strata, wind ripple lamination, and grainfall lamination (Hunter 1977). Deposition can occur from suspension, on sand surfaces that include particles from the reptation and saltation population as well as avalanches on the slipface of dunes. These depositional processes impose their own characteristics on the aeolian sediments, with fine-grained 'airfall' suspension deposits sometimes interbedded with inversely graded wind ripple lamination or avalanche deposits (see sedimentary structures below). Finally, diagenetic changes that occur after deposition which can include infiltration of fines, mixing by bioturbation can alter primary depositional textures.

- *Sandflow or avalanche strata:* When a pile of sand reaches a critical angle known as the angle of repose, usually between 32° and 34° for dry sand, the slope will fail and sand moves downslope as an avalanche, which is a mixture of sand grains in air where the interaction between the grains keeps the mass of sand moving downslope, often known as a slip face (Chapter 6). Sandflow or avalanche strata usually start high on the lee side of the dune and move towards the base of the slope where they wedge out against lower angle wind ripple and air-fall laminae (Figures 9.6 & 9.7). The deposits of sand avalanches, also known as grainflow or sandflow strata, form discrete tongues or inclined sheets of strata close to the angle of repose on the downwind side of a dune. The strata can be mm to a few cm in thickness, the sand is loosely packed and sometimes inversely graded, that is with coarser grains at the top.

Figure 9.7 An outcrop of aeolian sandstone with sedimentary structures produced by sand dunes in the Jurassic, Navajo Sandstone, Utah, USA. The pale layers are fine-grained wind ripple lamination, while the darker layers are coarser-grained avalanche cross-strata. The dip from left to right indicates that the dune migrated from left to right. The horizontal layer behind the seated figure is an interdune bounding surface. The repeated boomerang shaped packages of wind ripple lamination interbedded with the sets of cross-strata above the interdune surface, are interpreted as recording seasonal changes in wind direction in annual cycle (Hunter and Rubin 1983), probably associated with a monsoon climate (Loope et al. 2001). Beneath the seated figure the preserved dune deposit is dominated by toesets with wind ripple lamination.

- *Wind ripple lamination:* Wind ripples are very common on the surface of dunes and aeolian sand sheets and their deposits are widespread (see Chapter 2). Hunter (1977) defined two types of strata formed by wind ripples: climbing ripple lamination and translent strata. In climbing ripple lamination, ripple foresets can be preserved depending on the angle of climb, where the critical angle of climb is equal to the slope of the windward-side of the ripple. If the angle of climb is greater than the slope of the ripple stoss slope, then the ripple is said to be super-critically climbing. If the angle of climb is less than the slope of the stoss-side, then the ripple is said to be sub-critically climbing. When the ripple is sub-critically climbing and there is no preservation of foreset laminae, the strata are described as translent strata. Wind ripple lamination and translent strata produce layers with coarser particles above finer-grained particles and are said to be inversely graded.

- *Grain-fall lamination:* Grain-fall lamination is formed by the deposition of particles that are projected beyond the brink (pure airfall) as well as particles deposited from suspension in the separation zone on the lee side of a dune (Hunter 1977). Grainfall deposition is concentrated immediately downwind of the brink of the dune and contributes to the steepening of the slipface until it exceeds the angle of repose, resulting in slope failure and the formation of sandflow avalanches. Grainfall laminae are usually very thin, sometime less than 1 mm thickness, and drape the existing topography, covering wind ripple, and avalanche strata. In modern dune sands, airfall laminae tend to stand out in trenched sections because the fine particles are more cohesive and retain water by capillary pressure. In addition, airfall laminae may be differentiated by changes in colour. The interbedding of wind ripple translent strata and airfall laminae produces a characteristic paired lamination called pin-stripe lamination (Fryberger and Schenk 1988).

Although sand dunes are often characterised by large slipfaces with avalanches, the deposits preserved commonly include wind ripple lamination from the toesets because the lowest part of a dune has the greatest preservation potential (Figure 9.7).

9.4.2 Bounding Surfaces

Bounding surfaces are erosional structures that truncate sets of cross-strata in dune sands. Kocurek (1996) defined three types of bounding surfaces: (i) reactivation surfaces; (ii) superposition surfaces; and (iii) interdune surfaces. Reactivation surfaces are very common in dune sediments and mark changes in dune shape as a response to a change in wind direction. For example, if the wind direction reverses, the dune crest and slipface will be eroded. When the initial wind resumes, that erosion surface will be buried as the dune crest and slipface are reformed. Superposition surfaces are formed by one dune migrating over the lee side of another dune (Kocurek 1996). Interdune surfaces separate one set of dune cross-stratification from another and are formed where one dune migrates across another (Figure 9.8).

- *Reactivation surfaces:* Reactivation surfaces are formed when a sand dune is reshaped by a change in the wind direction or wind velocity which erodes sand from one part of a dune and deposits it on another. Changes in wind direction and wind velocity are very common and, as a consequence, reactivation surfaces are very common if not ubiquitous. Even in an area such as the Moroccan Atlantic Sahara, described as having some of the most stable wind conditions on Earth (Elbelrhiti and Douady 2011), barchan dunes have been observed to reverse and change shape in response to temporal changes in the wind regime. The morphological changes in dunes are recorded by the sedimentary structures within the dunes imaged by ground penetrating radar (GPR) profiles (Box 9.2) (Bristow

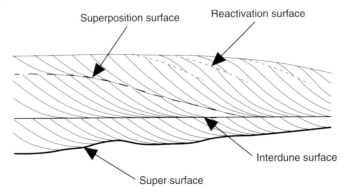

Figure 9.8 Bounding surfaces are erosional structures that truncate sets of cross-strata in dune sands. Gary Kocurek defined three types of bounding surfaces: Reactivation surfaces, superposition surfaces and interdune surfaces. Reactivation surfaces are very common in dune sediments and mark changes in dune shape as a response to a change in wind direction. For example, if the wind direction reverses the dune crest and slipface will be eroded. When the initial wind resumes, that erosion surface will be buried as the dune crest and slipface are reformed. Superposition surfaces are formed by one dune migrating over the lee side of another dune (Kocurek 1996). Interdune surfaces separate one set of dune cross-stratification from another and are formed where one dune migrates across another.

Box 9.2 Ground-Penetrating Radar (GPR)

Ground-penetrating radar (GPR) is a geophysical technique for imaging the shallow subsurface. GPR systems usually consist of a signal generator with transmitting and receiving antennas and a data logger or recording system. The signal generator produces a pulsed signal that is transmitted into the ground by the transmitting antenna. GPR antennas are not focused and the signal radiates out into the air as well as the ground. Some antennas are shielded to reduce signal transmission through the air. When the signal reaches a surface with a contrast in dielectric permittivity, part of the signal is reflected and can be recorded by the receiving antenna when it reaches the surface. The signal is recorded in two-way travel-time (twt), which is the time it takes for the signal to travel from the transmitter through the ground and back to the receiver at the surface, in nanoseconds (ns). The two-way travel-time depends on the depth to the reflector and the velocity of the signal through the ground. The velocity of the radar wave through sand typically varies between 0.15 and 0.17 m ns^{-1} in dry desert dune sands, through 0.12 m ns^{-1} in damp sand such as a coastal dune, to 0.06 m ns^{-1} in water-saturated sand beneath the water table.

By collecting multiple measurements at points spaced a few decimetres apart along a line on the surface, a profile of reflections from sub-surface reflectors can be put together to create an image of the shallow subsurface (Box Figure 9.1). By collecting multiple, closely-spaced parallel lines, or a grid of closely-spaced lines, it is possible to build up a three-dimensional image of radar reflections in the shallow subsurface (e.g. Grasmueck et al. 2005) (Bristow et al. 2007b) (Box Figure 9.2).

Sand dunes are usually very good environments for using GPR because the sands have a high resistivity, allowing good depth of signal penetration in the sub-surface. In addition, dune sands contain large sedimentary structures that can be imaged by GPR (Bristow 2009). Interpretation of GPR profiles of sand dunes has been used to reconstruct the relative chronology of dune deposits (Bristow et al 2005), and the selection of sites for dating, permitting the reconstruction of past rates of dune migration (Bristow et al 2005). However, there are some examples where the depth of penetration is poor, such as in clay-rich dunes of central Australia, where GPR was found not to be very effective at imaging dune stratigraphy (Bristow et al. 2007a).

In dune sands, reflections are most likely to come from changes in water content, which has a very strong control on dielectric permittivity. Very strong reflections are produced at the water table where there is a significant change from dry to damp and then saturated sands. Beneath the water table, where sand is saturated, reflections are observed from sedimentary structures such as sets of cross-strata. Above the water table reflections from sedimentary structures are usually attributed to changes in water content associated with changes in particle size. When sands are completely dry, which is rare even in desert dunes, the changes in grain-size, shape, and packing associated with cross-strata and bounding surfaces are sufficient to cause radar reflections (Guillemoteau et al. 2012).

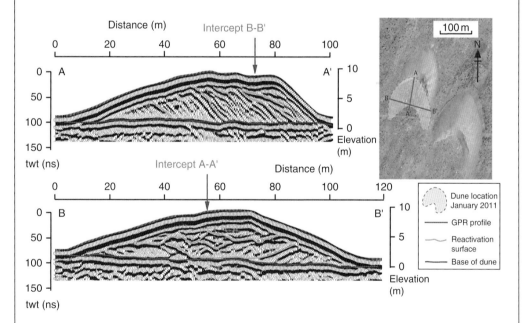

Box Figure 9.1 The internal structure of a barchan dune, near Tarfaya in southern Morocco, imaged with ground-penetrating radar (GPR). Section A-A' is parallel to the direction of the migration of the dune and shows steeply-dipping foresets that record the former position of the lee side of the slipface. The blue lines pick out reactivation surfaces where reflections are truncated marking erosion and reshaping of the dune due to changes in the wind direction. Section B-B' is roughly perpendicular to the direction of dune migration and shows a strike section with the preservation of the horns of the dune at the base, overlain by concave reflections from the trough-shaped dune slipface. The inset is a Google Earth™ image from May 2005 together with the position of the dune at the time of the survey in January 2011, showing that this 9 m high dune has moved 100 m downwind within six years. *Source:* From Bristow and Mountney (2012). See insert for colour representation of this figure.

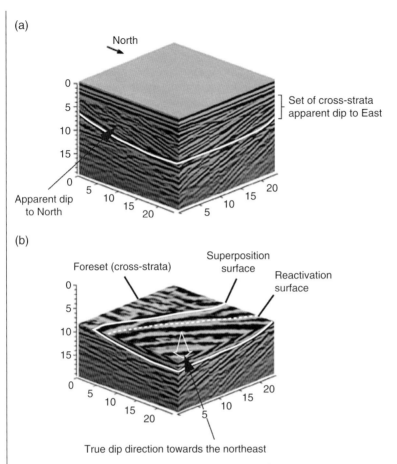

Box Figure 9.2 Three-dimensional cube of GPR data sliced horizontally to show changes in the orientation of sets of cross-strata and bounding surfaces, where a superimposed dune has been migrating along the flank of a linear dune, creating a superposition surface marked by a 70° change in the orientation of the strata and subsequent smaller change in the orientation of the strata, showing where the dune changed shape marked by a reactivation structure. *Source:* Modified from Bristow et al. (2007b). See insert for colour representation of this figure.

and Mountney 2012). The longitudinal profile A-A′ (Box Figure 9.1) shows strata dipping in the downwind direction with multiple reactivation surfaces formed during wind reversals. The transverse profile B-B′ shows undulating bounding surfaces due to the preservation of the horns, attributed to oblique migration, as well as truncation of strata due to the changes in dune morphology that are attributed to a combination of effects including: the presence of superimposed dunes, reshaping of the dune during reversing winds and oblique winds. Thus,

even a simple barchan in an area that is supposed to have some of the most stable wind conditions on Earth shows complex structure because the fluctuation of the wind is sufficient to perturb the shape of the dunes. These changes in dune morphology are preserved as reactivation surfaces. In addition, if the dune migration is not directly downwind but veers slightly sideways, then one of the horns will be preserved beneath the dune (Bristow and Mountney 2012).

- *Superposition surfaces:* Superposition surfaces are formed where one dune migrates over another dune. This is most likely to occur where small dunes migrate over larger dunes because small dunes migrate faster than larger dunes. In addition, small dunes may be superimposed on the flanks of larger dunes, for example, where transverse dunes migrate across larger dune bedforms (called megadunes or draa). In both cases the superimposed dunes truncate the underlying strata of earlier dunes and the resulting bounding surfaces and sets of cross-strata have been revealed by trenching (McKee 1966) and imaged using GPR (Bristow et al. 2007a; Bristow 2009). In the Namibian example (Box Figure 9.2), the superimposed dune is migrating perpendicular to the trend of the linear dune and there is a distinct change in dip direction of around 70° between the two sets of cross-strata.

- *Interdune surfaces:* Interdune surfaces are formed by the migration of dunes and aggradation, resulting in the preservation of sections of interdune sediments between sets or co-sets of dune cross-strata. The interdune deposits may be dry, e.g. a near horizontal wind-ripple-dominated sand sheet, or wet, e.g. an inland sabkha. The interdune surfaces may be inclined at very low angles (typically no more than a few tenths of a degree), with climb in a downwind direction only evident over distances of many hundreds of metres or even kilometres: such surfaces usually appear horizontal at outcrop but form important stratigraphic markers. Interdune surfaces may be constrained by a palaeo-water table which formed a downward limit to aeolian scour through the increased cohesion of damp or wet sand and through early cementation by evaporates precipitated in the sediments, as water evaporates at or close to the sand-air interface, see Fryberger et al. (1988) and Bristow and Mountney (2012).
- *Super surfaces:* Super surfaces are generated when aeolian accumulation ceases, possibly as a result of changes in climate or sea-level rise accompanied by a rise in the inland water table. They are of great lateral extent and continuity and typically bound entire aeolian accumulations

or large parts thereof. Some super surfaces may be correlated into adjacent non-aeolian, fluvial, or coastal sediments (Kocurek 1996).

9.4.3 Sedimentary Models for Dunes and Interdunes

Our knowledge of the internal structure of sand dunes owes much to the pioneering work of McKee, who excavated pits and trenches in sand dunes in order to photograph and sketch the sedimentary structures that were exposed in the trench walls (McKee 1957, 1966; McKee and Tibbits 1964; McKee and Bigarella 1972). The next major advance came through modelling strata conceptually and using computer simulations of migrating bedforms (Rubin 1987). Most recently, the use of GPR now allows researchers to image dune structure without resorting to the destructive use of trenches (e.g. Bristow et al. 2000; Bristow 2009).

The sedimentary deposits of different types of sand dunes are superficially similar, especially the barchanoid and transverse dune forms that have a single dominant migration direction. However, as the morphology and dynamics of dunes become more complex, so does the pattern of deposition recorded by the dune strata. Subtle differences in the dip and dip direction of sets of cross-stratification and bounding surfaces can be used to reconstruct past directions of dune migration. It has been suggested by Glennie (1972) that patterns of dip and dip direction could be used to reconstruct dune morphology. More sophisticated models of cross-strata can be constructed with computer models (Rubin 1987). Simple dunes such as barchan dunes, transverse, barchanoid, and parabolic dunes are relatively well understood. The influence of varying wind regimes and consequent changes in dune morphology introduce complexity such as bounding surfaces formed by the reshaping of dunes during wind reversals. The structure of linear dunes was much debated with different models depending upon the dune dynamics. The earliest model for linear dune structure presented by Bagnold (1941) assumed that the dunes remained static and built up vertically. It was later suggested that linear dunes might migrate laterally (Rubin and Hunter 1985) which would alter the way in which dune strata were preserved. GPR results from linear dunes in Namibia support the lateral migration hypothesis (Bristow et al. 2000; Bristow et al. 2007a). The strata of the tallest dunes, star dunes, which have multiple arms and are a common feature of the largest sand deserts on Earth are not well understood and require further investigation. Compound dunes which consist of two or more of the same type of dune (McKee 1979) and complex dunes that consist of two different basic types of dunes are less well understood and the intricacies of dune-dune interactions is a developing subject e.g. Ewing and Kocurek 2010, and Brothers et al. 2016).

Sedimentary structures in some coastal dunes can differ from other typically mobile dunes where sand accumulates on the downwind side because they can accrete upwind. Foredune ridges are commonly found running along the coast paralleling the shoreline. They derive their sediment from the beach and although occasionally cut back by coastal erosion during storm events, they can and do accrete towards the direction of sediment supply from the beach. This style of accumulation, termed foreslope accretion (Carter and Wilson 1990) is aided by vegetation such as Marram grass which traps windblown sand and stabilises the dune.

Sand dunes are not restricted to warm deserts, and coastal settings, but may also be found in cold climates at high latitudes or high elevations at mid-latitudes where ice and snow affect dune sediments. These niveo-aeolian sediments can include interbedded layers of sand and snow (Figure 9.9a). Seasonal melting of snow layers creates denivation structures (Dijkmans 1990), which include cracks, pits and mounds as the snow layers melt, deforming the interbedded sand layers (Figure 9.9b). Other indicators of cold climate dunes include ice and sand wedges. Following the last ice age, when vegetation was less extensive, there was more sediment available for aeolian transport so there were extensive dune fields across northern Europe and North America, many of which have been reactivated by changes in land use and vegetation (e.g.Tolksdorf and Kaiser 2012; Halfen and Johnson 2013).

- *Interdunes* are relatively flat areas found in between dunes. They are commonly classified into deflationary or depositional: and these in turn are informally subdivided into dry, wet and evaporative, based on water content (Ahlbrandt and Fryberger 1981). Deflationary interdunes commonly lack sandy sediments, exposing bedrock or non-aeolian sediments, such as lacustrine, fluvial or alluvial fan deposits. A lag of coarse grained sand or wind-polished ventifacts, pebbles, or pieces of rock shaped by sand-blasting, including three sided dreikanter pebbles may also be present. Dry interdunes' sediments usually include wind ripple laminated sands and layers of deflation marked by granule lags and ventifacts. On damp surfaces, windblown sand can stick to the surface, forming adhesion ripples. Wet interdunes are surprisingly common in coastal dune fields and deserts where the water table is close to the surface. In wet interdunes, temporary pools or shallow lakes can form after rainfall or become more extensive if the climate changes to become more humid. Wet interdune sediments can therefore include shallow lakes, ephemeral channels, and inland sabkhas. Warren (1989) defined sabkhas as marine and continental mudflats where displacive and replacive evaporite minerals are forming in the capillary zone above a saline water table. The displacive and replacive growth of evaporate minerals, salt (NaCl), and gypsum ($CaSO_4$ $2H_2O$), produces deformation of the sediment layers with small-scale folding and disruption of the beds known as haloturbation. When they dry out, muds deposited in temporary interdune ponds shrink, forming mudcracks and mud curls

(a) (b)

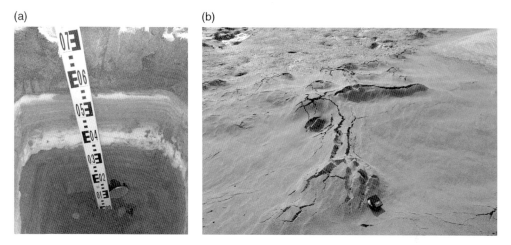

Figure 9.9 Niveo-aeolian sediments. (a) layers of interbedded sand and snow in Antarctica. (b) melting of the snow layers creates deformation of the sand layers, forming cracks, mounds, and pits on the surface, termed denivation structures.

(a) (b) (c)

Figure 9.10 Interdune sediments can be divided into deflationary or depositional. (a) A deflation surface on an alluvial fan at the margins of a dune field in the Draa Valley, Morocco (notebook is 20 cm across). Examples of depositional interdunes include ephemeral rivers and ponds formed during flash floods. (b) Desiccation cracks formed in an interdune pond after a flood in the Draa Valley, Morocco. (c) Mud curls formed in an ephemeral channel after a flash flood near Erg Chebbi, Morocco.

(Figure 9.10). The desiccated muds may then be reworked and broken by the wind, forming mud pellets and dust.

- *Sandsheets* are aeolian sand deposits that are often found around the margins of dune fields but lack definable dunes although nebkha dunes can occur. Nebkha dunes are formed where sand accumulates around

vegetation. Typically, sandsheets accumulate sand as sub-horizontal layers of wind ripples and grainfall deposits during sand storms (Fryberger et al. 1983). The association with vegetation, including xerophytic grasses and shrubs can result in root traces in the sediments along with burrows from insects and small vertebrates.

9.5 Eolianites

Eolianites are wind-driven subaerial accumulations of carbonate-dominated and carbonate-cemented sand (Brooke 2001). It is a term that was used by Sayles (1931) to describe carbonate dune deposits in Bermuda and has since come to be associated with carbonate aeolian dunes as distinct from siliciclastic (quartz) dominated aeolian sands (Brooke 2001). Carbonate dunes are mostly found in coastal areas, especially those with warm shallow seas, with high biogenic carbonate production coupled with low terrigenous input, onshore-directed trade winds and a seasonal water budget deficit (Brooke 2001). However, they are not restricted to coastal settings and can be found inland as well (Goudie and Sperling 1977). Carbonate dunes were widespread in the Quaternary (Brooke 2001) and appear to have good preservation potential due to early cementation. However, eolianites appear to be under-represented in the rock record which could be due to misinterpretation (Frébourg et al. 2008). Frébourg et al. (2008) noted that there is a lack of simple diagnostic features for eolianites at the micro-scale: although grainstone

(a granular sediment with the particles in contact and supporting one another) textures are ubiquitous, grainsize and sorting are variable. This might in part be due to differences in grain density between grains of quartz, and carbonate grains such as oolites, shell fragments or tests of foraminifera. Most eolianites are associated with large scale landward dipping cross-strata (see Figure 9.11a), grainflow and pinstripe lamination, slump scar features, footprints, and pedogenic alteration. The presence of pendant and meniscus cements typically indicate cementation occurred above the water table. A lack of micrite envelopes due to rapid export from the beach or nearshore and into backshore dunes is another potential indicator although micritisation of bioclasts and lithoclasts may be abundant in low-stand eolianites (Frébourg et al. 2008). The preservation of plant root fossils, and impressions of tree trunks and leaves (Figure 9.11b) also appear to be more common in eolianite than in quartz-rich dunes, presumably a signature of the relatively early diagenetic cementation of eolianite, although early cementation cannot be so early that it prevents landward migration of coastal dunes (Rowe and Bristow 2015).

Figure 9.11 (a) Cross-stratification in eolianite Bermuda; The height of the face is around 5 m. At the base of the section, partially covered by stone blocks, is a palaeosol horizon, across which the dune has migrated from left to right. (b) fossil mould of a palm tree preserved within eolianite, Bermuda. See insert for colour representation of this figure.

9.6   Conclusion

Dune sediments are mostly sand-sized particles of quartz, because sand sized particles are readily entrained by the wind and quartz is the most stable silicate mineral on the Earth's surface. However, there are natural variations largely due to changes in sediment sources (provenance), so that dunes can be made from volcanic rock fragments, clay pellets, or other sediments such as pieces of shell. The migration of dunes produces sets of cross-stratification that record the shape of the dune and bounding surfaces, which record changes in the shape of the dune that result from changes in the wind or interaction with other dunes. Preservation of cross-strata in the rock record permits reconstruction of dune morphodynamics.

Further Reading

A study of global sand seas by McKee (1979) and the chapters therein still offers a valuable source of information on sand dune morphology and sediments. Hunter (1977) described small-scale sedimentary structures in dune sediments. Reviews by Blott and Pye (2001, 2008) cover particle shape and size and Garzanti et al. (2012) also provided a case study on the use of particle size measures. Rubin (1987) simulated cross-strata by modelling dune bedforms and compares them with outcrops of aeolian sandstones. Bristow (2009) is a useful starting point on GPR

References

Ahlbrandt, T.S. (1979). Textural parameters of eolian deposits. In: *A Study of Global Sand Seas* (ed. E.D. McKee), 21–51, London: Geological Survey Professional Paper 1052.

Ahlbrandt, T.S. and Fryberger, S.G. (1981). Sedimentary features and significance of interdune deposits. In: *Recent and Ancient Nonmarine Depositional Environments: Models for Exploration* (ed. F.G. Ethridge and R.M. Flores), 293–314. Society of Economic Paleontologists and Mineralogists (SEPM), Special Publication 31.

Bagnold, R.A. (1941). *The Physics of Blown Sand and Sand Dunes.* London: Methuen.

Blackwelder, E. (1934). Yardangs. *Bulletin of the Geological Society of America* 24: 159–166.

Blott, S.J. and Pye, K. (2001). GRADISTAT: a grain size distribution and statistics package for the analysis of unconsolidated sediments. *Earth Surface Processes and Landforms* 26 (11): 1237–1248. doi: 10.1002/esp.261.

Blott, S.J. and Pye, K. (2008). Particle shape: a review and new methods of characterisation and classification. *Sedimentology* 55: 31–63.

Bowler, J.M. (1973). Clay dunes: their occurrence, formation and environmental significance. *Earth-Science Reviews* 9 (4): 315–338. doi: 10.1016/0012-8252(73)90001-9.

Bristow, C.S. (2009). Ground-penetrating radar in aeolian dune sands. In: *Ground Penetrating Radar: Theory and Applications* (ed. H.M. Jol), 273–297. Rotterdam: Elsevier Science.

Bristow, C.S., Bailey, S.D., and Lancaster, N. (2000). The sedimentary structure of linear sand dunes. *Nature* 406 (6791): 56–59. doi: 10.1038/35017536.

Bristow, C.S., Drake, N., and Armitage, S. (2009). Deflation in the dustiest place on Earth: the Bodélé Depression, Chad. *Geomorphology* 105 (1–2): 50–58. doi: 10.1016/j.geomorph.2007.12.014.

Bristow, C.S., Duller, G.A.T., and Lancaster, N. (2007b). Age and dynamics of linear dunes in the Namib Desert. *Geology* 35 (6): 555–558. doi: 10.1130/G23369A.1.

Bristow, C.S., Jones, B.G., Nanson, G.C., et al. (2007a). GPR surveys of vegetated linear dune stratigraphy in central Australia: Evidence for linear dune extension with vertical and lateral migration. In: *Stratigraphic Analysis Using GPR* (ed. G.S. Baker and H.M. Jol), 19–34. Geological Society of America Special Paper 432, doi:10.1130/2007.2432(02).

Bristow, C.S., Lancaster, N., and Duller, G.A.T. (2005). Combining ground penetrating radar surveys and optical dating to determine dune migration in Namibia. *Journal of the Geological Society* 162 (2): 315–321. doi: 10.1144/0016-764903-120.

Bristow, C. and Mountney, N. (2012). Eolian landscapes: eolian stratigraphy. In: *Treatise on Geomorphology*, vol. 11 (ed. J. Shroder and N. Lancaster), 246–268. San Diego, CA: Academic Press.

Brooke, B.P. (2001). The distribution of carbonate eolianite. *Earth-Science Reviews* 55 (1–2): 135–164. doi: 10.1016/S0012-8252(01)00054-X.

Brothers, S.C., Kocurek, G.A., Brothers, T.C., and Buynevich, I.V. (2006). Stratigraphic architecture resulting from dune interactions: White Sands Dune Field, New Mexico. *Sedimentology* 64: 686–713. doi: 10.1111/sed.12320.

Buckley, R.C. (1989). Grain-size characteristics of linear dunes in Central Australia. *Journal of Arid Environments* 16 (1): 23–28. doi: 10.1007/s40333-015-0005-4.

Carter, R.W.G. and Wilson, P. (1990). The geomorphological, ecological and pedological development of coastal dunes at Magilligan Point, Northern Ireland. In: *Coastal Dunes: Form and Process* (ed. K.F. Nordstrom, N.P. Psuty and R.W.G. Carter), 129–157. Chichester: Wiley.

Chen, X.Y., Chappell, J., and Murray, A.S. (1995). High (ground) water levels and dune development in Central Australia: TL dates from gypsum and quartz dunes around Lake Lewis (Napperby), Northern-Territory. *Geomorphology* 11 (4): 311–322. doi: 10.1016/0169-555X(94)00072-Y.

Churchward, H.M. (1961). Soil studies at Swan Hill Victoria Australia: I. Soil layering. *Journal of Soil Science* 12: 73–86.

Coffey, G.N. (1909). Clay dunes. *The Journal of Geology* 17 (8): 754–755. doi: 10.1086/621681.

Crocker, R.L. (1946). The Simpson Desert expedition, 1939, scientific reports: 8 – the soils and vegetation of the Simpson Desert and its borders. *Transactions of the Royal Society of South Australia* 70: 235–258.

Dare-Edwards, A.J. (1983). Loessic clays of south Eastern Australia, *Loess Letter*, Supplement 2.

Dijkmans, J.W.A. (1990). Niveo-aeolian sedimentation and resulting sedimentary structures; Søndre Strømfjord area, Western Greenland. *Permafrost and Periglacial Processes* 1: 83–96.

Dorn, R.I. and Oberlander, T.M. (1981). Microbial origin of desert varnish. *Science* 213 (4513): 1245–1247. doi: 10.1126/science.213.4513.1245.

Drake, N.A. (1997). Recent aeolian origin of surficial, gypsum crusts in southern Tunisia: geomorphological, archaeological and remote sensing evidence. *Earth Surface Processes and Landforms* 22 (7): 641–656. doi: 10.1002/(SICI)1096-9837(199707)22:7<641:AID-ESP737>3.0.CO;2-R.

Edgett, K.S. and Lancaster, N. (1993). Volcaniclastic aeolian dunes: terrestrial examples and application to Martian sands. *Journal of Arid Environments* 25 (3): 271–297. doi: 10.1006/jare.1993.1061.

Elbelrhiti, H. and Douady, S. (2011). Equilibrium versus disequilibrium of barchan dunes. *Geomorphology* 125 (4): 558–568. doi: 10.1016/j.geomorph.2010.10.025.

Ewing, R.C. and Kocurek, G.A. (2010). Aeolain dune interactions and dune field pattern formation: White Sands Dune Field, New Mexico. *Sedimentology* 57: 1199–1219. doi: 10.1111/j.1365-3091.2009.01143.x.

Folk, R.L. and Ward, W.C. (1957). Brazos River bar: a study in the significance of grain size parameters. *Journal of Sedimentary Research* 27 (1): 3–26. doi: 10.1306/74D70646-2B21-11D7-8648000102C1865D.

Frébourg, G., Hasler, C.-A., Davaud, E., and de Guern, P. (2008). Facies characteristics and diversity in carbonate eolianites. *Facies* 54 (2): 175–191. doi: 10.1007/s10347-008-0134-8.

Fryberger, S.G., Al-Sari, A.M., and Clisham, T.J. (1983). Eolian dune, interdune, sand sheet, and siliciclastic sabkha sediments of an offshore prograding sand sea, Dhahran area, Saudi Arabia. *AAPG Bulletin* 67: 280–312.

Fryberger, S.G. and Schenk, C. (1988). Pin stripe lamination: a distinctive feature of modern and ancient eolian sediments. *Sedimentary Geology* 55 (1–2): 1–16. doi: 10.1016/0037-0738(88)90087-5.

Fryberger, S.G., Schenk, C., and Krystinik, L.F. (1988). Stokes surfaces and the effects of near-surface groundwater-table on aeolian deposition. *Sedimentology* 35 (1): 21–41. doi: 10.1111/j.1365-3091.1988.tb00903.x.

Garzanti, E., Andò, S., Vezzoli, G. et al. (2012). Petrology of the Namib Sand Sea: long-distance transport and compositional variability in the wind-displaced Orange Delta. *Earth-Science Reviews* 112 (3–4): 173–189. doi: 10.1016/j.earscirev.2012.02.008.

Garzanti, E., Resentini, A., Ando, S. et al. (2015). Physical controls on sand composition and relative durability of detrital minerals during ultra-long distance littoral and aeolian transport (Namibia and southern Angola). *Sedimentology* 62 (4): 971–996. doi: 10.1111/sed.12169.

Glennie, K.W. (1972). Permian Rotliegendes of Northwest Europe interpreted in the light of modern desert sedimentation studies. *The American Association of Petroleum Geologists Bulletin* 56 (6): 1048–1071.

Goudie, A.S. and Sperling, C.H.B. (1977). Long distance transport of formaniferal tests by wind in the Thar Desert, northwest India. *Journal of Sedimentary Research* 47 (2): 630–633. doi: 10.1306/212F71FD-2B24-11D7-8648000102C1865D.

Goudie, A.S. and Thomas, D.S.G. (1985). Pans in southern Africa with particular reference to South Africa and Zimbabwe. *Zeitschrift für Geomorphologie* 29: 1–19.

Goudie, A.S. and Watson, A. (1981). The shape of desert sand dune grains. *Journal of Arid Environments* 4: 185–190.

Goudie, A.S., Warren, A., Jones, D.K.C., and Cooke, R.U. (1987). The character and possible origins of the aeolian sediments of the Wahiba Sand Sea, Oman. *The Geographical Journal* 153 (2): 231–256. doi: 10.2307/634875.

Grasmueck, M., Weger, R., and Horstmeyer, H. (2005). Full resolution 3D GPR imaging. *Geophysics* 70: k12–k19.

Greene, R.S.B. and Nettleton, W.D. (1996). Soil genesis in a longitudinal dune swale landscape New South Wales, Australia. *AGSO Journal of Australian Geology and Geophysics* 16: 277–287.

Guillemoteau, J., Bano, M., and Dujardin, J.-R. (2012). Influence of grainsize, shape and compaction on georadar waves: example of an aeolian dune. *Geophysical Journal International* 190: 1455–1463.

Halfen, A.F. and Johnson, W.C. (2013). A review of Great Plains dune field chronologies. *Aeolian Research* 10: 135–160.

Hartmann, D. (2007). From reality to model: operationalism and the value chain of particle-size analysis of natural sediments. *Sedimentary Geology* 2002 (3): 383–401. doi: 10.1016/j.sedgeo.2007.03.013.

Hesse, P. (2011). Sticky dunes in a wet desert: formation, stabilisation and modification of the Australian desert dunefields. *Geomorphology* 134: 309–325.

Hunter, R.E. (1977). Basic types of stratification in small eolian dunes. *Sedimentology* 24 (3): 361–387. doi: 10.1111/j.1365-3091.1977.tb00128.x.

Hunter, R.E. and Rubin, D.M. (1983). Interpreting cyclic crossbedding, with an example from the Navajo Sandstone. *Developments in Sedimentology* 38: 429–454.

Jones, D.J. (1953). Gypsum-oolite dunes, Great Salt Lake desert, Utah. *American*

Association of Petroleum Geologists Bulletin 37 (1): 2530–2538.

Kocurek, G. (1996). Desert aeolian systems. In: *Sedimentary Environments and Facies*, 3e (ed. H.G. Reading), 125–153. Oxford: Blackwell.

Kocurek, G., Carr, M., Ewing, R. et al. (2007). White Sands dune field, New Mexico: age, dune dynamics and recent accumulations. *Sedimentary Geology* 197 (3–4): 313–331. doi: 10.1016/j.sedgeo.2006.10.006.

Krinsley, D.H. and Donahue, J. (1968). Environmental interpretation of sand grain surface textures by electron microscopy. *Geological Society of America Bulletin* 79 (6): 743–748. doi: 10.1130/0016-7606(1968)79[743:EIOSGS]2.0.CO;2.

Krinsley, D.H. and Doornkamp, J.C. (1973). *Atlas of Quartz Surface Textures by Electron Microscopy*. Cambridge: Cambridge University Press.

Krinsley, D.H. and Trusty, P. (1985). Environmental interpretation of quartz grain surface textures. In: *Provenance of Arenites* (ed. G.G. Suffer), 213–229. Dordrecht: Reidel.

Krumbein, W.C. (1938). Size frequency distributions and the normal phi curve. *Journal of Sedimentary Petrology* 8: 84–90. doi: 10.1306/D4269008-2B26-11D7-8648000102C1865D.

Lancaster, N. (1986). Grain-size characteristics of the linear dunes of the southwestern Kalahari. *Journal of Sedimentary Petrology* 56: 395–400.

Lancaster, N. (1989). *The Namib Sand Sea: Dune Forms, Processes, and Sediments*. Rotterdam: A.A. Balkema.

Leys, J., McTainsh, G., Koen, T. et al. (2005). Testing a statistical curve-fitting procedure for quantifying sediment populations within multi-modal particle-size distributions. *Earth Surface Processes and Landforms* 30 (5): 579–590. doi: 10.1002/esp.1159.

Livingstone, I. (1987). Grain-size variation on a 'complex' linear dune in the Namib Desert. In: *Desert Sediments: Ancient and Modern*,

vol. 35 (ed. L.E. Frostick and I. Reid), 281–291. Geological Society of London, Special Publication.

Livingstone, I., Bullard, J.E., Wiggs, G.F.S., and Thomas, D.S.G. (1999). Grain-size variation on dunes in the Southwest Kalahari, Southern Africa. *Journal of Sedimentary Research* 69 (3): 546–552. doi: 10.2110/jsr.69.546.

Livingstone, I. and Warren, A. (1996). *Aeolian Geomorphology, an Introduction*. Harlow: Addison-Wesley Longman.

Loope, D.B., Rowe, C.M., and Joeckel, R.M. (2001). Annual monsoon rains recorded by Jurassic dunes. *Nature* 412: 64–66. doi: 10.1038/35083554.

Macdonald, P.D.M. and Green, P.E.J. (1988). *User's Guide to Program MIX: An Interactive Program for Fitting Mixtures of Distributions*. Hamilton, Ontario: Ichthus Data Systems.

McKee, E.D. (1957). Primary structures in some recent sediments. *Bulletin of the American Association of Petroleum Geologists* 41 (8): 1704–1742.

McKee, E.D. (1966). Structures of dunes at White Sands National Monument, New Mexico (and a comparison with structures of dunes from other selected areas). *Sedimentology* 7 (1): 1–69. doi: 10.1111/j.1365-3091.1966.tb01579.x.

McKee, E.D. (1979). *A Study of Global Sand Seas*. Geological Survey Professional Paper 1052. Washington, DC: U.S. Government Printing Office.

McKee, E.D. and Bigarella, J.J. (1972). Deformational structures in Brazilian coastal dunes. *Journal of Sedimentary Research* 42 (3): 670–681. doi: 10.1306/74D725F4-2B21-11D7-8648000102C1865D.

McKee, E.D. and Tibbitts, G.C. Jr. (1964). Primary structures of a seif dune and associated deposits in Libya. *Journal of Sedimentary Research* 34 (1): 5–17. doi: 10.1306/74D70FBA-2B21-11D7-8648000102C1865D.

McTainsh, G.H., Livingstone, I., and Strong, C. (2013). Fundamentals of aeolian sediment transport. In: *Treatise on Geomorphology*, vol. 11 (ed. J. Shroder and N. Lancaster), 23–42. Oxford: Academic Press.

Mohamedou, A.O., Aventurier, A., Barbiero, L. et al. (1999). Geochemistry of clay dunes and associated pan in the Senagal Delta, Mauritania. *Arid Soil Research and Rehabilitation* 13 (3): 265–280. doi: org/10.1080/089030699263302.

Potter, R.M. and Rossman, G.R. (1977). Desert varnish: the importance of clay minerals. *Science* 196 (4297): 1446–1448. doi: 10.1126/science.196.4297.1446.

Rowe, M.P. and Bristow, C.S. (2015). Landward-advancing Quaternary eolianites of Bermuda. *Aeolian Research* 19: 235–249.

Rubin, D.M. (1987). Cross-bedding, bedforms and palaeocurrents. Society of Economic Paleontologists and Mineralogists (SEPM) 1.

Rubin, D.M. and Hunter, R.E. (1985). Why deposits of longitudinal dunes are rarely recognized in the geologic record. *Sedimentology* 32 (1): 147–157. doi: 10.1111/j.1365-3091.1985.tb00498.x.

Sayles, R.W. (1931). Bermuda during the Ice Age. *Proceedings of the American Academy of Arts and Sciences* LXVI (11): 382–467.

Shi, S. (1995). Observational and theoretical aspects of tsunami sedimentation. Unpublished PhD thesis. Coventry University.

Sneh, A. and Weissbrod, T. (1983). Size-frequency distribution on longitudinal dune ripple flank sands compared to that of slipface sands of various dune types. *Sedimentology* 30 (5): 717–726. doi: 10.1111/j.1365-3091.1983.tb00705.x.

Stapor, F.W., May, J.P., and Barwis, J. (1983). Eolian shape-sorting and aerodynamic traction equivalence in the coastal dunes of Hout Bay, Republic of South Africa. In: *Eolian Sediments and Processes, Developments in Sedimentology* 38 (ed. M.E. Brookfield, and T.S. Ahlbrandt), 149–164. Amsterdam: Elsevier.

Thomas, D.S.G. (1988). Analysis of linear dune sediment-form relationships in the Kalahari dune desert. *Earth Surface Processes and Landforms* 13 (6): 545–553. doi: 10.1002/esp.3290130608.

Tolksdorf, J.F. and Kaiser, K. (2012). Holocene aeolian dynamics in the European sand-belt as indicated by geochronological data. *Boreas* 41: 408–421. doi: 10.1111/j.1502-3885.2012.00247.x.

Udden, J.A. (1914). The mechanical composition of clastic sediments. *Bulletin of the Geological Society of America* 25 (4): 655–744. doi: 10.1130/GSAB-25-65.

Vermeesch, P. and Drake, N. (2008). Remotely sensed dune celerity and sand flux measurements of the world's fastest barchans (Bodélé, Chad). *Geophysical Research Letters* 35: L24404. doi: 10.1029/2008GL035921.

Vermeesch, P., Fenton, C.R., Kober, F. et al. (2010). Sand residence times of one million years in the Namib Sand Sea from cosmogenic nuclides. *Nature Geoscience* 3 (12): 862–865. doi: 10.1038/ngeo985.

Vos, K., Vandenberghe, N., and Elsen, J. (2014). Surface textural analysis of quartz grains by scanning electron microscopy (SEM): from sample preparation to environmental interpretation. *Earth-Science Reviews* 128: 93–104. doi: 10.1016/j.earscirev.2013.10.013.

Walker, T.R. (1979). Red color in dune sand. In: *A Study of Global Sand Seas* (ed. E.D. McKee), 61–81), United States Department of the Interior, U.S. Geological Survey, Professional Paper 1052.

Wang, X., Dong, Z., Zhang, J. et al. (2003). Grain size characteristics of dune sands in the central Taklimakan Sand Sea. *Sedimentary Geology* 161: 1–14.

Warren, J.K. (1989). *Evaporite Sedimentology: importance in hydrocarbon accumulation*. Englewood Cliffs, NJ: Prentice Hall.

Wasson, R.J. (1983). Dune sediment type, sand colour, sediment provenance and hydrology in the Strzelecki –Simpson Dunefield, Australia. In: *Eolian Sediments and Processes, Developments in Sedimentology*, vol. 38 (ed. M.E. Brookfield and T.S. Ahlbrandt), 165–196. Amsterdam: Elsevier.

Watson, A. (1986). Grain-size variations on a longitudinal dune and a barchan dune.

Sedimentary Geology 46 (1): 49–66. doi: 10.1016/0037-0738(86)90005-9.

Wentworth, C.K. (1922). A scale of grade and class terms for clastic sediments. *The Journal of Geology* 30: 377–392.

White, K., Walden, J., and Gurney, S.D. (2007). Spectral properties, iron oxide content and provenance of Namib dune sands. *Geomorphology* 86 (3–4): 219–229. doi: 10.1016/j.geomorph.2006.08.014.

Willetts, B.B., Rice, M.A., and Swaine, S.E. (1982). Shape effects in aeolian grain transport. *Sedimentology* 29 (3): 409–417. doi: 10.1111/j.1365-3091.1982.tb01803.x.

10

Dune Palaeoenvironments

David S.G. Thomas

University of Oxford, Oxford, UK

10.1   Introduction

Aeolian sedimentary structures and dune bedding are well represented within the geological record, as in the Devonian Old Red Sandstone in the British Isles (Nichols 2005), the Permian Rotliegendes Beds of northern Europe (Glennie 1983), the Jurassic Navajo Sandstone in North America (Loope et al. 2012), and the Tertiary Tsondab Sandstone in Namibia (Kocurek et al. 1999). This demonstrates that dune environments have long been part of the earth system but also that the distributions of dune activity and desert dune systems have changed through time. The dunes present in today's landscapes can also be palimpsests of past conditions. Large dunes may need a long time to form, and are likely to have a long reconstitution time, embedded in the concept of 'dune memory' (Warren and Kay 1987).

This chapter focuses on dunes and dunefields that are found in degraded, eroded, or heavily vegetated states, outside regions where aeolian processes operate effectively today. It is likely that these dunes contain a record of the timing of past periods of aeolian activity. Since the pioneering work of Singhvi et al. (1982), the capacity has existed to unlock the record of past events in these landscapes, through the application of luminescence dating (Box 5.2), which can establish the timing of deposition of sand from within dune bodies. Dune bodies can also contain within them evidence of times when aeolian processes did not operate. The presence of palaeosols, units of sediment deposited by other processes, or even datable fossils can all indicate when dunes were inactive, or when processes other than aeolian activity affected dune landscapes.

Together, characteristics of these deposits (long-term accumulation, changed locations of dune formation, and the cessation of aeolian activity) have the potential to allow dune records to contribute usefully to the reconstruction of environmental and climate changes during the Quaternary Period. As more is known about the controls on dune development, and as more dated sequences of dune accumulation become available, it becomes increasingly clear that dune palaeoenvironments, and the data available to reconstruct them, are much more complex than has previously been recognised. Focusing on the Quaternary Period and on continental (as opposed to coastal) dunes and dunefields, this chapter considers: how dune palaeoenvironments can be chronometrically defined; how dunes can be defined as inactive, and therefore regarded as evidence of past dune-forming periods; what environmental and climatic parameters can be deduced from old dune sediments and landforms, and how dunes fit into Quaternary histories of climate and environmental change. It is not the intention in this chapter to present a review of regional chronologies of Quaternary dune activity. These exist elsewhere (e.g. Singhvi and Porat 2008; Thomas and Burrough 2016; Hesse 2014), are constantly evolving as new data and interpretations develop, and can be contested, though sometimes based on a misunderstanding of what luminescence dating can deliver (see e.g. Miller (2014) and Thomas and Burrough (2012) for consideration of the issues concerned).

10.2 A Date with Dunes

It has long been recognised that not all desert dunes are active today: for example, Passarge (1904) identified inactive dunes in the 'Middle Kalahari' of southern Africa, a semi-desert region where aeolian activity is not prevalent today, and regarded them as evidence of more arid conditions in the past. But knowing *when* such dunes accumulated was problematic for a long time, because the general absence of organic material from dune sediments meant that the benefits much of Quaternary science received from the development of radiocarbon dating from the 1950s onwards did not extend to quartz-rich dune sands. Thus, even as recently as the 1970s and 1980s, the development of dunes considered inactive today was commonly ascribed either to gaps in dated records of wetter conditions or by reference to overriding general theories of global climate change. For example,

Sarnthein (1978), in a classic paper, inferred that areas of presently inactive desert dunes, world-wide, were attributable to aridity during the last glacial maximum (Figure 10.1), basing his view on inferred correlation with aeolian sand and dust units in off-shore marine cores that were chronologically tied by bracketing dates from organic sediments.

This situation changed when Singhvi et al. (1982) made the first application of thermoluminescence (TL) dating to dune sands, in a study from the Thar Desert in India. The history and technique of luminescence dating are explained in Chapter 5, Box 5.2.

It is clear from Figure 10.2 that luminescence dating is a technique that has proved popular, even essential, in studies of dune palaeoenvironments. Almost 1000 ages have been published in dune studies from North America (Halfen et al. 2015), almost 700 from Australia (Hesse 2014), and over 600 from southern Africa (Thomas and Burrough 2016), with many more from other dune systems, not just in the low latitudes and on the fringes of deserts, but from mid- (e.g. Bateman et al. 1999) and high- (Murton et al. 2007) latitudes, including Antarctica (Bristow et al. 2011), as well as complex coastal dune systems (e.g. Hu Fangen et al. 2013). What is evident is how far dune palaeoenvironmental studies have advanced because of the capacity of luminescence dating to deliver chronologies. Singhvi and Porat (2008) rightly noted that it was down to this technique that simple text-book models of globally-synchronous expansion and contractions of deserts and dunefields had been debunked. This does not mean, however, that it has become any simpler to establish controls on the dynamics of ancient dunes. Indeed, the complexities of long-term dune accumulation and associated palaeoenvironmental conditions that have emerged from many regions have arguably made the task even more challenging.

Today

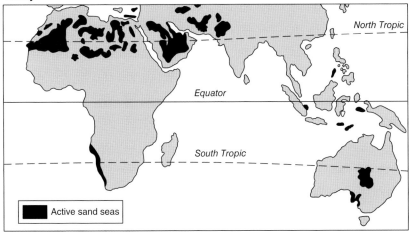

At Last Glacial Maximum

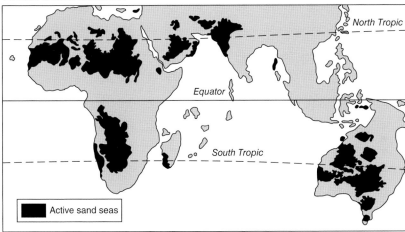

Figure 10.1 Sarnthein's (1978) model of the extent of active desert dunes during the Last Glacial Maximum and today. This map is important as it brought the potential value of palaeo-dune studies to the attention of the wider scientific community. It has spawned much subsequent detailed regional research that has shown a more temporally-complex history of Quaternary dunefield activity than was initially thought. *Source:* Often reproduced (this version from Thomas (2011)).

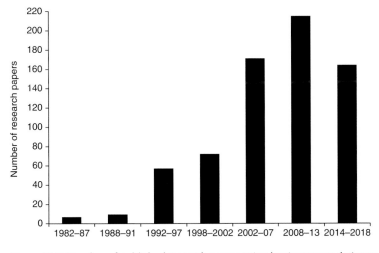

Figure 10.2 Number of published research papers using luminescence dating to provide chronometric data of dune accumulation histories.

10.3 A Challenge: Inactive Versus Active Dunes

It once seemed quite simple to designate dunes as active or inactive. Inactive dunes could be identified by being 'drowned' in lake waters or lake sediments (e.g. Singh et al. (1972) in the Thar Desert, India; and Grove (1958) in the Chad basin in Nigeria), or by sea level rise (e.g. Jennings (1975) in northern Australia). Dune patterns have also been disrupted by tectonic processes, indicating inactive aeolian processes since tectonic displacement occurred (McFarlane and Eckardt 2007). Well-developed surface soils (Smith 1963 in North Africa), surface calcification (Allchin et al. 1978, in the Thar), cementation as with aeolianites (cemented dune sands, often retaining bedding structures; e.g. Sperling and Goudie (1975) in India; and Gardner (1988) in Oman), and highly degraded morphologies (Flint and Bond (1968) in Zimbabwe) also indicate that dune sediments or dunes are not being shaped by aeolian processes today. It is the presence of live vegetation, however, that is by far the primary indicator of dune inactivity in palaeoenvironmental studies, sometimes in conjunction with other criteria as shown by Flint and Bond (1968), but often vegetation, on its own, particularly where initial identification of dune forms has been by aerial photographs (e.g. Grove (1969) in Botswana) or satellite imagery (May 2013). The use of vegetation cover information to infer changes in precipitation since the time of dune development has also occurred in studies where, regionally, dunes are found along a rainfall gradient where both 'active' and 'inactive' forms are found (e.g. Thomas (1984) in southern Africa).

Today the ability to designate a dune 'inactive', as well as to identify the environmental parameters that have changed since dune formation, is seen to be much more complex. This is down to a number of factors, which include an improved understanding of the environmental controls on dune dynamics (Chapter 6), and the implications of the controls for dune preservation (Munyikwa 2005); realisation that the relationship between dunes and vegetation cover is neither simple nor the same for all dunes or vegetation types (Livingstone and Thomas 1993); issues in deciphering which environmental parameter(s) (e.g. precipitation, wind energy, sediment supply) have changed (Lancaster and Tchakarian 1996; Chase 2009); and, ironically, because the application of luminescence dating has shown that dune evolution and development have been far more complex through time that was previously thought to be the case.

10.3.1 Vegetation

Dune systems under dense savannah vegetation, for example, in western Zambia (O'Connor and Thomas 1999), or under British woodland (e.g. Bateman et al. 1999) are clearly not experiencing aeolian activity today. Many such dunes have also been degraded, by runoff and other slope processes.

Dunes with a sparser vegetation cover, in environments subject to high levels of interannual rainfall variability and drought, are less clearly inactive. This is the case in many dune systems in subtropical and tropical locations, some of which have been regarded as palimpsests of palaeo-aridity. In many instances, for example, on the margins of the Sahara and in the Kalahari and in Australia, these dune systems are dominated by linear forms. Even fully active linear dunes are relatively immobile compared to transverse forms (Thomas and Shaw 1991; and see Chapter 6), and commonly support an element of plant cover even under arid conditions (Tsoar and Møller 1986), because of the moisture-holding capacity of sand. It is therefore important to consider dune type when the role of vegetation as evidence of inactivity is considered. This applies not only to linear dunes, but to forms such as parabolic dunes, on which a vegetated surface and its partial disturbance form an intrinsic part of dune development.

It is now widely accepted, therefore, that partially vegetated dunes cannot necessarily be termed 'inactive' (Hesse and Simpson 2006). There is not, in present semi-arid or arid environments, a distinct threshold between aeolian activity and inactivity, but rather different levels of activity (Livingstone and Thomas 1993; Mason et al. 2009), which might include anything from limited surface change through to more extensive activity including slip-face development. Other factors (such as wind energy) being equal, a gradual and progressive decrease occurs in sand movement as the density of vegetation increases (Ash and Wasson 1983; Wiggs et al. 1995). One implication here is that dunes found across a spatial gradient of vegetation are likely to have differing palaeoclimatic significance, in the sense that 'more has to change' to reactivate/build dunes in areas under closed woodland compared to those under sparse vegetation cover. This issue has been explored in the context of linear dunes in the Kalahari by Thomas and Burrough (2012), where it is argued that dunes in the drier, arid south-west of that area are less likely to record a strong climate (precipitation change) signal than those in the wetter, tropical, north-east.

The conceptual gradient of activity is not simply spatial. The temporal dimension should also be considered. Activity is modulated by factors including the seasonality of rainfall and the recurrence of droughts (Bullard et al. 1997), the legacy of which may last for several years in terms of vegetation cover and sand movement. This is evident in dated dune sediments from Mali, which demonstrate reactivation as a function of the great Sahel drought of the late twentieth century (Stokes et al. 2004).

10.3.2 Interacting Environmental Controls

Subject to the energy of the wind, the transport of significant quantities of sand may occur on dunes where vegetation covers up to 35% of the surface (Ash and Wasson 1983). Many dunefields that have been regarded as evidence of palaeoclimatic, particularly precipitation, change (e.g. in southern Africa, interior Australia and in north-west India) have relatively low, and temporally variable, levels of cover. Some authors (e.g. Singh et al. 1990; Tsoar 2005; Chase 2009) have concluded that present-day dune inactivity is not necessarily a function of increased effective precipitation (erodibility), but may be due to a decline in windiness (or erosivity). However, after analysis of a range of vegetational variables along the interior Australian dunefield climate gradient, Hesse and Simpson (2006) concluded that, over millennial time scales, it is vegetation that is the primary control both on the activity of a linear dunefield, and on the rate of the growth of a dune, rather than the strength of the wind. A study of the relationship between wind energy and dune activity in northern China (Mason et al. 2009) also suggested wind strength is not a primary control of activity.

Recognising the complexity of the roles of controlling climate variables, several approaches have attempted to integrate the interactions between changes in precipitation (affecting erodibility, primarily through vegetation cover) and erosivity (wind energy) in explaining dunefield inactivity and the degree of climatic changes since dunes were active. Talbot (1984), Wasson (1984), Lancaster (1988), Chase and Brewer (2009) and others have all applied various 'dune mobility indices' to the problem of identifying climate conditions at the time of palaeodune activity (Table 10.1).

Climate is not the only dynamic element that can change and influence a dune's palaeoenvironment. Sediment supply is an important and variable factor that can have an impact both on the development of a dune

Table 10.1 Dune activity/mobility indices used to assess changes in climate since Quaternary dunes were active.

Index	Source	Terms	Other information	Initial test area
$M = V^3/Mo^2$	Talbot (1984)	M = mobility value V = mean wind speed Mo = a moisture value based on monthly precipitation and temperature	Where M: <1, dunes stable. 5–10 episodically active. >10 dunes fully active.	Sahel
$M = W/(P{:}PE)$	Lancaster (1988)	M = mobility value W = % time wind speed above threshold for sand transport P = monthly precipitation, mm PE = potential evapotranspiration, mm/month	Where M: <50, dunes inactive. >50, <100 crests active. >100, <200 crest and flanks active. >200 dunes fully active.	South-west Kalahari, used to assess changes needed from present climate to reactive dunes
$A_{pGCM} = \bar{U}^3 (P_{lag}/Ep_{lag} + P_{rainy}/Ep_{rainy})$	Thomas et al. (2005)	A_{pGCM} = dune activity index, [GCM refers to data inputs from GCM runs] $\bar{U}^3$ = cube of mean wind speed P = precipitation E = potential evapotranspiration P_{lag}/Ep_{lag} = residual effect of recent P and Ep, where current and previous months are included P_{rainy}/Ep_{rainy} = effect of rainy season P and Ep	Where A_{pGCM} > 700, highly dynamic dune landscapes, bare dune bodies and sparsely vegetated interdunes. 160–700, significant dune activity, dunes crests bare, dune flanks rippled but moderate interdune vegetation cover. 70–160, dune activity limited to crests, dune flanks vegetated, interdunes well vegetated. < 70 indicates vegetation cover across the whole dune, dunes inactive.	Kalahari, used to assess future activity through application of GCM data to model. Used by Chase and Brewer (2009) to calculate activity at the LGM.

and the palaeoenvironmental conclusions drawn from studies of dune development. This is well exemplified by dunes both in Sudan and central Australia, where over three decades ago Williams (1985) noted that many presently inactive dunes did not represent records of extreme aridity in the past, because they were constructed from sediments deflated by winds from channel systems with seasonally-varying flow regimes. If a component of precipitation change were recorded at all in the dunes referred to by Williams, it was not one representing conditions in the dunefield itself, but instead related to precipitation in the river catchment from which dune sediments were sourced. Seasonality of river flow may also be inferred in some cases, for example, in the series of small dunefields recently identified in Bolivia (May 2013). In parts of north-west India, stabilised dunes are found *within* the channel systems of major braided rivers, overlying alluvial sediments (Saini and Mujtaba 2012). In Australia, fluvially-influenced dune forms are commonly referred to as source-bordering dunes, and have been distinguished from more spatially-extensive dune systems in the Simpson Desert by the properties of the sediment (Nanson et al. 1995). In the Strzelecki Desert the luminescence ages of source bordering dunes are distinct from ages derived from forms in the main dunefield (Hesse 2014). These sediments peak in age at the Last Glacial Maximum (LGM), but source-bordering dunes peak at 15–10 ka, when monsoonal rains in northern tropical catchments were reinstated. There was a conspicuous absence in this area of dune with ages corresponding to the Last Glacial Maximum.

The supply of sediment is also an important consideration in other palaeodune contexts. Dunes (sometimes called lunettes) on the margins of small closed dryland depressions (also known as playas or pans) have been utilised in palaeoenvironmental studies from the western USA (Rich 2013), southern Africa (Holmes et al. 2008), and Australia (Fitzsimmons et al. 2014). Their formation is supply-limited, although there has been debate in the literature as to whether they form by deflation from the desiccated floors of the lake basins (Lancaster 1978), or from sediments brought to the downwind end of basins by wave action in flooded basins (Bowler 1973), analogous to the formation of many coastal dunes. Investigation of lunettes across the hydrological gradient in Australia by Bowler (1986) provided a still-relevant model of differential formation according to hydrological context (Figure 10.3) which can be used as an ergodic analogue (in which space is substituted for time) for interpreting developments in the past.

Locations where topography interacts with sand transport (Figure 10.4) provide the potential for the storage of aeolian sediments and, in some cases, the accumulation of landforms that may preserve, in a stratigraphically-integrated manner, evidence of periods of aeolian activity and the operation of other processes, including slope runoff and talus flows. These features were first studied in North America, and were termed sand ramps (Lancaster and Tchakerian 1996). Studies of sand ramps have also contributed to palaeoenvironmental reconstruction in Iran (Thomas et al. 1997), West Africa (Rendell et al. 2003), southern Africa (Thomas et al. 2002; Telfer et al. 2012), and in the Mohave Desert (Rendell et al. 1994; Bateman et al. 2012). In many cases sand has been supplied from a desiccated lake, a lake shoreline or a river system, so, like source bordering dunes, the timing of aeolian accumulation may not simply affect local aridity (Rowell et al. 2017).

In many cases, the non-aeolian sedimentary units within a sand ramp may provide evidence of enhanced runoff from the topographic barrier, or even to colder conditions that contributed to the development of the talus (Thomas et al. 1997). The ramps may also incorporate palaeosols, which suggest periods of ramp stability (Lancaster and Tchakarian 1996). Some ramps appear to accumulate rapidly, possibly in a single period of a few thousand years (e.g. Bateman et al. 2012), but others record a more complex and

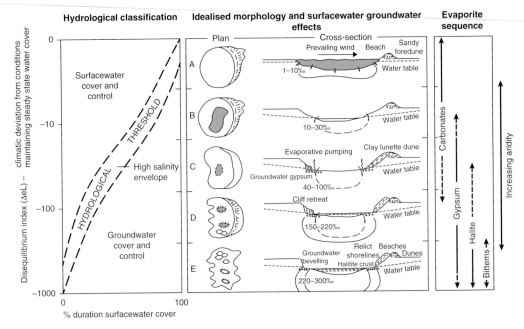

Figure 10.3 Bowler's (1986) schematic diagram of differences in lunette dune formation and pan/playa morphologies across the Australian hydrological gradient. The model embraces spatially changing relationships between basin hydrological state, morphology, fringing lunette dunes, and sediment characteristics. It integrates ideas that sandy lunettes form where deflation occurs from pan-margin beach deposits, while a higher fraction of fine (clay pellet) sediments may be present in lunettes if deflation has occurred from a dry pan floor, where clay has previously accumulated. The model provides a helpful tool for interpreting lunette forms and sediments in the Quaternary record.

Figure 10.4 A large sand ramp that has accumulated on the eastern margin of the Namib Desert, southern Africa. The ramp of aeolian sediment extends from left (east) to right (west) of the photograph and is over 2 km in length. See insert for colour representation of this figure.

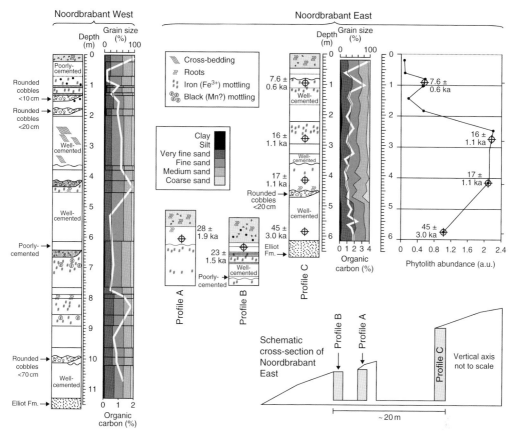

Figure 10.5 Example of the complex and variable stratigraphy that can be recorded within the body of a sand ramp, in this case from Noordbrabant in eastern South Africa. Three sections of the ramp (A, B, C) have been dated using OSL from three sections exposed in gullies, producing a record of fluctuating environmental conditions over 45 ka, including those that created the slope deposits (cobbles). The longer Noordbrabant West section is undated. The aeolian accumulations at Noordbrabant have been interpreted as a function of enhanced deflation during cold periglacial conditions in this upland location. *Source:* Telfer et al. (2012).

longer record of accumulation, which may have lasted over 45 000 years (Telfer et al. 2012; Figure 10.5). As such, the potential for analysis of sand ramps to contribute to palaeoenvironmental analyses is considerable, but they are also complex in terms of applying a single theory to the significance of their development.

10.4 Complex Dune Development

Before luminescence-dating provided chronologies of aeolian accumulation, it was widely assumed that most of the dune systems that are inactive today recorded relatively simple histories of accumulation, which were widely associated in low latitudes with enhanced former aridity (e.g. the classic paper by Sarnthein 1978). In mid- and higher latitudes, the development of now-relict dunes has often been viewed as a function of glacial retreat that had resulted in large volumes of outwash sediment being available in glacial-foreland areas, which also experienced strong winds, and luminescence-dating has often confirmed this to be the case (e.g. Bateman and Murton 2006). While a limited cover of vegetation may have played a role in promoting aeolian activity, especially in the higher latitudes, in some contexts, vegetation may also have played an important role in trapping sand (Koster et al. 1993).

In lower latitudes, complexity in accumulated records was inferred in some studies. For example, spatially distinct phases of accumulation in the Wahiba Sands dunefield, Oman, were attributed to variations in sediment supply from the continental shelf (Goudie et al. 1987). In the Kalahari Desert, in southern Africa, phases of linear dune development, each recorded in a separate dunefield, were differentiated by their general orientation, and were thought to result from shifts in atmospheric circulation patterns (Lancaster 1981). In the same region, paired pan-margin lunette dunes, with different mean azimuth orientations (Lancaster 1978), also suggested formation under different wind regimes. In all these examples, the distinctions within dune system formation and timing were based on spatial patterns of inactive dunes within the landscape.

10.4.1 Complex Dune Profile Records

Most luminescence dating studies of dune systems have employed the collection of sediment samples from within dunes, either in natural exposures, sample pits or by drilling through dunes, the latter allowing the longest sequences of dune records to be achieved (Figure 10.6). In one study of the Wahiba Sands, drilling and sampling for dating attained depths of over 200 m (Preusser et al. 2002). An important general consequence of this approach has been the recognition that most dune systems, and many individual dunes, preserve within their sediments

Figure 10.6 Hand or mechanical auguring of dune sediments using light-tight sampling heads has allowed the timing of dune accumulation to be established using luminescence dating. This example is from UAE. Age datasets for dunefields can be analysed using a range of statistical methods to allow integrated histories of sand sea development to be produced (Thomas and Bailey 2017).

records of multiple periods of accumulation, often over tens of thousands of years. Significantly, this applies not only to dune systems that are widely regarded as largely inactive today, in Australia, southern Africa, and parts of the USA, for example, but to dunefields that are active and in presently hyperarid or very arid areas, such as the Namib (Bristow et al. 2007), the Rub' al Khali, Arabia (e.g. Atkinson et al. 2011), and the Gran Desierto dunefield, Mexico (Beveridge et al. 2006).

Complex histories of accumulation are perhaps not surprising, even in active sand seas today, given that large dunes do not form instantaneously and that aeolian processes may only affect the upper parts of large dunes. Besler (1982), Glennie (1983) and others long suspected that the largest dunes in modern-day desert sand seas were formed in the past; and luminescence data give some support to this hypothesis, at least as concerns moderately recent dunes. The apparent antiquity of large dunes simply reflects the notion of dune memory and the time that it takes for bedform reconstitution to occur (e.g. Warren and Kay 1987; Warren and Allison 1998). An implication for the large body of dated dune sands that now exists from around the world, is that the distinction between active and inactive dunes is even more blurred, and that even demonstrably-active dune systems can preserve a palaeoenvironmental record.

10.5 Interpreting Dated Dune Records

The first directly-dated studies of dune systems led to the proposal that there had been distinct phases of dune accumulation. For example, the first 50 age determinations from the Kalahari dune systems resulted in the identification of four accumulation phases, at 115-95, 46-41, 26–20, and 16-9 ka (Stokes et al. 1997a), with some spatial variations in the intensity of accumulation along the region's north-east–south-west rainfall gradient. These data led to the inference that there had been 'punctuated' Late Quaternary aridity, taking account of central ages and their 1 sigma errors. Similar interpretations of episodic activity and inferred aridity emerged from other dunefields, for example, in the Simpson Desert, Australia (Nanson et al. 1995) and the Algodones in the SW USA (Stokes et al. 1997b). In some dune systems, such as the Australian Strezlecki, the occurrence of features interpreted as weak palaeosols has aided the recognition of hiatuses in accumulation and affirmed the notion of punctuated dune development (e.g. Fitzsimmons et al. 2007).

In some, but not all, sand seas, dunes have patterns that can be inferred to represent different generations of overlapping dunes. An inference to be drawn from these is that environmental conditions, particularly atmospheric circulation and resultant sand transport directions, have changed over time, and, because sand supply has not been limited, generations of dunes have successively developed without the destruction of older, underlying forms. In western Mauritania, Lancaster et al. (2002) applied OSL dating to sediments from three overlapping linear dune patterns, with respective ages of 24-15, 13-10, and younger than 5 ka for each dune 'generation' (see also Section 8.8.2). A comparable study from the eastern Sahara (Bubenzer and Bolten 2008) drew similar conclusions regarding the formation of successive dune generations. Again, these studies suggested that dune palaeoenvironments experienced punctuated activity in the late Quaternary period. Differences in dune size have also sometimes been related to age differences (Warren and Allison 1998).

As more age-data has been produced from dune sands, so earlier interpretations of dune chronologies have been questioned and revised. Most records from individual dunefields are generated by the combination of data from many sampling locations. The complexity of the dune palaeoenvironmental records that has emerged is illustrated for southern Africa (Figure 10.7) and Australia

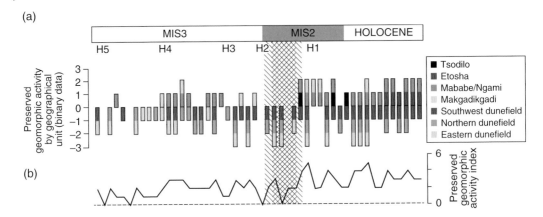

Figure 10.7 The ages of linear dunes in the Kalahari Desert (after Thomas and Burrough 2012). There are many ways of presenting data on the age of dunes, and there has been much discussion about the best way to do so without introducing biases in interpreted peaks and gaps in records. In this diagram the record of luminescence age, for the last 50 ka, taken from studies of linear dunes in three regions of the Kalahari (south-west, eastern, and northern dunefields) is shown (a) along with luminescence records from four lake-shoreline sequences in the northern and middle Kalahari (Tsodilo, Etosha, Mababe/Ngami, and Makgadikgadi). Ages are presented as binary plots in 1000-year time slices, taking central ages, i.e. not including 1 sigma errors (see Thomas and Burrough (2012) for an additional diagram that compares this method with the raw ages plus errors). Sampling bias is avoided in this method by representing the dated deposits in each 1000-year 'bin' as a value of 1, regardless of the number of ages that actually occur. Zero occurs where no age exists for a 1000-year bin. In (b) all the data for each bin are summed as a measure of millennial scale geomorphic activity in the region. Three observations can be made. (i) Dune activity has been widespread and frequent throughout the last 50 ka; few 1000-year periods showing no dated activity. (ii) For many millennia, both lake and dune activity are recorded. (iii) This shows a dynamic landscape, without simple 'dry' and 'wet' phases, at least in terms of the millennial intervals over which the data are shown (sub-millennial variability may occur but is not represented in the dataset). The grey bars show the Last Glacial Maximum as represented by Gasse et al. (2008) (dark grey) and Chase and Meadows (2007) (wider, light grey, bar), this being the period in which Sarnthein (1978) proposed there to have been the maximum extension of dunefields. H1, H2, etc. On the diagram it shows the timing of Heinrich events. See Thomas and Burrough (2012; 2016) for detailed discussion of issues surrounding dune age data representation and interpretation.

(Figure 10.8), two of the most intensively dated dune environments. In the Kalahari, the emergence of over 140 ages from linear dunes in the southern and western dunefields, many from samples from drilled full-dune profiles, has resulted in both an extension back in time of the period in which accumulation is recorded (from < 30 ka in Stokes et al. 1997a, to almost 200 ka in Stone and Thomas 2008) and a 'filling in' of the gaps in the 'punctuated aridity' model that was first proposed (Figure 10.7).

In Australia, Hesse (2014) expressed concern about the dominance of shallow samples in the chronometric records from many dunefields, possibly introducing a potential sampling bias in the available data (Figure 10.8).

An overall interpretation of that record was seen to comprise a low, level, background of sand transport through most of the last glacial cycle, with evidence for an activity/accumulation 'peak' at the LGM only present in the record for the Malee dunefield, in SE Australia. While many studies continue to represent dune-age data on a frequency basis (either histograms or probability density plots), the bias that this can introduce to interpretations is not lost (Hesse 2014). Thus, Thomas and Burrough (2012) used binary age data plots (Figure 10.7) to reduce the effects of sampling bias. While this results in some loss of information, it does result in the complexity (or simplicity!) of dunefield accumulation records being displayed.

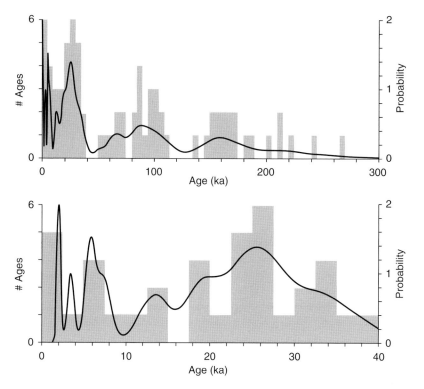

Figure 10.8 OSL ages from the Malee dunefield, Australia. This figure presents ages in a different manner than Figure 10.7, and together the two figures illustrate some of the difficulties of interpreting the data that dating provides for understanding dunefield palaeoenvironmental histories. 'A' shows the total data set, spanning 300 ka and using 5 ka bins to group age data; the last 40 ka are shown in more detail in 'B' with 2.5 ka bins. The grey columns are histograms of total age numbers in each bin, the black lines are probability density curves, in which the height of peaks is influenced by the number of ages. *Source:* Adapted from Hesse (2014).

The emergence of larger data sets has also had impacts on records from dunefields with multiple, overlapping dune generations. Atkinson et al. (2011, 2012) showed that, in the northeast Rub 'al-Khali, a set of basal in the ages of samples from 'upper' dunes overlaps significantly with the ages of samples from the underlying megadune ridges. Thus, there are the two generations of dunes, each with a different mean orientation, but they do not represent distinct phases of formation. This finding has wider implications for inferences about changes in the atmospheric circulation that have been based on overlapping dune patterns, especially in studies that are based on a small number of samples that appear to demonstrate distinct periods if development (e.g. Lancaster et al. 2002). Given recent experimental investigations on the controls of dune orientation, which suggest that multiple dune orientations can emerge within single wind regimes (Courrech du Pont et al. 2014; Lü Ping et al. 2014), this may not be unexpected.

10.6 A Schematic View of Interpreting Dune Palaeoenvironmental Records

The preceding discussion demonstrates that it is not straightforward to extract meaningful palaeoenvironmental interpretations from sedimentary records of dunes. Dunes in different contexts can respond to different forcing factors, with these differences including dune type, dune location within the environment, and whether a dune is an isolated form or part of a more extensive sand sea. Thus, we have moved beyond the assumption that palaeodunes are simply a record of past aridity. A further issue relates to the interpretation of emerging datasets as a sample of a dunefield record as a whole. The manner in which dated dune records have evolved during the past 30 years has been somewhat piecemeal. In many analyses, sampling has been opportunistic, or has only resulted in the generation of age data from a small (sometimes very small) number of dunes from a dunefield comprising hundreds or more individual forms. The issue of representativeness therefore arises (Hesse 2014). It cannot be assumed that all dunes in a dunefield preserve the same sedimentary record: observation of the complex surface dynamics of today's active dunes makes this very evident: the same record *should not* be expected to occur everywhere within a dunefield or even within an individual dune, whether that dune is an isolated feature or part of a more extensive dunefield. This advice is borne out in studies that have attempted to sample individual dunes intensively (Telfer and Thomas 2006; Telfer 2012), and through a study of the effect of sampling patterns on the resultant dated record (Stone and Thomas 2008).

Figure 10.9 is useful when considering the complexity of accumulated dune records. Based on longstanding analyses of dune sedimentary records, in both the rock record and in Quaternary systems, Kocurek (1998) illustrated the sedimentary and climatic interactions that can influence the deposition of dune sediments in a context of a simple climatic shift from humid to arid. In particular, Kocurek's analysis showed that the accumulated record at any point on a dune can be influenced by climatic controls on sediment production, supply to the dune, and by a possible reduction of supply over time during the period of potential maximum sediment transport, which can result in erosion. This means that if the energy to transport sediment on a dune exists, but sediment supply (the arrival of new sediment) is reduced, then

erosion is likely to occur, rather than accumulation. The important points here are that: (i) individual records from a dune are unlikely to represent the full environmental history; and (ii) incompleteness may be manifested in identifiable sedimentary features (see Leighton et al. 2013a, 2013b; Chapter 9), but not all of these features will have the same chronometric meaning (see Figure 10.10), and, in any case, many dunes may lack strong internal stratigraphy (Stone and Thomas 2008).

These complexities have been explored in recent models of dune development (Telfer et al. 2012; Bailey and Thomas 2014). They show that dune sediments are filtered records of changes in the environmental conditions (Figure 10.11). It is therefore not surprising that dated dune records, of which there are now many, are proving difficult clearly to interpret as palaeoenvironmental archives. A switch between erosion and deposition at any location in a dunefield or even on an individual dune in a dunefield is likely to have occurred, so that discontinuous and highly variable stratigraphic sequences are not surprising, even at locations close to each other (Bailey and Thomas 2014). If a reliable palaeoenvironmental record for a region, or dunefield, is to be achieved, it must be based on many sampling locations and depths.

10.6.1 Accumulation: A Better Way to Look at Dune Palaeoenvironmental Sequences?

In the light of such findings, the making of the palaeoenvironmental interpretations of dune ages needs a new approach: the generation of ages alone is insufficient to achieve a meaningful proxy of past conditions. One approach is to use ages to identify changes in the rate of accumulation of dune sands over time, rather than to treat the ages themselves as the proxy for an undefined environmental parameter (Bailey and Thomas 2014). The variability of accumulation rates has been assessed by Hesse (2014) for the Australian linear dune age

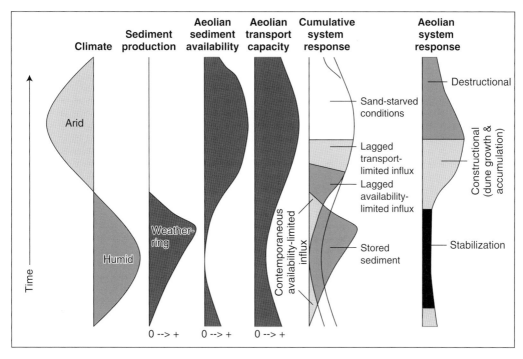

Figure 10.9 A model of dune system responses to climatic changes, from Kocurek (1998). The model assumes a simple transition from humid conditions, when dunes were inactive but future sediment for deflation was being created by other geomorphic processes, to arid conditions, when dunes were active. The model hypothesises that in arid phases, dunes do not simply accumulate, but their behaviour varies through time. As aridity increases, vegetation cover declines, and sediment availability increases, including new material derived from weathering in the humid-phase. The availability of sediment is at first slow to increase, due to lagged supply effects (as vegetation dies off in the source area, sand gradually becomes available to the wind). As the arid phase proceeds, the availability of sediment may fall, so that accumulation of aeolian sediment in dunes may decline, as the ability of the wind to move sediment persists, but the supply of new sediment declines as the source is exhausted, and erosion, rather than deposition, occurs on dunes. The model is not meant to be indicative of all dune environments, nor all dune types; its importance is that it shows that the behaviour of a dune system can be complex in the face of changing climates and other environmental factors, even in a simulation where the climate transition is simple.

dataset, who found maximum rates of $1\,\mathrm{m\,ka^{-1}}$. Using luminescence ages for the last 30 ka from the eastern Rub' al Khali and Wahiba sand seas in Arabia, Leighton et al. (2014); Figure 10.12 calculated rates of up to $15\,\mathrm{m\,ka^{-1}}$.

Leighton et al. (2014) also compared the variability of accumulation in the age data set with changes in accumulation rates modelled using the procedure of Bailey and Thomas (2014), and driven by changes through time in sediment supply, sediment availability, and transport capacity. This follows the framework of 'aeolian system state'

of Kocurek and Lancaster (1999), in which data for these variables was obtained from other regional proxy records of climate change. Figure 10.12b shows how moisture (rainfall), windspeed and sediment supply varied through time, in a manner compatible with the recorded phases of dune accumulation (Figure 10.12a). These initial studies of changes in the accumulation rate over time offer an opportunity to relate dune dynamics to a wider suite of controlling variables, contributing both to paleoenvironmental investigations and to a deeper understanding of dune dynamics.

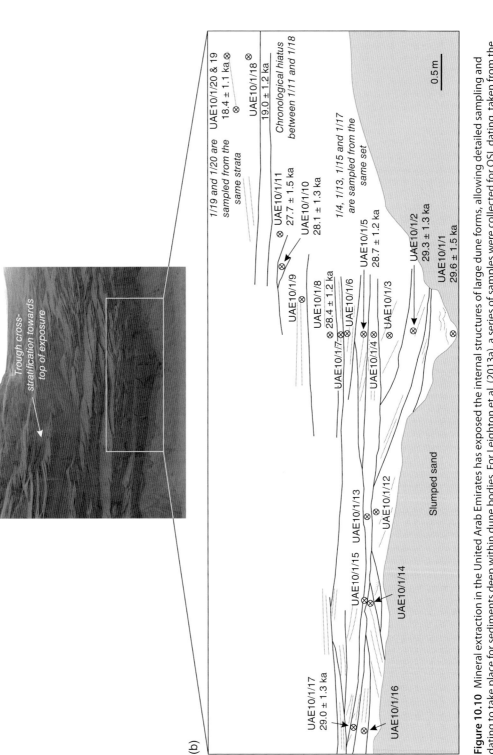

Figure 10.10 Mineral extraction in the United Arab Emirates has exposed the internal structures of large dune forms, allowing detailed sampling and dating to take place for sediments deep within dune bodies. For Leighton et al. (2013a), a series of samples were collected for OSL dating, taken from the exposed sedimentary structures in several dunes. 'Bounding surfaces' are apparent stratigraphical breaks in a dune, separating sediments that have accumulated at different times. In this example, two bounding surfaces, shown as AB1 and AB2, separate sediments below that were dated as 27 000 years or older, from those above that were 19 000 years and younger. However, other surfaces, within the older lower sediments in the dune, do not have equivalent significance as features demarcating long breaks in deposition. See insert for colour representation of this figure.

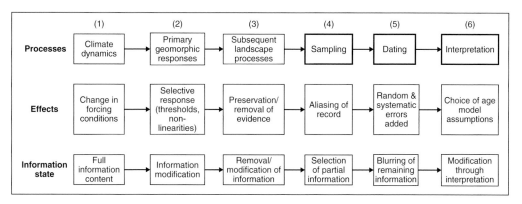

Figure 10.11 Schematic diagram to illustrate why the samples that are dated by OSL in dune studies may not preserve fully the past environmental histories of a dune. Six stages are shown, and for each, there is a 'process', an 'effect' of that process on the system state, and an 'information state' that represents knowledge of the process. The diagram moves, from left to right, from the initial climate of a period of time in the past (i), to the landscape response to that climate (ii) and the processes that ensue (iii), to the sampling (iv) of the landforms that are undertaken to collect samples to date in the laboratory (v), and finally how the ages generated from dating are influenced by the actual processes used in the laboratory (vi). Thus, moving from (i) through to (vi), there are progressive modifications, and losses of information, from the past climate that occurred through to the ages that we generate through analysis. The diagram shows that the information available to us to interpret the past climate is constantly being modified both by the geomorphic processes that have occurred and by what we do as scientists in our attempts to extract data from sediments and landforms. Boxes with thick outlines represent processes that to a degree can be controlled by analytical activities; factors in other boxes cannot be controlled and represent the realities of a preserved dune palaeoenvironmental record. *Source:* From Bailey and Thomas (2014).

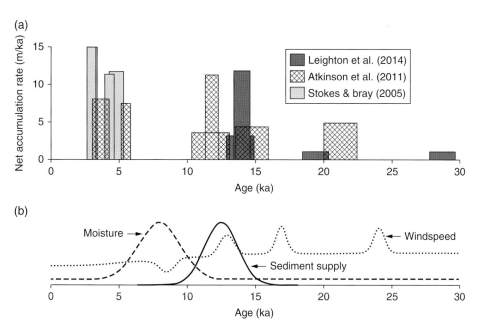

Figure 10.12 (a) Net dune sand accumulation rates derived from some of the dated dune records from the Arabian Peninsula. Rates are determined from records that have two or more (paired) ages in stratigraphic succession, with the time periods on the x axis derived from the paired OSL dates and the accumulation rate on the y axis derived from the thickness of sediment (in metres) between each pair of ages, divided by the time interval between the ages. For example, between 18.5 and 20.5 ka, sediment accumulated in a dune at the rate of ~1 m per 1000 years, while another sampled dune recorded accumulation between 20 and 22.5 ka at a rate of almost 5 m ka^{-1}. No sediment accumulation is recorded between 22.5 and 28 ka. (b) schematically shows the independent records of windspeed, moisture (rainfall) and sediment supply for the region, and illustrates a compatibility between the timing of phases of dune accumulation and the changes in key controlling variables. *Source:* Simplified from Leighton et al. (2014).

10.7 Conclusion

The study of dune palaeoenvironments has moved a long way since early studies that recognised that dunes could persist in the landscape as both active forms and as palimpsests of the past. The interpretation of dune palaeoenvironments has proved more complex, however, as it has met challenges, ranging from a better understanding of dune-forming processes, to dating issues, identified in recent overviews (Thomas 2013; Telfer and Hesse 2013).

The extent of ancient dunefields is well established in Africa, India, Australia, North America, Europe, and to a lesser extent in South America. They are significant elements of landscapes in low-, mid- and high-latitudes. Luminescence dating has provided both an opportunity and a challenge to establishing the periodicities of development through the Late Quaternary, and the environmental and climatic factors that have driven changes in their accumulation and preservation. There has been a tendency for investigations to focus on geographically extensive dunefields, but studies of these landscapes have not always generated age and sedimentary data that are necessarily representative of dune systems that cover many hundreds and thousands of square kilometres. Isolated dunes, including those that are topographically controlled, offer the opportunity to better constrain interpretations of development, both because of their more constrained and confined nature, and because of the identifiable interactions between aeolian and other processes that have affected their development. What is clear is that, whether dealing with dunefields or isolated dunes, there is no single, simple explanation of the timings of their development, nor or the factors that have driven their accumulation.

Further Reading

The following provide useful additional reading on the main topics in this chapter: Leighton et al. (2014); Singhvi and Porat (2008); Stone and Thomas (2008); Telfer and Hesse (2013); Thomas (2013); and Thomas and Burrough (2013).

References

Allchin, B., Goudie, A., and Hegde, K.T.M. (1978). *The Prehistory and Paleogeography of the Great Indian Desert*. London: Academic Press.

Ash, J.E. and Wasson, R.J. (1983). Vegetation and sand mobility in the Australian desert dunefield. *Zeitschrift für Geomorphologie, Supplementband* 45: 7–25.

Atkinson, O.A.C., Thomas, D.S.G., Goudie, A.S., and Bailey, R.M. (2011). Late Quaternary chronology of major dune ridge development in the northeast Rub' al-Khali, United Arab Emirates. *Quaternary Research* 76 (1): 93–105. doi: 10.1016/j.yqres.2011.04.003.

Atkinson, O.A.C., Thomas, D.S.G., Goudie, A.S., and Parker, A. (2012). Holocene development of multiple dune generations in the northeast Rub' al-Khali, United Arab Emirates. *The Holocene* 22 (2): 179–189. doi: 10.1177/0959683611414927.

Bailey, R.M. and Thomas, D.S.G. (2014). A quantitative approach to understanding dated dune stratigraphies. *Earth Surface Processes and Landforms* 39 (5): 614–631. doi: 10.1002/esp.3471.

Bateman, M.D., Bryant, R.G., Foster, I.D.L. et al. (2012). On the formation of sand ramps: a case study from the Mojave Desert.

Geomorphology 161–162: 93–109. doi: 10.1016/j.geomorph.2012.04.004.

Bateman, M.D., Hannam, J., and Livingstone, I. (1999). Late Quaternary dunes at Twigmoor Woods, Lincolnshire, UK: a preliminary investigation. *Zeitschrift für Geomorphologie Supplementband* 116: 131–146.

Bateman, M.D. and Murton, J.B. (2006). The chronostratigraphy of late pleistocene glacial and periglacial aeolian activity in the Tuktoyaktuk Coastlands, NWT, Canada. *Quaternary Science Reviews* 25: 2552–2568. doi: 10.1016/j.quascirev.2005.07.023.

Besler, H. (1982). The north-eastern Rub' al Khali within the borders of the United Arab Emirates. *Zeitschrift für Geomorphologie* 26 (4): 495–504.

Beveridge, C., Kocurek, G., Ewing, R.C. et al. (2006). Development of spatially diverse and complex dune-field patterns: Gran Desierto Dune Field, Sonora, Mexico. *Sedimentology* 53 (6): 1391–1409. doi: 10.1111/j.1365-3091.2006.00814.x.

Bowler, J.M. (1973). Clay dunes: their occurrence, formation and environmental significance. *Earth Science Reviews* 9: 315–338.

Bowler, J.M. (1986). Spatial variability and hydrologic evolution of Australian lake basins: analog for Pleistocene hydrologic change and evaporite formation. *Palaeogeography, Palaeoclimatology, Palaeoecology* 54: 21–41. doi: 10.1016/0031-0182(86)90116-1.

Bristow, C.S., Augustinus, P., Rhodes, E.J. et al. (2011). Is climate change affecting rates of dune migration in Antarctica? *Geology* 39 (3): 831–834. doi: 10.1130/G32212.1.

Bristow, C.S., Duller, G.A.T., and Lancaster, N. (2007). Age and dynamics of linear dunes in the Namib Desert. *Geology* 35: 555–558. doi: 10.1130/G23369A.1.

Bubenzer, O. and Bolten, A. (2008). The use of new elevation data (SRTM/ASTER) for the detection and morphometric quantification of Pleistocene megadunes (draa) in the eastern Sahara and the southern Namib. *Geomorphology* 10 (2): 221–231. doi: 10.1016/j.geomorph.2008.05.003.

Bullard, J.E., Thomas, D.S.G., Livingstone, I., and Wiggs, G.F.S. (1997). Dunefield activity and interactions with climatic variability in the southwest Kalahari Desert. *Earth Surface Processes and Landforms* 22 (2): 165–174. doi: 10.1002/(SICI)1096-9837 (199702)22:2<165::AID-ESP687>3.0.CO;2-9.

Chase, B.M. (2009). Evaluating the use of dune sediments as a proxy for palaeo-aridity: a southern African case study. *Earth-Science Reviews* 9 (1–2): 31–45. doi: 10.1016/j.earscirev.2008.12.004.

Chase, B.M. and Brewer, S. (2009). Last Glacial Maximum dune activity in the Kalahari Desert of southern Africa: observations and simulations. *Quaternary Science Reviews* 28 (3–4): 301–307. doi: 10.1016/j.quascirev.2008.10.008.

Chase, B.M. and Meadows, M.E. (2007). Late Quaternary dynamics of southern Africa's winter rainfall zone. *Earth-Science Reviews* 84 (3–4): 103–138. doi: 10.1016/j.quascirev.2008.10.008.

Courrech du Pont, S., Narteau, C., and Gao, X. (2014). Two modes for dune orientation. *Geology* 4 (9): 743–746. doi: 10.1130/G35657.1.

Fitzsimmons, K.E., Rhodes, E.J., Magee, J.W., and Barrows, T.T. (2007). The timing of linear dune activity in the Strzelecki and Tirari Deserts, Australia. *Quaternary Science Reviews* 2 (39): 2598–2616. doi: 10.1016/j.quascirev.2007.06.010.

Fitzsimmons, K.E., Stern, N., and Murray-Wallace, C.V. (2014). Depositional history and archaeology of the Central Lake Mungo lunette, Willandra Lakes, southeast Australia. *Journal of Archaeological Science* 41: 349–364. doi: 10.1016/j.jas.2013.08.004.

Flint, R.F. and Bond, G. (1968). Pleistocene sand ridges and pans in western Rhodesia. *Bulletin of the Geological Society of America* 79 (3): 299–314. doi: 10.1130/0016-7606 (1968)79[299:PSRAPI]2.0.CO;2.

Gardner, R.A.M. (1988). Aeolianties and marine deposits of the Wahiba Sands: character and palaeoenvironments. *Journal of Oman Studies*, Special Report 3: 75–94.

Gasse, F., Chalie, F.G., Vincens, A. et al. (2008). Climatic patterns in equatorial and southern Africa from 30,000 to 10,000 years ago reconstructed from terrestrial and near-shore proxy data. *Quaternary Science Reviews* 27 (25–26): 2316–2340. doi: 10.1016/j.quascirev.2008.08.027.

Glennie, K.W. (1983). Early Permian (Rotliegendes) palaeowinds of the North Sea. *Sedimentary Geology* 34 (2–3): 245–265. doi: 10.1016/0037-0738(83)90088-X.

Goudie, A.S., Warren, A., Jones, D.K.C., and Cooke, R.U. (1987). The character and possible origins of the aeolian sediments of the Wahiba sand sea, Oman. *Geographical Journal* 153: 231–256. doi: 10.2307/634875.

Grove, A.T. (1958). The ancient erg of Hausaland and similar formations on the south side of the Sahara. *Geographical Journal* 124 (2): 528–533. doi: 10.2307/1790942.

Grove, A.T. (1969). Landforms and climate change in the Kalahari and Ngamiland. *Geographical Journal* 135 (2): 191–212. doi: 10.2307/179682.

Halfen, A.F., Lancaster, N., and Wolfe, S. (2015). Boundary conditions for aeolian activity in North American dune fields. adsabs.harvard.edu/abs/2014AGUFMEP43B3560H

Hesse, P.P. (2014). How do longitudinal dunes respond to climate forcing? Insights from 25 years of luminescence dating of the Australian desert dunefields. *Quaternary International* doi: 10.1016/j.quaint.2014.02.020.

Hesse, P.P. and Simpson, R.L. (2006). Variable vegetation cover and episodic sand movement on longitudinal desert sand dunes. *Geomorphology* 81 (3–4): 276–291. doi: 10.1016/j.geomorph.2006.04.012.

Holmes, P.J., Bateman, M.D., Thomas, D.S.G. et al. (2008). A Holocene-late Pleistocene aeolian record from lunette dunes of the western Free State panfield South Africa. *The Holocene* 1 (8): 1193–1205. doi: 10.1177/0959683608095577.

Hu, F., Li, Z., Jin, J. et al. (2013). Coastal environment evolution record from Anshan coastal aeolian sand of Jinjiang, Fujian Province, based on the OSL dating. *Acta Geographica Sinica* 68: 343–356.

Jennings, J.N. (1975). Desert dunes and estuarine fill in the Fitzroy estuary (North-Western Australia). *Catena* 2 (3–4): 215–262. doi: 10.1016/S0341-8162(75)80015-4.

Kocurek, G. (1998). Aeolian system response to external forcing factors – a sequence stratigraphic view of the Saharan region. In: *Quaternary Deserts and Climatic Change* (ed. A.S. Alsharhan, K.W. Glennie, G.L. Whittle and C.G.S.C. Kendall), 327–337. Rotterdam: Balkema.

Kocurek, G. and Lancaster, N. (1999). Aeolian system sediment state: theory and Mojave Desert Kelso dune field example. *Sedimentology* 46: 505–515. doi: 10.1046/j.1365-3091.1999.00227.x.

Kocurek, G., Lancaster, N., Carr, M., and Frank, A. (1999). Tertiary Tsondab sandstone formation: preliminary bedform reconstruction and comparison to modern Namib Sand Sea dunes. *Journal of African Earth Sciences* 29 (4): 629–642. doi: 10.1016/S0899-5362(99)00120-7.

Koster, E.A., Castel, I.I.Y., and Nap, R.L. (1993). Genesis and sedimentary structures of late Holocene aeolian drift sands in Northwest Europe. In: *The Dynamics and Environmental Context of Aeolian Sedimentary Systems*, vol. 72 (ed. K. Pye), 247–267. London: Geological Society Special Publication. doi: 10.1144/GSL.SP.1993.072.01.20.

Lancaster, I.N. (1978). Composition and formation of southern Kalahari pan margin dunes. *Zeitschrift für Geomorphologie* 22: 148–169.

Lancaster, N. (1981). Palaeoenvironmental implications of fixed dune systems in Southern Africa. *Palaeogeography, Palaeoclimatology, Palaeoecology* 33: 327–346. doi: 10.1016/0031-0182(81)90025-0.

Lancaster, N. (1988). The development of linear dunes in the southwestern Kalahari. *Journal of Arid Environments* 14 (3): 233–244. doi: 10.1016/0037-0738(88)90090-5.

Lancaster, N., Kocurek, G., Singhvi, A. et al. (2002). Late Pleistocene and Holocene dune activity and wind regimes in the Western Sahara Desert of Mauritania. *Geology* 30 (11): 991–994. doi: 10.1130/0091-7613-31.1.e18.

Lancaster, N. and Tchakerian, V.P. (1996). Geomorphology and sediments of sand ramps in the Mojave Desert. *Geomorphology* 17 (1–3): 151–165. doi: 10.1016/0169-555X (95)00101-A.

Leighton, C.L., Bailey, R.M., and Thomas, D.S.G. (2013a). The utility of desert sand dunes as Quaternary chronostratigraphic archives: evidence from the northern Rub' al Khali. *Quaternary Science Reviews* 78: 303–318. doi: 10.1016/j. quascirev.2013.04.016.

Leighton, C.L., Thomas, D.S.G., and Bailey, R.M. (2013b). Allostratigraphy and Quaternary dune sediments: Not all bounding surfaces are the same. *Aeolian Research* 11: 55–60. doi: 10.1016/j.aeolia.2013.09.001.

Leighton, C.L., Bailey, R.M., and Thomas, D.S.G. (2014). Interpreting and modelling late Quaternary dune accumulation in the southern Arabian Peninsula. *Quaternary Science Reviews* 102: 1–13. doi: 10.1016/j. quascirev.2014.08.002.

Livingstone, I.P. and Thomas, D.S.G. (1993). Modes of dune activity and their palaeoenvironmental significance: an evaluation with reference to southern African examples. In: *The Dynamics and Environmental Context of Aeolian Sedimentary Systems*, vol. 72 (ed. K. Pye) Geological Society Special Publication, 91–101. doi: 10.1144/GSL.SP.1993.072.01.10.

Loope, D.B., Elder, J.F., and Sweeney, M.R. (2012). Downslope coarsening in aeolian grainflows of the Navajo Sandstone. *Sedimentary Geology* 265–266: 156–162. doi: 10.1016/j.sedgeo.2012.04.005.

Lu, P., Narteau, C., Dong, Z. et al. (2014). Emergence of oblique dunes in a landscape-scale experiment. *Nature Geoscience* 7: 99–103. doi: 10.1038/ngeo204.

Mason, J.A., Lu, H., Zhou, Y. et al. (2009). Dune mobility and aridity at the desert margin of northern China at a time of peak monsoon strength. *Geology* 37 (10): 947–950. doi: 10.1130/G30240A.

May, J.-H. (2013). Dunes and dunefields in the Bolivian Chaco as potential records of environmental change. *Aeolian Research* 10: 89–102. doi: 10.1016/j.aeolia.2013.04.002.

McFarlane, M.J. and Eckardt, F.D. (2007). Palaeodune morphology associated with the Gumare fault of the Okavango graben in the Botswana/Namibia borderland: a new model of tectonic influence. *South African Journal of Geology* 110 (4): 535–542. doi: 10.2113/gssajg.110.4.535.

Miller, R.M. (2014). Evidence for the evolution of the Kalahari dunes from the Auob River, southeastern Namibia. *Transactions of the Royal Society of South Africa* 69 (3): doi: 10.1080/0035919X.2014.955555.

Munyikwa, K. (2005). The role of dune morphogenetic history in the interpretation of linear dune luminescence chronologies: a review of linear dune dynamics. *Progress in Physical Geography* 29 (3): 317–336. doi: 10.1191/0309133305pp451ra.

Murton, J.B., Frechen, M., and Maddy, D. (2007). Luminescence dating of mid- to Late Wisconsinan aeolian sand as a constraint on the last advance of the Laurentide Ice Sheet across the Tuktoyaktuk Coastlands, western Arctic Canada. *Canadian Journal of Earth Sciences* 44 (6): 857–869. doi: 10.1139/e07-015.

Nanson, G.C., Chen, X.Y., and Price, D.M. (1995). Aeolian and fluvial evidence of changing climate and wind patterns during the past 100 ka in the western Simpson Desert, Australia. *Palaeogeography, Palaeoclimatology, Palaeoecology* 113 (1): 87–102. doi: 10.1016/0031-0182(95)00064-S.

Nichols, G.J. (2005). Sedimentary evolution of the Lower Clair Group, Devonian, West of Shetland: climate and sediment supply controls on fluvial, aeolian and lacustrine deposition. *Petroleum Geology Conference Proceedings* 6: 957–967.

O'Connor, P.W. and Thomas, D.S.G. (1999). The timing and environmental significance of late Quaternary linear dune development in western Zambia.

Quaternary Research 52 (1): 44–55. doi: 10.1006/qres.1999.2042.

Passarge, S. (1904). *Die Kalahari*. Berlin: Dietrich Riemer.

Preusser, F., Radies, D., and Matter, A. (2002). A 160,000-year record of dune development and atmospheric circulation in Southern Arabia. *Science* 296: 2018–2020. doi: 10.1016/j.crte.2009.02.003.

Rendell, H.M., Clarke, M.L., Warren, A., and Chappell, A. (2003). The timing of climbing dune formation in southwestern Niger: Fluvio-aeolian interactions and the rôle of sand supply. *Quaternary Science Reviews* 22 (10–13): 1059–1065. doi: 10.1016/S0277-3791(03)00026-X.

Rendell, H.M., Lancaster, N., and Tchakerian, V.P. (1994). Luminescence dating of late quaternary aeolian deposits at Dale Lake and Cronese Mountains, Mojave Desert, California. *Quaternary Science Reviews* 13 (5–7): 417–422. doi: 10.1016/0277-3791(94)90052-3.

Rich, J. (2013). A 250,000-year record of lunette dune accumulation on the Southern High Plains, USA and implications for past climates. *Quaternary Science Reviews* 62: 1–20. doi: 10.1016/j.quascirev.2012.11.015.

Rowell, A., Thomas, D., Bailey, R. et al. (2017). Controls on sand ramp formation in southern Namibia. *Earth Surface Processes and Landforms* doi: 10.1002/esp4159.

Saini, H.S. and Mujtaba, S.A.I. (2012). Depositional history and palaeoclimatic variations at the northeastern fringe of Thar Desert, Haryana plains, India. *Quaternary International* 250: 37–48. doi: 10.1016/j.quaint.2011.06.002.

Sarnthein, M. (1978). Sand deserts during the last glacial maximum and climatic optimum. *Nature* 272 (5648): 43–46. doi: 10.1038/272043a0.

Singh, G., Joshi, R.D., and Singh, A.P. (1972). Stratigraphic and radiocarbon evidence for the age and development of three salt lake deposits in Rajasthan, India. *Quaternary Research* 2 (4): 496–505. doi: 10.1016/0033-5894(72)90088-9.

Singh, G., Wasson, R.J., and Agrawal, D.P. (1990). Vegetational and seasonal climatic changes since the last full glacial in the Thar Desert, Northwest India. *Review of Palaeobotany and Palynology* 64 (1–4): 351–358. doi: 10.1016/0034-6667(90)90151-8.

Singhvi, A.K. and Porat, N. (2008). Impact of luminescence dating on geomorphological and palaeoclimate research in drylands. *Boreas* 37 (4): 536–558. doi: 10.1111/j.1502-3885.2008.00058.x.

Singhvi, A.K., Sharma, Y.P., and Agrawal, D.P. (1982). Thermo-luminescence dating of sand dunes in Rajasthan, India. *Nature* 295 (5847): 313–315. doi: 10.1038/295313a0.

Smith, H.T.U. (1963). Eolian geomorphology, wind direction and climatic change in North Africa.US Air Force, Cambridge Research Laboratories, Report AF19 (628).

Sperling, C. and Goudie, A.S. (1975). The miliolite of western India: a discussion of aeolian and marine hypotheses. *Sedimentary Geology* 13 (1): 71–75. doi: 10.1016/0037-0738(75)90052-4.

Stokes, S., Bailey, R.M., Fedoroff, N., and O'Marah, K.E. (2004). Optical dating of aeolian dynamism on the West African Sahelian margin. *Geomorphology* 59 (1–4): 281–291. doi: 10.1016/j.geomorph.2003.07.021.

Stokes, S., Kocurek, G., Pye, K., and Winspear, N.R. (1997b). New evidence for the timing of aeolian sand supply to the Algodones dunefield and East Mesa area, southeastern California, USA. *Palaeogeography, Palaeoclimatology, Palaeoecology* 128 (1–4): 63–75. doi: 10.1016/S0031-0182(96)00048-X.

Stokes, S., Thomas, D.S.G., and Washington, R. (1997a). Multiple episodes of aridity in southern Africa since the last interglacial period. *Nature* 388 (6638): 154–158. doi: 10.1038/40596.

Stone, A.E.C. and Thomas, D.S.G. (2008). Linear dune accumulation chronologies from the southwest Kalahari, Namibia: challenges of reconstructing late Quaternary palaeoenvironments from aeolian landforms. *Quaternary Science Reviews* 27 (17–18): 1667–1681. doi: 10.1016/j.quascirev.2008.06.008.

Talbot, M.R. (1984). Late Pleistocene rainfall and dune building in the Sahel. *Palaeoecology of Africa* 16: 203–214.

Telfer, M.W. (2012). Growth by extension, and reworking, of a south-western Kalahari linear dune. *Earth Surface Processes and Landforms* 36: 1125–1135. doi: 10.1002/esp.2140.

Telfer, M.W. and Hesse, P.P. (2013). Palaeoenvironmental reconstructions from linear dunefields: recent progress, current challenges and future directions. *Quaternary Science Reviews* 78: 1–21. doi: 10.1016/j.quascirev.2013.07.007.

Telfer, M. and Thomas, D.S.G. (2006). Spatial and temporal complexity of lunette dune development, Witpan, South Africa: implications for palaeoclimate and models of pan development in arid regions. *Geology* 34 (10): 853–856. doi: 10.1130/G22791.1.

Telfer, M.W., Thomas, Z.A., and Breman, E. (2012). Sand ramps in the Golden Gate Highlands National Park, South Africa: evidence of periglacial aeolian activity during the last glacial. *Palaeogeography, Palaeoclimatology, Palaeoecology* 313–314: 59–69. doi: 10.1016/j.palaeo.2011.10.008.

Thomas, D.S.G. (1984). Ancient ergs of the former arid zones of Zimbabwe, Zambia and Angola. *Transactions of the Institute of British Geographers* 9 (1): 75–88. doi: 10.2307/621868.

Thomas, D.S.G. (2011). Aeolian landscapes and bedforms. In: *Arid Zone Geomorphology: Process, Form and Change in Drylands* (ed. D.S.G. Thomas), 427–453. Chichester: Wiley.

Thomas, D.S.G. (2013). Reconstructing paleoenvironments and palaeoclimates in drylands: what can landform analysis contribute? *Earth Surface Processes and Landforms* 3 (1): 8): 3-16. doi: 10.1002/esp.3190.

Thomas, D.S.G. and Bailey, R.M. (2017). Is there evidence for global-scale forcing of Southern Hemisphere Quaternary desert dune accumulation?. A quantitative method for testing hypotheses of dune system development. *Earth Surface Processes*

and Landforms 42 (1): 2280–2294. doi: 10.1002/esp4183.

Thomas, D.S.G., Bateman, M.D., Mehrshahi, D., and O'Hara, S.L. (1997). Development and environmental significance of an eolian sand ramp of last-glacial age, Central Iran. *Quaternary Research* 48 (2): 155–161. doi: 10.1006/qres.1997.1923.

Thomas, D.S.G. and Burrough, S.L. (2012). Interpreting geo-proxies of late Quaternary climate change in African drylands: implications for understanding environmental and early human behaviour. *Quaternary International* 253: 5–17. doi: 10.1016/j.quaint.2010.11.001.

Thomas, D.S.G. and Burrough, S.L. (2016). Luminescence-based dune chronologies in southern Africa: analysis and interpretation of dune database records across the subcontinent. *Quaternary International* 410, Part B: 30–45. doi: org/10.1016/j.quaint.2013.09.008.

Thomas, D.S.G., Holmes, P.J., Bateman, M.D., and Marker, M.E. (2002). Geomorphic evidence for late Quaternary environmental change from the eastern Great Karoo margin, South Africa. *Quaternary International* 89 (1): 151–164. doi: 10.1016/S1040-6182(01)00086-6.

Thomas, D.S.G., Knight, M., and Wiggs, G.F.S. (2005). Remobilization of southern African desert dune systems by twenty-first century global warming. *Nature* 435 (7046): 1218–1221. doi: 10.1038/nature03717.

Thomas, D.S.G. and Shaw, P.A. (1991). Relict desert dune systems: interpretations and problems. *Journal of Arid Environments* 20 (1): 1–14.

Tsoar, H. (2005). Sand dunes mobility and stability in relation to climate. *Physica A: Statistical Mechanics and its Applications* 357 (1): 50–56. doi: 10.1016/j.physa.2005.05.067.

Tsoar, H. and Møller, J.T. (1986). The role of vegetation in the formation of linear dunes. In: *Aeolian Geomorphology*, vol. 17 (ed. W.G. Nickling), 75–95. Binghampton Symposia in geomorphology, International series. Boston: Allen and Unwin.

Warren, A. and Allison, D. (1998). The palaeoenvironmental significance of dune size hierarchies. *Palaeogeography, Palaeoclimatology, Palaeoecology* 137 (3–4): 289–303. doi: 10.1016/S0031-0182 (97)00110-7.

Warren, A. and Kay, S. (1987). Dune networks. In: *Desert Sediments: Ancient and Modern* (ed. L.E. Frostick and I. Reid), 205-212, Geological Society of London, Special Publication 35, doi:10.1144/GSL. SP.1987.035.01.14.

Wasson, R.J. (1984). Late Quaternary palaeoenvironments in the desert dunefields of Australia. In: *Late Cainozoic*

Palaeoclimates of the Southern Hemisphere (ed. J.C. Vogel), 419–432. Rotterdam: Balkema.

Wiggs, G.F.S., Thomas, D.S.G., Bullard, J.E., and Livingstone, I. (1995). Dune mobility and vegetation cover in the southwest Kalahari Desert. *Earth Surface Processes and Landforms* 20 (6): 515–530. doi: 10.1002/ esp.3290200604.

Williams, M.A.J. (1985). Pleistocene aridity in tropical Africa, Australia and Asia. In: *Environmental Change and Tropical Geomorphology* (ed. I. Douglas and T. Spencer), 219–235. London: Allen and Unwin.

11

Planetary Aeolian Geomorphology

Mary C. Bourke[1], Matthew Balme[2], Stephen Lewis[2], Ralph D. Lorenz[3], and Eric Parteli[4]

[1] Trinity College, Dublin, Ireland
[2] The Open University, Milton Keynes, UK
[3] Johns Hopkins University, Laurel, Maryland, USA
[4] University of Cologne, Cologne, Germany

11.1 Introduction

Aeolian processes play an essential role not only in the dynamics of beaches and deserts on the Earth, but also contribute to surface landforms on several bodies in our solar system.

11.2 Planetary Atmospheres

There are at least four bodies in our solar system that have sufficient enough atmosphere to sustain winds that can transport sediment: Mars, Titan, Venus, and Earth. These atmospheres interact with geological processes and influence the morphology and composition of surfaces (Grotzinger et al. 2013).

11.2.1 Mars

Landforms and geochemical signatures suggest that the atmosphere of Mars was significantly thicker in its past, even as Mars today has the thinnest atmosphere of all the bodies that contain confirmed aeolian features. On Mars. the orbital parameters of

eccentricity, obliquity, and season of perihelion are more variable than on Earth, leading to a strongly forced global climate that has varied significantly through time (Laskar et al. 2004).

As on Earth, atmospheric circulation patterns on Mars are generated by air flows at many different scales, varying spatially, and seasonally. Martian global wind patterns are derived from Hadley cells. Also, baroclinic instability on Mars is produced by an obliquity and rotation similar to Earth (see Table 11.1). Air rises at intertropical convergence zones (ITCZ) and sinks at the poles. Unlike Earth, however, the Hadley cells are of unequal size in different seasons as the Martian ITCZ shifts latitudinally due to the rapid thermal response of the surface. For example, at the solstices, the summer hemisphere Hadley cell crosses the equator and sinks in the winter hemisphere. The resultant surface winds are, like Earth, deflected to the west as they approach the equator by a Coriolis effect. The significantly higher elevations in the southern hemisphere together with severe eccentricities (see Table 11.1), in which the perihelion occurs during the southern summer, contribute to a stronger

Table 11.1 Properties of planetary atmospheres and surfaces.

Body	Venus	Earth	Mars	Titan
Mean distance from sun (AU)	0.72	1	1.52	9.58
Diameter (Km)	12 104	12 742	6779	5150
Gravitational acceleration (m s^{-2})	8.87	9.81	3.71	1.35
Rotation period (sidereal day)	223	1	1.04	16
Mean surface atmospheric pressure (bar)	91	1.01	0.007	1.47
Mean surface air temperature (°C)	464	15	−63	−178
Dominant atmospheric gas	CO_2	N2, O_2	CO_2	N_2, CH_4
Atmospheric density (kg m^{-3})	64	1.25	0.02	5.4
Dynamic viscosity (10^{-6} Pa-s)	35	17	13	6
Planetary boundary layer (km)	0.2?	0.3–3	>10	2–3
Total topographic relief (km)	13.7	19.8	29.4	~0.5
Minimum threshold friction speed (ms^{-1})[a]	0.02	0.2	2.0	0.04
Dune field cover (x10^6 km^2)[b]	0.0183	5	0.9	10
Dune field cover (% land and ocean surface area)[c]	0.004	0.98	0.62	12.5

[a]Greeley and Iverson (1985).
[b]Estimates from Greeley et al. (1995), Livingstone and Warren (1996), Fenton and Hayward (2010), Le Gall et al. (2011).
[c]*Source:* After Fenton et al. (2013).

Hadley circulation during the southern summer than in the northern. Thus, the largest dust storms on Mars occur during the southern summer.

However, it should be noted that there is also a significant variability of atmospheric pressure on Mars at the seasonal scale. During the Martian winter, atmospheric temperatures move below the condensation point of CO_2. This results in a drop of atmospheric pressure by 30% over a Mars year as the seasonal CO_2 ice cap forms. These periods of low pressure depress the conditions for saltation globally while the aeolian deposits are sequestered under their seasonal ice caps at the poles. The seasonal ice caps are deposited down to approximately 55°N and 50°S. Wintertime jet streams that generate westerly winds and fronts are produced by the strong thermal contrasts at the edges of these polar caps. These are particularly effective in areas of low topographic roughness such as the northern plains. During the springtime sublimation of these seasonal caps, off-cap katabatic winds are generated.

There is significant difference in the elevation of the southern hemisphere on Mars to the lower (by several kilometres) northern hemisphere. Atmospheric pressure and density vary considerable across these topographic heights on Mars. For example, atmospheric pressure at the base of the Tharsis Volcanoes is 6mb, and this drops by a factor of almost 6 at the 22 km high summit of the volcanoes. This is reflected in the larger wavelength of bedforms on the volcanic peaks (Lorenz et al. 2014). Topographic forcing at canyon walls, crater rims, and volcanic peaks produces geomorphologically significant slope winds that can generate net upslope flow during the day and downslope flow during the night. The enhanced wind velocities generated by topographic forcing may create areas of enhanced sediment transport on Mars (Bourke et al. 2004; Jackson et al. 2015). On a diurnal scale, surface heating leads to convective generation of wind gusts and dust devils (see Section 11.6.1).

11.2.2 Titan

Saturn dominates the global circulation patterns, length of year, and seasons on its moon, Titan. A year on Titan is 29.46 terrestrial years, and the small rotation rate of Titan (1 Titan day = 15.95 Earth days) inhibits mid and high-latitude baroclinic instability. The eccentric orbit (0.056) induces shorter and more intense southern hemisphere summers (similar to Mars). Titan's atmosphere super-rotates at altitude but this flow does not dominate surface wind patterns. Near-surface circulation on Titan is dominated by solar heating that induces Hadley cell formation, with the winds generated by the return flow of each cell. The cells rise either side of the ITCZ (Flasar et al. 2010) and extend to high latitudes. At each solstice a single Hadley cell rises in the high latitudes in the summer hemisphere and sinks at the winter pole. These produce northerly surface winds during the southern summer and southerly winds during the northern summer. The Coriolis force causes westerly components in the mid-latitudes and an easterly deflection close to the equator. A modelled strong westerly $1–1.5\,\mathrm{m\,s^{-1}}$ at the equinoxes may orient the linear equatorial dunefields (Tokano 2010).

11.2.3 Venus

The atmosphere on Venus is the densest of the four bodies (Table 11.1). As a consequence, an extreme greenhouse effect traps heat and reduces spatial and temporal surface temperature variability. Hadley cell circulation rises at the equator and descends at the poles, with southern hemisphere winds deflected to the west by topography (as planetary rotation rate is very low (1 Venus day = 116.8 Earth days) (Table 11.1). Global circulation is generally towards the equator. Very little data is available on the atmospheric circulation of Venus. Similar to Mars, slope winds may be important on Venus.

11.3 Planetary Sediment Transport (Mars, Titan, Venus)

Sediment transport is responsible for a broad range of geophysical phenomena, including dust storms on Earth and Mars and the formation and migration of dunes on both planetary bodies, as well as on Venus and Titan (Bourke et al. 2010; Kok et al. 2012). In particular, the shape and migration of dunes provide excellent proxies for the characteristics of sediment transport across the surface of planetary bodies (Lorenz and Zimbelman 2014). This section aims briefly to discuss these characteristics, as well as presenting open questions on planetary sediment transport that need to be addressed to improve our understanding of planetary geomorphology, with emphasis on dune formation and migration.

Indeed, there are important differences in the scale and dynamics between dunes on Earth and their extra-terrestrial counterparts. A considerable body of research in the last two decades has been devoted to the understanding of these differences. This has helped us to identify characteristics of sediment transport that are particular to each distinct physical environment, and has pushed forward our understanding of the sediment transport on our own planet (Kok et al. 2012).

As discussed in Chapter 2, the transport of sediments along the surface occurs mainly through saltation, which starts when the wind shear velocity u_* exceeds a minimal threshold $u_{*\mathrm{ft}}$ – called the *fluid threshold* – for initiating transport (Bagnold 1941; Iverson and White 1982). Once transport begins, particles move in nearly ballistic trajectories close to the ground thereby ejecting new particles upon collisions with the bed (splash). Because of the contribution of splash to grain entrainment, saltation can be sustained at a wind velocity that is less than the one required to lift sand grains. The associated threshold shear velocity is called the *impact threshold* u_{*t}, which for aeolian transport on Earth is about 80% of u_{*ft}. However, the relation between u_{*t} and u_{*ft} is not the same for Earth, Mars, Venus, and Titan, as the characteristics of sediment transport differ substantially among the four planetary bodies as shown in Table 11.2.

11.3.1 Mars

Because the atmosphere of Mars is thin (nearly 80 times less dense than the Earth's), sediment transport begins at wind speeds one order of magnitude larger than those required to entrain sand on Earth. The higher entrainment wind speeds, the lower gravity, and thinner atmosphere of Mars lead to much longer and higher saltation trajectories than those of terrestrial sand grains (see Table 11.2). A critical consequence of the larger saltation trajectories on Mars is that Martian grains hit the surface with much higher energy than saltating particles on Earth, which causes the efficiency of the splash in ejecting grains from the surface on Mars to be higher than on Earth (Marshall and Stratton 1999; Parteli and Herrmann 2007).

The puzzle is this: wind shear velocity values on Mars estimated from *in situ* measurements of the wind velocity profile by Martian Landers are typically *below* $1.0\,\mathrm{m\,s^{-1}}$, which equates to estimated average values of u_* on Mars in the range between 0.3 and $0.6\,\mathrm{m\,s^{-1}}$ (Sutton et al. 1978; Sullivan et al. 2000; Holstein-Rathlou et al. 2010). Yet such values are well below the fluid threshold u_{*ft} on Mars (see Table 11.2). That sediment moved at all on Mars was intriguing.

Observations by early missions to Mars provided only occasional evidence for aeolian surface changes on loose surfaces on Mars (Kok et al. 2012). In particular, the first *in situ* evidence for aeolian transport on the surface of an extraterrestrial planet was registered only after the third winter of the Viking 1 mission, that is, after 2.5 Martian years or five Earth years of mission observations. Arvidson et al. (1983) reported enthusiastically:

> (...) The movement of the rock, the alternations of the conical piles, clods, trenches, and other features, and the increase in scene contrast, demonstrate that an erosion event or events of substantial magnitude occurred during the third winter season, probably between Sols 1720 and 1757.

In fact, in sol 1742 (a mean Martian solar day, or 'sol', is about 24 hours, 39 minutes and 35.244 seconds), an image was acquired, that showed an enormous storm in progress, which has been named 'The Martian Dust Storm of Sol 1742' (Moore 1985). The changes in the surface during that event included significant erosion of sand piles constructed by the lander at the landing site and the formation of a ripple-like bed with wavelengths of several centimetres (Arvidson et al. 1983). Thus, the bed modifications represented the first evidence that saltation could occur under present atmospheric condition of Mars.

In the three subsequent decades, the general view of Martian aeolian processes was that saltation transport occurs in short gusts of strong (but rare) aeolian activity (Sullivan et al. 2005; Parteli and Herrmann 2007; Almeida et al. 2008).

However, advances in image modelling and processing, made in the last decade, have permitted the measurement of relative displacements between images to the sub-pixel level, and led to a revolution in our knowledge of Martian dune dynamics (Bridges et al. 2012b; Silvestro et al. 2013; Ayoub et al. 2014). In particular, the migration of dunes and ripples on Mars has been found to be ubiquitous: large sand dunes migrate at several centimetres a year (Bridges et al. 2012a). Such high migration rates come as a surprise considering that the fluid threshold shear velocities on Mars are much larger than the average wind shear velocities on the Red Planet (Kok et al. 2012).

Table 11.2 Characteristics of saltation transport on Earth, Mars, Venus, and Titan.

Planetary body	Gravity (g) $(\mathrm{m\,s^{-2}})$	Air density $(\mathrm{kg\,m^{-3}})$	Particle density $(\mathrm{kg\,m^{-3}})$	Typical particle size (μm)	Fluid threshold u_{*ft} $(\mathrm{m\,s^{-1}})^a$	Ratio of impact to fluid threshold, $u_{*t}/u_{*ft}{}^b$	Typical saltation height $(\mathrm{cm})^b$	Typical saltation length $(\mathrm{cm})^b$	Dune scale (m)
Earth	9.81	1.2	2650	150–250	~0.2	~0.8	~3	~30	5–10
Mars	3.71	0.02	3000	100–500	~1.5	~0.1	~10	~100	~100
Venus	8.87	66	3000	Unknown	~0.02	>1	~0.2	~1	0.1–0.3
Titan	1.352	5.1	1000	Unknown	~0.04	>1	~0.8	~8	1–2

[a] computed with the equations of Iversen and White (1982).
[b] obtained from numerical simulations by Kok et al. (2012).

The puzzle has been somewhat resolved by computer simulations (Almeida et al. 2008; Kok 2010a) that demonstrated the effect of the Martian splash. Subsequent entrainment of saltating particles could cause the ratio u_{*t}/u_{*ft} on Mars to be substantially lower than on Earth. Values of this ratio as small as 10% have been predicted, which means that sand transport on Mars might ensue even if the wind strength decreases down to Earth-like values. Although there is little evidence for a lower u_{*t}/u_{*ft} on Mars than the corresponding value on Earth (Ayoub et al. 2014), this difference could be the explanation.

The new picture of Martian sediment transport proposed by Almeida et al. (2008) and Kok (2010b) offers an explanation for such sand fluxes in spite of the high Martian threshold for aerodynamic entrainment. Specifically, a plausible explanation for the Earth-like fluxes on Mars is the low Martian impact threshold u_{*t}, which allows sand to be transported at normal Martian wind speeds (Almeida et al. 2008; Kok 2010a). This explanation still needs to be supported by wind tunnel measurements that confirm the predicted value of u_{*t}. Nevertheless there are deeper consequences for a broad range of aeolian processes on the surface of the Red Planet.

Much progress has been achieved in the development of reliable theoretical expressions for computing the saturation length L_{sat}, and saturated flux, Q, of sediment transport as a function of flow conditions and the attributes of sediment and fluid. By estimating these quantities from the size and migration speed of planetary dunes, valuable information can be gained indirectly about grain size, wind velocity and duration of the sand moving wind gusts on planetary surfaces (Kok et al. 2012; Pähtz et al. 2013).

11.3.2 Venus and Titan

As shown in Table 11.2, the threshold shear velocities for direct entrainment on Venus and Titan differ from those on Mars, and are an order of magnitude smaller than on Earth.

The much lower wind speeds required to mobilise sand grains and the large atmospheric density relative to the Earth's lead to smaller saltation trajectories, with particles jumping to heights lower than about a centimetre. Moreover, the lower trajectories mean that particles acquire low downwind velocities, which substantially decreases the strength of their impact on the surface. In other words, the splash on Venus and Titan is of negligible importance in sand entrainment compared to the direct entrainment processes by aerodynamic forces. Therefore, saltation on Venus and Titan is probably much more akin to saltation under water, and cannot be sustained below the threshold for aerodynamic entrainment u_{*ft} (Kok et al. 2012). The difference in the modes of sediment transport on Venus and Titan also have important consequences for determining the scale of dunes on these planetary bodies, as discussed below.

11.4 From Sediment Transport to Aeolian Bedforms

Much of our knowledge about sand flux, wind regimes, and attributes of sediment in Martian dune fields has been gained from the study of dynamics and scale of ripples and dunes detected from satellite coverage.

The main characteristics of the mechanism leading to dune formation are depicted in Figure 11.1. A topographic feature is an obstacle in the wind, which causes a reduction in the wind velocity at its upwind front and an upward deflection of the flow. At the crest, a negative pressure perturbation keeps the flow attached to the surface. This leads to a compression of the flow streamlines and thus to a higher velocity gradient (and consequently a larger shear velocity) near the crest than in the adjacent lower topography. However, the position of maximum shear velocity is not at the crest. The upward deflection of the flow at the windward side is

Figure 11.1 Streamlines of wind flow over a hill of Gaussian profile in the along-wind direction (solid surface). L_{sat} gives the spatial lag between the positions of maximal shear stress (τ_{max}) and maximal flux (q_{max}).

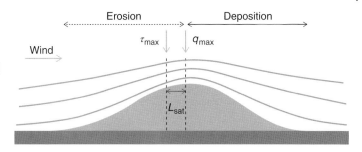

counteracted by the inertia of the turbulent velocity fluctuations at the downwind side of the hill, such that the maximal shear velocity at the surface is shifted upwind with respect to the profile of the terrain (Kroy et al. 2002; Claudin et al. 2013). That is, if the sand flux responded with no delay to wind velocity, the maximum flux would always occur upwind of the crest, and thus sand would be always deposited on the bump thus leading to dune growth.

However, there is a finite time (or equivalently a saturation length, L_{sat}) for the sand flux to react to a change in flow conditions. The bump grows and evolves into a dune only if its size is large enough such that the flux is maximal upwind of the bump's crest. In other words, the saturation length defines a minimal length scale below which the bump is eroded and a dune cannot form (Kroy et al. 2002). Based on empirical observations of the minimal dune size in air and under water, the following approximate relation has been proposed (Hersen et al. 2002):

$$L_{sat} \approx 2d\,\rho_{grain}/\rho_{fluid} \qquad (11.1)$$

where d is the average particle diameter, ρ_{fluid} is the fluid density and ρ_{grain} is the particle density. This scaling relation has been improved by a theoretical model for L_{sat} (Pähtz et al. 2013), which further reproduces the dependence of L_{sat} on u_* observed for sediment transport in the subaqueous regime (which dictates dune formation on Venus and Titan) and is consistent with measurements of the saturation length over at least 5 orders of magnitude in the ratio of fluid and particle density (Figure 11.2).

Ripples are smaller than L_{sat}. The origin of ripple formation is a 'screening' instability of the sand surface exposed to the impact of saltating particles (Anderson 1987; Fourrière et al. 2010). Saltating grains colliding obliquely onto a sand bed eject many *reptating* grains (see Chapter 2), thereby generating small depressions on the surface and leading to a chain of small-scale undulations of asymmetric profile. The windward side of the perturbations is less steep than the lee side, and is the side most exposed to impacts of the saltating particles. In contrast, the lee side is less susceptible to erosion by grain impact such that reptating grains falling in the lee accumulate there, thus leading to ripple growth. While dune migration velocity gives a proxy for saltation flux, ripple migration rates are controlled by the flux of reptating particles.

11.4.1 Deriving Sediment Transport Characteristics from Planetary Bedform Characteristics

The main quantities controlling the characteristics of sediment transport, that is the steady-state flux (or saturated flux; see Chapter 2) and the saturation length, can be obtained indirectly from the dynamics and scale of planetary dunes.

Specifically, the saturated (bulk) flux Q (in units of area per unit time) can be obtained by measuring the displacements of planetary barchan dunes. Indeed, the migration velocity of a barchan can be estimated for a dune in steady state by applying mass conservation at the dune's crest (Bagnold 1941), which yields a scaling of $v_m \approx aQ/W$,

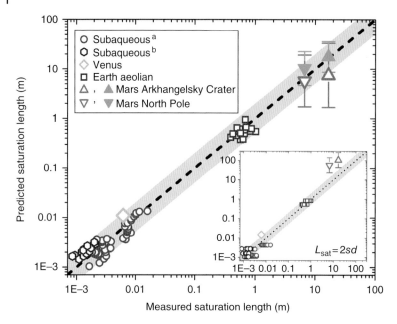

Figure 11.2 Values of L_{sat} estimated from measurements and predicted from theory (Pähtz et al. 2013). Measurements of L_{sat} for subaqueous sediment transport were based on experimental data of Fourrière et al. (2010) and Franklin and Charru (2011) (data sets 'a' and 'b', respectively). For Venus, L_{sat} was estimated from the crest-to-crest distance of elementary dunes (Greeley et al. 1992). Field and wind-tunnel measurements of L_{sat} for aeolian transport on Earth were provided by Andreotti et al. (2010). For Mars, L_{sat} was derived from the minimal size of barchans (Lorenz and Zimbelman 2014; Parteli et al. 2007), and plotted against L_{sat} predicted using average particle size $d = 100$–600 μm (Bourke et al. 2010). The filled symbols use $u_* = u_{*ft}$, while the open symbols use u_* obtained from Viking 2 wind speed measurements using a theoretical model (Pähtz et al. 2013). The inset shows measured and predicted values of L_{sat} for the same conditions as in the main plot, but using a different model proposed by Claudin and Andreotti (2006). According to this model, $L_{sat} = 2sd$, where s is the grain to fluid density ratio. *Source:* After Pähtz et al. (2013).

where $a \approx 50$ (Durán et al. 2010), and, equivalently, an approximate scaling of v_m with Q/H, with H standing for the dune height. Therefore, the saturated flux follows the scaling $Q \sim HD/T$, where D is the distance covered by a barchan within the observation time span T (Bridges et al. 2012b). Note that Q is related to the average, height-integrated steady-state mass flux q of transported particles, in units of mass per unit length and per unit time, through the equation $Q = q/\rho_{bed}$, where ρ_{bed} is the average density of the sediment bulk, which for loose sand is typically around 62% of ρ_{grain} (Bagnold 1941; Durán et al. 2010; Kok et al. 2012).

The saturation length L_{sat} can also be estimated indirectly from the minimal crosswind width W of barchan dunes through the approximate relation $W \approx 12\,L_{sat}$ (Parteli et al. 2007). Barchans smaller than the minimal size do not display limbs or a slip face and have a dome-like shape. The largest dome which has neither slip face nor limbs and is smaller than the smallest barchan dune indicates the minimal dune size. This method has been used to estimate values of L_{sat} in different dune fields on Mars from the scale of barchan dunes (Parteli and Herrmann 2007), and the values obtained by this method were reproduced by theoretical calculations adopting plausible grain sizes and average wind shear velocities valid for Mars (Pähtz et al. 2013).

Yet another method of estimating L_{sat} is based on a theoretical model which relates the saturation length to the wavelength or crest-to-crest distance of 'elementary dunes', i.e. the smallest superimposed dunes occurring on a flat surface or on top of large barchan dunes (Andreotti et al. 2010; Fourrière et al. 2010). This method was used to obtain L_{sat} from the crest-to-crest distance of Venusian microdunes (10–30 cm) produced in wind tunnel experiments under atmospheric conditions valid for Venus (Greeley et al. 1992). Finally, by computing L_{sat} under atmospheric conditions valid for Titan, the predicted wavelength of elementary dunes on Titan is about 1.5 m (Claudin and Andreotti 2006), which, unfortunately, is well below the resolution of Cassini's radar.

11.5 Planetary Aeolian Deposition Features: Dunes, Sand Seas, Sediment Source

11.5.1 Mars

The geomorphic processes that generate sediment for aeolian entrainment have changed through time on Mars. Early in Mars's history (the Noachian and Hesperian: 4.1–3.0 Ga) surface processes were dominated by impact-cratering processes that broke apart the primordial crust to form a megaregolith (a range of grain sizes from megaclasts to silt and clay). It was also a time of extensive volcanic activity and a period where a thicker atmosphere permitted a phase of chemical weathering (leading to extensive deposits of clay minerals), active fluvial processes, catastrophic flooding, and the formation of lakes and potentially an ocean. It was also a time of aeolian activity.

Stratigraphic evidence exposed in the walls of small craters suggests that the history of windblown dune sediment is long on Mars. The Opportunity Rover imaged thick sequences of aeolian cross-bedded sandstone that probably date back 3.5 Ga (Grotzinger et al. 2005). These are the oldest known aeolian sandstones in our solar system.

We know, from several decades of surface and orbiting observations, that Mars's

surface displays aeolian dune forms with a wide range of sizes and morphologies (typically transverse). Other, smaller aeolian bedforms, such as ripple-like features, have now been observed by various Rover missions (Greeley et al. 2004).

There are approximately 2000 dunefields on Mars that cover an area of ~904 000 km^2 (Hayward et al. 2007; Fenton and Hayward 2010). The majority of these dunes are located in large dunefields that circle the North Polar ice cap. The remainder of the dunefields are located in topographic traps on the floor of impact craters and inter-crater plains. The dunefields are found both in the deepest topographic lows on Mars (e.g. Hellas basin −7 km) and at elevations of 10–21 km above datum.

The lithology of the majority of grains on Mars is basalt. Grain density is 3.0 g cm^{-3} as compared to quartz (2.65 g cm^{-3}). Dunes may contain some more locally-derived minerals. For example, dunes in the Olympiae Undae have gypsum concentrations that may be ablated from the polar cap (Massé et al. 2010).

Many of the modern extensive dune fields have local sediment sources. The North Polar dunefields sediment is likely sourced from the adjacent polar layered deposits (Edgett et al. 2003). The sources of several southern hemisphere dunefields are thought to be locally derived from exposures of sand sources in crater walls and pits in the crater floor (Tirsch et al. 2011).

Data from space-borne platforms at Mars suggest that the grain size of dune sand is medium- to coarse-grained sand (Edgett and Christensen 1991). Lander and Rover platforms suggest finer sand with sizes of 200–300 μm sand (Sullivan et al. 2008). Sites that may contain sand-sized aggregates of dust arranged in 5–30 m wide reticulate bedforms (Bridges et al. 2010) have also been identified. The NASA Mars Exploration Rovers examined several ripple-like landforms, each a few metres long and 5–10 cm or so in height, and were able to examine their interiors by 'trenching' through them with their wheels. They found that although

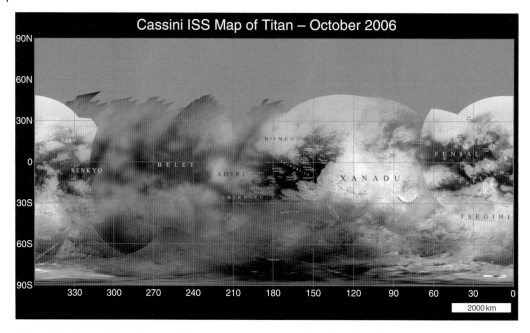

Figure 11.3 A near-infrared map of Titan, showing major named bright and dark regions around the equator: the dark regions are sand seas.

their surfaces were covered by sand and sometimes granule-sized grains, their interiors included many grains too small to be resolved by the microscopic imager (<30 μm in diameter) as well as ~50–100 μm diameter fine sand-sized material (Sullivan et al. 2005).

11.5.2 Titan

The likelihood of dunes on Titan was noted, as soon as it was determined by the Voyager 1 encounter in 1980 that it had a thick atmosphere (Greeley and Iversen 1985). But while a dense atmosphere and low gravity mean the wind speed threshold to move particles is small, it also means that the solar-driven buoyant forces lead to proportionately weak winds. The expectation of hydrocarbon seas on Titan also conditioned expectations, since such seas might act as sand traps. However, these cautions were confounded by Titan's diversity (Lorenz et al. 1995). Because Titan rotates slowly, its Hadley circulation is more extensive and its lower latitudes become desiccated and methane moisture is transported to the North Pole. It turned out that the dark

regions girding Titan's equator (Figure 11.3) are vast seas of giant linear dunes, covering some ~15% of Titan's surface (Lorenz et al. 2006). A couple of low, dark dunes are visible in the mosaic of images made by the Huygens probe during its parachute descent in 2005, but the principal means of surveying Titan's dunes has been the Cassini radar mapper.

The dunes appear (Figure 11.4) to have a morphology and size very similar to those of the largest linear dunes on Earth, in the Namib and Arabian deserts, with spacing of around 3 km and lengths of tens to hundreds of km (Radebaugh et al. 2010). The dune heights have been inferred from radar measurements to be up to about 150 m. Despite Titan's atmosphere being dense, its gravity being $1.35\,ms^{-2}$ (like the Earth's moon) and the particle composition being very different from Earth, it seems that the controlling parameter on the scale of mature dunes (Andreotti et al. 2009) is the thickness of the atmospheric boundary layer, which is about 3 km on both Titan and Earth (Lorenz et al. 2010), Titan's denser atmosphere is compensated for by its longer day-night cycle.

Figure 11.4 A Cassini radar image (about 100 × 300km) of the Belet sand sea on Titan. The linear features are sand dunes – in other places simply as dark streaks, but here with bright glints on their north flanks (the radar illumination is from the north/above) indicating their positive relief of about 100 m.

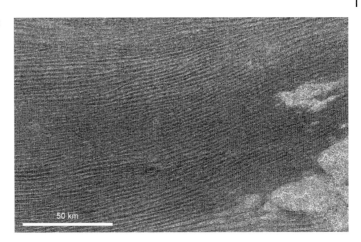

Although there may be a few barely-resolved hooked barchans, and a few transverse dunes where topographic barriers 'straighten' seasonally-varying winds, the dominance of linear dunes on Titan (e.g. Lorenz et al. 2006) attests to a bidirectional wind regime (Chapter 6; see also Reffet et al. 2010). This regime was first thought to be caused by atmospheric tides, but it now appears to be due to a seasonal north–south Hadley circulation (Tokano 2008). For some years atmospheric modellers struggled to explain the clearly West–East direction of dune propagation on Titan, because, for most of the year, the low-latitude winds should be blowing in an East-to-West direction. It may be that the dunes reflect only the brief but fastest winds around the stormy spring equinox, when winds are in fact West–East; another is that haboob-type outflows from methane rainstorms preferentially cause eastwards transport (Charnay et al. 2015). The topic remains an area of active research (e.g. Lucas et al. 2014). Circulation models do, at least, satisfactorily predict that a band around the equator should dry out, occasional methane rainstorms notwithstanding, which is where the dunefields are found (Mitchell 2008). The ±30° band of dunes makes an interesting contrast – owing mainly to Titan's slow rotation which determines the Hadley circulation– to the two latitude belts where most of the terrestrial deserts are found. The dunefields (the major ones are Belet, Shangri-La, Fensal, and Aztlan – see Figure 11.3) form a dark equatorial belt that is visible (even with the largest telescopes) from Earth. This belt is broken by a large bright and mountainous region named Xanadu – it is not yet clear whether the sand circumnavigates Titan; if so, it must somehow be transported through or around Xanadu (Barnes et al. 2015).

Titan's dunes are of an optically-dark material with a low dielectric constant, consistent with an organic composition, and an organic, possibly aromatic, character is indicated by some Cassini spectroscopic data. Thus, the sands are presumably derived ultimately from the photochemical processing of methane in the atmosphere, which also causes an organic-rich haze which makes Titan's atmosphere optically-opaque. However, the process by which 1-µm aerosol particles (which may be a sticky mix of refractory organics as well as liquid ethane) become ~200 µm saltating particles is not known – it may be related to hydrologic erosion of massive deposits of organics, or it may be connected with climate cycles on Titan wherein the seas dry out at alternate poles over periods of tens of thousands of years. Even at Titan's low temperature of 94 K, simple organics like benzene are white, waxy solids: a more complex composition appears to be needed, perhaps polycyclic aromatic hydrocarbons (PAHs), such as pyrene and anthracene, or even some nitrogen-containing versions of these. There is at present relatively little understanding of the behaviour of such materials in saltation in Titan's conditions – would sand made of such material be rapidly ground into dust?

Assuming that Titan sands are not appreciably stickier than terrestrial sands, the optimum in the saltation curve (see Lorenz et al. 2014) should correspond roughly to $1\,\mathrm{ms}^{-1}$ windspeeds and particle sizes of 200–300 μm, i.e. not too different from terrestrial sand. Some recent wind tunnel measurements suggest saltation thresholds may be somewhat higher, $2–3\,\mathrm{ms}^{-1}$. Such wind speeds are rather larger than those measured at the surface by the Huygens probe ($\sim 0.3\,\mathrm{ms}^{-1}$) but models predict that such winds – also strong enough to generate waves on Titan's hydrocarbon seas – may occur in polar summer or equatorial equinox, and storm winds (Charnay et al. 2015) especially can exceed this (Burr et al. 2015).

11.5.3 Venus

Although Venus's thick atmosphere should make aeolian sediment transport relatively easy, near-surface winds may be somewhat gentle. However, a more important consideration may be the limited availability of sediment: there are no freeze–thaw cycles to generate sediment, nor is there a (hydrological) cycle, such as the methane cycle on Titan. Thus, the main source of sand-sized sediment may be ejecta from impact craters and there do appear to be some streaks of material associated with small volcanoes (Greeley et al. 1992).

The Magellan mission which mapped Venus in the early 1990s had a best radar resolution of only ~100 m, so only the largest dune features are visible. They form two dune fields, Algaonice and Fortuna-Meshkenet (Greeley et al. 1992). There may be rather widespread 'microdunes' suggested by wind tunnel experiments under Venus

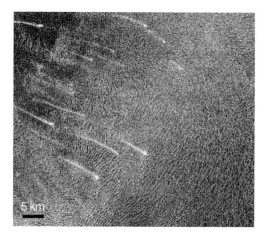

Figure 11.5 Magellan radar mosaic of the Fortuna-Mesknet dunefield on Venus. These dunes appear to be rather more extensive than the Algaonice dunes. Like the Algaonice, these dunes appear to be transverse, orthogonal to the prominent windstreaks.

conditions which show bedforms with slip-faces, but with spacings of only a few tens of centimetres, detectable indirectly. Future missions, with instruments that have higher resolution, may reveal more abundant dune forms.

The Algaonice dunes (also known as Menat Undae) on Venus at 25°S, 340°E cover some $1300\,\mathrm{km}^2$, are 0.5–5 km in length, and are quite bright, likely because there are slip faces of ~30° oriented towards the radar illumination. The dunes lie at the end of the ejecta outflow channel from the Algaonice impact crater of that name, but because the dunes are only barely resolved, there is little more information.

A more northern dunefield, Fortuna-Meshkenet (also known as Al-Uzza Undae), lies at 67°N, 91°E in a valley between Ishtar Terra and Meshkenet Tessera. The dunes are 0.5–10 km long, 0.2–0.5 km wide and spaced by an average of 0.5 km. They appear (see Figure 11.5) to be transverse dunes, in that there are several bright wind streaks visible in the region, which seem generally to be orthogonal to the dunes (Greeley et al. 1997; Lorenz and Zimbelman 2014).

11.6 Aeolian Dust

Aeolian dust is usually defined as sediment that travels in suspension in the atmosphere, rather than in saltation or as bedload (creep or reptation, see Chapter 2). Dust is therefore distinct from larger sand- and granule-sized material, but there is no specific size at which dust becomes sand; differences in particle density and shape can produce materials with transitional behaviours. In fact, on planetary bodies with different gravity and atmospheric compositions and densities, this transition between suspension and saltation can span very different size ranges. In general, though, a commonly applied distinction is that sand-grade material is >62.5 μm in diameter, with smaller particles being defined as dust. This size definition equates dust with what a sedimentologist would define as 'clay and silt'. In the current terrestrial atmosphere, two populations of aeolian dust exist: fine dust <15 μm in diameter, which can remain airborne in suspension almost indefinitely, and coarse dust >50 μm in diameter, which is transported in suspension for only tens or a few hundreds of kilometres (see Chapter 4).

The dustiness of the Earth's atmosphere is, of course, not constant on short or long timescales: dust storms, lasting days or weeks, can raise local atmospheric dust concentration by orders of magnitude, and enormous dust deposits (loess), found across the globe, and sometimes hundreds of metres thick, are testament to long-duration enhancements of atmospheric dustiness in the past. For example, many of the thickest loess deposits are dated to the Pleistocene, during which there may have been long periods when the atmosphere was up to 40 times dustier than today (Taylor et al. 1993) (see Chapters 4 and 5).

Aeolian dust in a planetary context is, by definition, limited to those bodies with an atmosphere capable of transporting sediments: Earth, Mars, Venus, and Titan. Although there are observations of movement of sand-sized material on Mars, the limited in-situ and remote sensing data available for Venus and Titan means that no observations of *dust* lifting or deposition have been made for those bodies. Hence, the significance of dust as a geomorphic process on Venus and Titan is unknown. On Mars, though, the story is quite different, and there is extensive evidence for dust in the atmosphere, on the surface and an active exchange between the two (e.g. Read and Lewis, 2004).

11.6.1 Dust in the Martian Atmosphere

On Mars, dust is most clearly evident in the atmosphere (Figure 11.6). Planet-encircling dust storms, in which all or part of the surface is obscured from astronomical observation, have been recognised for more than a century, and Mars's atmosphere maintains a dusty 'haze' even in relatively clear periods. Large dust storms generally occur during the Martian southern hemisphere spring or summer. In any given Martian year there is about a one-in-three chance of a planet-encircling dust storm, and about an 80% chance of regional dust storms (Zurek and Martin 1993).

Dust plays a critical role in the behaviour of the Martian atmosphere. This is because dust is crucial to an important atmospheric feedback process. When airborne dust absorbs solar and infrared radiation, it heats the atmosphere and creates temperature contrasts between different parts of the atmosphere. These contrasts create spatial differences in density, and these are linked to winds. The winds carry airborne dust and might also lift dust from the surface, which in turn changes patterns of heating in the atmosphere. This feedback is crucial to Martian climate variability (e.g. Mulholland et al. 2013). Even during relatively clear periods, normally during the northern hemisphere summer, when Mars is furthest from the Sun, the atmosphere has a significant haze of dust particles with effective radii of

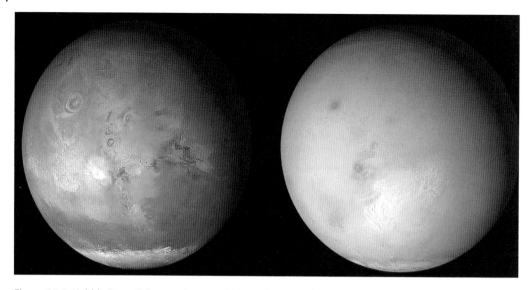

Figure 11.6 Hubble Space Telescope images of Mars taken on (left) 10 July, 2001, and (right) 31 July, 2001. The second picture shows the same hemisphere (note the circular feature that is the large volcano, Olympus Mons at upper left) as the first, but after the beginning of the great global dust storm of 2001. *Source:* NASA/Hubble Heritage Team.

about 1.0–1.5 μm (e.g. Wolff 2003; Wolff et al. 2006), which absorb up to 10% of the incoming sunlight. Slightly larger suspended particles have been detected during the more severe dust storms, during which more than half of the visible sunlight is absorbed in the atmosphere well above the surface. This means that less sunlight reaches the surface and so the surface becomes cooler.

Detailed models of the Martian atmosphere (Newman et al. 2002) show that the background level of atmospheric dust on Mars is probably maintained by dust devils (Box 11.1). An active process is required to support the haze, because the settling rate, as measured by the appearance of calibration targets on the Mars Exploration Rovers (Kinch et al. 2007), is of the order of several μm every Martian year, so that without such a process as dust devils, the

---

Box 11.1 Dust Devils

Dust devils are convective vortices, powered by insolation, and made visible by the dust and debris they have entrained (Balme and Greeley 2006). They are common in arid environments on Earth, forming especially frequently during dry, hot, late summers, but have also been seen in arid polar or high elevation environments. They are mostly of the order of a metre to a few tens of metres in diameter, and tens to hundreds of metres high. Dust devils are also common on Mars, where they can be an order of magnitude larger. Dust devils on Earth are often thought to be 'nuisance-level' phenomena, but they have caused aviation accidents (Lorenz and Myers 2005) and can contribute to poor air quality (Gillette and Sinclair 1990). On Mars, though, dust devils are both agents for geomorphological change, and play a significant role in the climate, as their movement across the surface can change the surface albedo by forming extensive 'tracks', as well as being responsible for maintaining the ongoing haziness of the atmosphere (Box Figure 11.1).

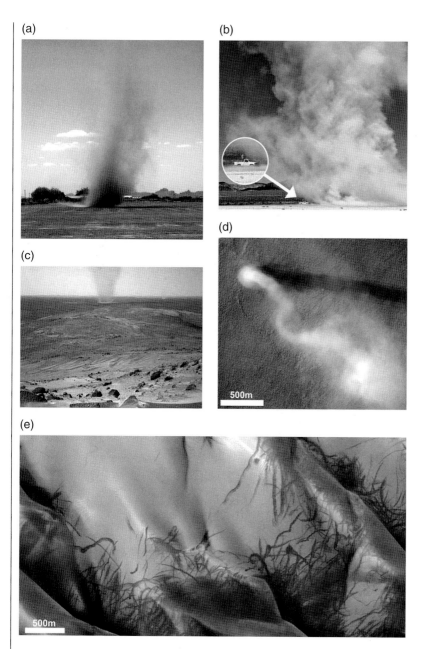

Box Figure 11.1 (a) A small (~5m diameter) dust devil on Earth. (b) A much larger dust devil on Earth. (c) Dust devils on Mars, observed from the surface by the NASA Mars Exploration Rover 'Spirit'. (d) A large Martian dust devil observed from orbit. Note the distinctive shadow and the sinuous column. Part of HiRISE NASA image ESP_026394_2160. (e) Dust devil tracks on top of Martian sand dunes as observed from orbit. Part of NASA HiRISE image ESP_014426_2070.

atmosphere would clear more quickly between dust storms than is observed. Dust devils are most active when there is a high surface (warm) to air (cool) thermal contrast, which occurs mostly during local summer, and are suppressed when the amount of dust in the atmosphere rises (i.e. during periods with significant local, regional or global dust storms).

11.6.2 Sources of Martian Dust

On Earth, dust is produced by a variety of process including comminution (the physical grinding of particles in glacial or fluvial systems), attrition (the breakdown of saltating grains), and weathering of rocks by chemical or physical means. Minor or sporadic sources include micrometeorites and volcanic ash (see Chapter 4). On Mars today there have been no observations of ongoing active volcanism and the surface is generally too cold and dry for significant fluvial or glacial processes, although limited thaw has occurred in geologically recent times (Balme et al. 2013). Hence, the production of large amounts of dust by comminution seems unlikely under present conditions. Observations of the active migration of aeolian bedforms demonstrate that saltation is occurring on Mars (Bridges et al. 2012b), so that some attrition is likely to be happening. The prevalence of dust devils and dust storms shows that dust is constantly being lifted, but it seems likely that much of this dust today is being reworked from older surface deposits, and is not being actively created in large amounts. Weathering is also likely to have been a much more significant process in Mars's ancient past than during the recent, more arid periods.

11.6.3 Landforms Associated with Aeolian Dust on Mars

The Mars Exploration Rovers have observed many grains too small to be resolved by the microscopic imager (< 30 μm in diameter) inside ripple bedforms and in dust aggregates coating rocks (Sullivan et al. 2008), leading to the inference that dust contributes to the formation of macro-forms like ripples. The small, ripple-like features observed by the various Mars Rovers transition into larger bedforms known as Transverse Aeolian Ridges (Bourke et al. 2003) that are extremely common in the Martian equatorial and mid-latitude regions (Balme et al. 2008). If these all contain significant amounts of sub-30 μm diameter materials, then globally these bedforms retain large volumes of dust. It has also been suggested that some Transverse Aeolian Ridges could be primarily formed from larger-scale loess-like dust deposits, which have been sculpted into their current ripple-like form by many thousands or millions of years of wind erosion (Geissler 2014).

As well as being present in small bedforms, dust has been a key component in the formation of global-scale Martian landscapes. For example, as first seen in Mariner 9 images, both Mars's northern and southern polar regions contain extensive layered deposits (e.g. Cutts 1973) thought to be formed of multiple horizontal to sub-horizontal alternating dark-bright layers of dust and ice (Blasius et al. 1982; Figure 11.7). It is not known whether the dark 'dust' in these layers is truly fine material, as opposed to sand-grade sediments, but their extensive nature, rhythmic layering, and topographic setting suggest that they were deposited by airfall, and so are likely to be 'dust'. The polar layered deposits are each more than a million square kilometres in spatial extent (Tanaka 2000) and are 2–3 km thick, so, even if they only contain a few percent of lithic material, tens of thousands of cubic kilometres of dust could be sequestered within them.

A second, dust-related landscape of global scale is the extensive blanketing terrain known as the 'latitude-dependent mantle' (LDM) (Figure 11.8). The LDM blankets the mid to high latitudes in both Martian hemispheres, and is geologically young, to judge by the paucity of superposing impact craters. The LDM is believed to comprise a mixture of dust and ice (Mustard et al. 2001; Head et al. 2003), both because it corresponds spatially to regions that have massive ground ice in their upper tens of centimetres, as measured by gamma ray spectrometry remote sensing (Feldman et al. 2004), and because it drapes the topography, and hence is likely to be of airfall origin. It covers nearly a quarter of the planet's surface and has an estimated depth of about 10 m (Kreslavsky and Head 2002). Although recent measurements suggest it is mainly ice (50–95% by

Figure 11.7 Dust/ice layering in Mars's Polar regions. This image shows a roughly east–west trending trough in the North Polar layered deposits. The dark, rough band, just below the middle of the image, is the lowest point, and contains a round structure that is an impact crater. At the top and bottom of the image are the higher edges of the trough, with layers visible in the trough walls as distinctive light and dark bands. The layers are particularly well exposed on the northern trough wall. These layers are thought to have been formed by differential erosion of variable mixtures of ice and dust. HiRISE image PSP_001462_2630. Image centred at 83.01°N, 94.82°E. *Source:* NASA/JPL/UofA.

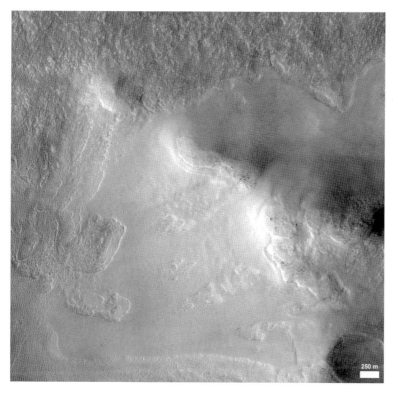

Figure 11.8 Latitude Dependent Mantle (LDM). The smoother parts of this scene are a terrain type thought to be an ice-rich, dusty 'mantle' or LDM. The lighting is from the left, revealing that the mantling material is partially 'pasted onto' the topographically higher terrain in the centre of the image, and appears to be draped over the relief in many places. Such material covers much of Mars's mid- to high-latitudes, and it is this draping property that allows the inference that the material was deposited by airfall, and hence is partly composed of dust. At the top and left of the image, the mantle is degraded, revealing the rougher, underlying surfaces. Further north (this image is at a latitude of about 50°N), the mantle becomes much less degraded and is essentially continuous. Part of HiRISE image ESP_034615_231, centred at 50.769°N, 35.252°E. North is towards the top in this image. *Source:* NASA/JPL/UofA.

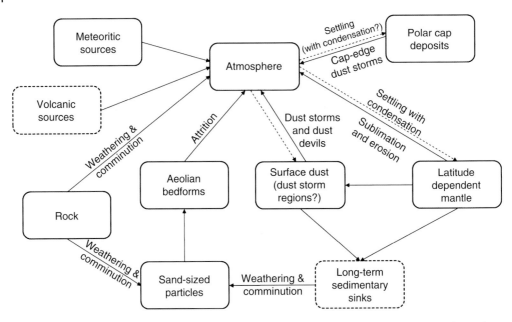

Figure 11.9 Links and feedbacks in the Martian dust cycle. Boxes represent reservoirs. Arrows indicate dust transport or formation pathway. Dashed arrows indicate gravitational settling from the atmosphere. Dashed boxes indicate sources mainly related to ancient epochs, rather than being 'active'. *Source:* Modified from figure in Kahn et al. (1992).

volume; Conway and Balme 2014), thousands of cubic kilometres of dust are probably contained within these deposits.

While the LDM is geologically young, having formed less than a few million years ago, traces of similar, older deposits have also been identified. Skinner and colleagues (2012) mapped extensive loess-like sediments thought to be about a billion years old. These deposits have a mean thickness of 32 m, and extend >3.1 × 10^6 km^2, so they account for tens of thousands of cubic kilometres of dust, assuming that they have only a small component of ice at the present day. Importantly, Skinner and colleagues (2012) suggest that they might have been formed by the degradation or erosion of the basal components of the north polar layered deposits – presumably mediated by transport of fine grained materials and perhaps deposition into or alongside ice. This demonstrates how the complex interactions between climate change and surficial deposits of dust and ice have been ongoing throughout Mars's history and are likely still happening today.

Mars, like Earth, in its present and past, has had an active dust cycle, the intensity and character of which are controlled by climate (Figure 11.9). Sources and sinks of dust exist on Mars that are similar to those on Earth – with the main exception of the terrestrial oceans. Dust certainly plays a key role in Mars's climate, both by absorbing incoming solar radiation when airborne, and by changing the albedo and roughness of the surface as it is removed or deposited. It is unknown whether dust plays such a central role on other planets, but new high-resolution imaging capabilities or new in-situ exploration missions are likely to be required before this question can be answered.

11.7 Planetary Wind-Eroded Landforms

In the current climate, Mars has an almost totally inactive hydrosphere, notwithstanding observations of small amounts of possible salty liquids at the surface (Ojha et al. 2015).

Figure 11.10 Image of etched sandstone taken by the Mastcam on the Curiosity Rover at Gale crater. *Source:* NASA/JPL-Caltech/MSSS.

The absence of stable liquid water results in low rates of chemical weathering (which were significant in Mars's past). The role of wind combined with sand to effect aeolian abrasion is perhaps one of the most important agents of landscape modification on Mars today. Although the Martian gravity and atmospheric density are significantly lower than Earth's (see Table 11.1), it does have: (i) a basaltic sand supply, and (ii) winds above the sediment transport threshold. Consequently, landforms that result from aeolian abrasion (ventifacts and yardangs) are observed on Mars's surface.

11.7.1 Ventifacts

Rocks that have been abraded by windborne particles (ventifacts, Chapter 3) are common on the Martian surface and have been observed at four Rover landing sites (Pathfinder, Spirit, Opportunity, and Curiosity) (Bridges et al. 1999; Sullivan et al. 2005; Thomson et al. 2008).

The atmospheric density and gravity on Mars suggest that abrasion will take at least 10–100 times longer than on Earth (Bridges and Laity 2013). Despite these differences, ventifact

textures similar to those on Earth have been described and include facets, elongated pits, grooves, flutes, and etching (Figure 11.10).

Ventifacts provide useful proxies of former Martian sand-transporting winds where, for example, analysis of ventifacts in images taken by the Spirit Rover as it traversed the undulating terrain of the Columbia hills show the direction of funnelled winds along troughs, thus providing a useful record of past climate and landscape modification (Thomson et al. 2008).

11.7.2 Yardangs

Yardangs were first identified on Mars during the Mariner 9 flyby (McCauley, 1973). The later Mars Orbiter Camera (MOC) camera on the Mars Global Surveyor Mission revealed a significant surface area displaying extensively eroded sedimentary rock (yardangs, Chapter 3). The most highly concentrated distribution is in the ash/ignimbrite Medussa Fossae formation (MFF) (see example in Figure 11.11) where yardangs occur as extensive fields, some with areas of up to $40\,000\,\mathrm{km^2}$ (Mandt et al. 2008).

Figure 11.11 Yardangs on Mars. Section of HiRISE image ESP_023051_1865. North is towards top of image. See insert for colour representation of this figure.

Yardangs on Mars are larger in scale than those on Earth. Ward (1979) identified individual ridges in the equatorial region of Mars that were tens of kilometres long, with valleys nearly 1 km wide. These mega-yardangs cover approximately 7% of the MFF. Their width ranges from 1–5 km and they range in height from 5–700 m, with most being in excess of 100 m tall (Mandt et al. 2008). Mega-yardangs on Earth may be tens of metres high and several kilometres long (e.g. Cooke et al. 1993). On Earth, the presence of an upper indurated facies can result in large, high aspect ratio (1:20–1:40) mega-yardangs that form 100 m tall ridges with steep to vertical walls (de Silva et al. 2010). Strong

unidirectional winds and a lithology with a vertical induration profile are also proposed to result in the mega-yardang formation in the MFF (Mandt et al. 2009). Yardangs on Mars are also complex in pattern as they can be both parallel and bidirectional. Resistant layers and jointing, in addition to prevailing wind directions, are thought to influence their orientation (Bradley et al. 2002).

Only one possible yardang field has been identified on Venus (Greeley et al. 1992; Greeley et al. 1997) at 9°N, 60.5°E, about 300 km south-east of the crater Mead. Because yardangs require an easily-erodable substrate such as lake sediments or ignimbrites, it may be worth considering that such substrates may not be widespread on Venus – with no hydrology for the last several billion years, lake deposits seem unlikely, and the high pressure on Venus makes gassy explosive eruptions on Venus difficult. As with the relative paucity of sand dunes on Venus, the transport of appropriately-sized sediment may not be the limiting factor, but rather the availability of abrading or abradable material. It should also be noted that most terrestrial yardangs would be invisible to Magellan's rather coarse imaging resolution of ~100 m.

On Titan, a handful of locales beyond the main dunefields have been identified with linear features with positive relief. These may be fossil dunes (they have a different orientation from the main equatorial dunes, and a different radar texture) although yardangs cannot be excluded (Paillou et al. 2016).

11.8 Conclusion

Evidence for the wind abrasion of rocks and the generation of soil particles has been documented on several bodies in our solar system. In particular, sand dunes with forms very similar to those on Earth have been confirmed on Mars, Venus, and Titan, and bedforms have been proposed on the comet 67P/Churyumov–Gerasimenko and Pluto (Jia et al. 2017, Telfer et al. 2018). In all these places aeolian deposits provide a proxy for regional wind regimes and the availability of sediment, largely because of the understanding we have of aeolian landscapes on Earth (see Chapter 6). As international space programmes extend our exploration to the edge of, and beyond our solar system, we can expect to expand our perception of aeolian geomorphology. The challenge is to continue to develop image and analytical tools that allow us to understand the full range of controls that permit aeolian geomorphology to operate in environments that are different to those on Earth.

Further Reading

A good introduction to dune systems on planetary bodies in our solar system will be found in Bourke et al. (2010). This review was updated by Zimbelman et al. (2013) and Fenton et al. (2013). A more recent review by Diniega et al. (2017) includes other planetary bodies such as Pluto and Io. For further reading on sediment transport that includes a planetary perspective, we recommend Kok et al. (2012). For specific topics, such as Martian dune migration, we recommend Bridges et al. (2013). A recent book on the topic of planetary dunes that includes analogue research and methods is Lorenz and Zimbelman (2014).

References

Almeida, M.P., Parteli, E.J.R., Andrade, J.S., and Herrmann, H.J. (2008). Giant saltation on Mars. *Proceedings of the National Academy of Sciences* 105 (17): 6222–6226. doi: 6210.1073/pnas.0800202105.

Anderson, R.S. (1987). A theoretical model for aeolian impact ripples. *Sedimentology* 34 (5): 943–956. doi: 10.1111/j.1365-3091.1987. tb00814.x.

Andreotti, B., Claudin, P., and Pouliquen, O. (2010). Measurements of the aeolian sand transport saturation length. *Geomorphology* 123 (3–4): 343–348. doi: 10.1016/j. geomorph.2010.08.002.

Andreotti, B., Fourriere, A., Ould-Kaddour, F. et al. (2009). Giant aeolian dune size determined by the average depth of the atmospheric boundary layer. *Nature* 457 (7233): 1120–1123. doi: 1110.1038/ nature07787.

Arvidson, R.E., Guinness, E.A., Moore, H.J. et al. (1983). Three Mars years: Viking Lander 1 imaging observations. *Science* 222 (4623): 463–468. doi: 10.1126/science.222. 4623.463.

Ayoub, F., Avouac, J.P., Newman, C.E. et al. (2014). Threshold for sand mobility on Mars calibrated from seasonal variations of sand flux. *Nature Communications* 5: 5096. doi: 10.1038/ncomms6096.

Bagnold, R.A. (1941). *The Physics of Blown Sand and Desert Dunes*. London: Methuen.

Balme, M.R., Berman, D., Bourke, M.C., and Zimbelman, J.R. (2008). Transverse aeolian ridges (TARs) on Mars. *Geomorphology* 101 (4): 703–720. doi: 710.1016/j.geomorph. 2008.1003.1011.

Balme, M.R., Gallagher, C.J., and Hauber, E. (2013). Morphological evidence for geologically young thaw of ice on Mars: a review of recent studies using high-resolution imaging data. *Progress in Physical Geography* 37 (3): 289–324. doi: 10.1177/0309133313477123.

Balme, M.R. and Greeley, R. (2006). Dust devils on Earth and Mars. *AGU100*.

Barnes, J., Lorenz, R., Radebaugh, J. et al. (2015). Production and global transport of Titan's sand particles. *Planetary Science* 4 (1): doi: 10.1186/s13535-015-9-0004-y.

Blasius, K.R., Cutts, J.A., and Howard, A.D. (1982). Topography and stratigraphy of Martian polar layered deposits. *Icarus* 50 (2–3): 140–160. doi: 10.1016/0019-1035 (82)90122-1.

Bourke, M.C., Bullard, J., and Barnouin-Jha, O. (2004). Aeolian sediment transport pathways and aerodynamics at troughs on Mars. *Journal of Geophysical Research (Planets)* 109 (E7): E07005. doi: 10.1029/2003JE002155.

Bourke, M.C., Lancaster, N., Fenton, L.K. et al. (2010). Extraterrestrial dunes: an introduction to the special issue on planetary dune systems. *Geomorphology* 121 (1–2): 1–14. doi: 10.1016/j. geomorph.2010.1004.1007.

Bourke, M.C., Wilson, A.A., and Zimbelman, J.R. (2003). The Variability of Transverse Aeolian Ridges in Troughs on Mars. Lunar and Planetary Science Conference, p. 2090.

Bradley, B.A., Sakimoto, S.E.H., Frey, H., and Zimbelman, J.R. (2002). Medusae Fossae Formation: new perspectives from Mars Global Surveyor. *Journal of Geophysical Research* 107 (E8): 5058. doi: 10.1029/ 2001JE001537.

Bridges, N.T., Ayoub, F., Avouac, J.P. et al. (2012a). Earth-like sand fluxes on Mars. *Nature* 485 (7398): 339–342. doi: 10.1038/nature11022.

Bridges, N.T., Banks, M.E., Beyer, R.A. et al. (2010). Aeolian bedforms, yardangs, and indurated surfaces in the Tharsis Montes as seen by the HiRISE camera: evidence for dust aggregates. *Icarus* 205 (1): 165–182. doi: 10.1016/j.icarus.2009.05.017.

Bridges, N.T., Bourke, M.C., Geissler, P.E. et al. (2012b). Planet-wide sand motion on Mars. *Geology* 40 (1): 31–34. doi: 10.1130/ G32373.1.

Bridges, N.T., Geissler, F., Silvestro, P., and Banks, M. (2013). Bedform migration on Mars: current results and future plans. *Aeolian Research* 9: 133–151. doi: 10.1016/j. aeolia.2013.02.004.

Bridges, N.T., Greeley, R., Haldemann, A.F.C. et al. (1999). Ventifacts at the Pathfinder landing site. *Journal of Geophysical Research* 104 (E4): 8595–8615. doi: 10.1029/98JE02550.

Bridges, N.T. and Laity, J.E. (2013). Fundamentals of Aeolian sediment transport: Aeolian abrasion. In: *Treatise on Geomorphology* (ed. J.F. Shroder), 134–148. San Diego: Academic Press.

Burr, D., Bridges, N., Marshall, J. et al. (2015). Higher-than-predicted saltation threshold wind speeds on Titan. *Nature* 517: 60–63.

Charnay, B., Barth, E., Rafkin, S. et al. (2015). Methane storms as a driver of Titan's dune orientation. *Nature Geoscience* 8 (5): 362–366. doi: 10.1038/ngeo2406.

Claudin, P. and Andreotti, B. (2006). A scaling law for aeolian dunes on Mars, Venus, Earth, and for subaqueous ripples. *Earth and Planetary Science Letters* 252 (1–2): 30–44. doi: 0.1016/j.epsl.2006.09.004.

Claudin, P., Wiggs, G.F.S., and Andreotti, B. (2013). Field evidence for the upwind velocity shift at the crest of low dunes. *Boundary-Layer Meteorology* 148: 195–206. doi: 10.1007/s10546-013-9804-3.

Conway, S.J. and Balme, M.R. (2014). Decameter thick remnant glacial ice deposits on Mars. *Geophysical Research Letters* 41 (15): 5402–5409. doi: 10.1002/2014GL060314.

Cooke, R.U., Warren, A., and Goudie, A.S. (1993). *Desert Geomorphology*. London: UCL Press Limited.

Cutts, J.A. (1973). Nature and origin of layered deposits of the Martian polar regions. *Journal of Geophysical Research* 78 (20): 4231–4249. doi: 10.1029/JB078i020p04231.

De Silva, S.L., Bailey, J.E., Mandt, K.E., and Viramonte, J.M. (2010). Yardangs in terrestrial ignimbrites: synergistic remote and field observations on Earth with applications to Mars. *Planetary and Space Science* 58 (4): 459–471. doi: 10.1016/j. pss.2009.10.002.

Diniega, S., Kreslavsky, M., Radebaugh, J. et al. (2017). Our evolving understanding of aeolian bedforms, based on observation of dunes on different worlds. *Aeolian Research* doi: 10.1016/j.aeolia.2016.10.001.

Durán, O., Parteli, E.J.R., and Herrmann, H.J. (2010). A continuous model for sand dunes: review, new developments and application to barchan dunes and barchan dune fields. *Earth Surface Processes and Landforms* 35 (13): 1591–1600. doi: 10.1002/esp.2070.

Edgett, K.S. and Christensen, P.R. (1991). The particle size of Martian aeolian dunes. *Journal of Geophysical Research, Planets* 96 (E5): 22765–22776. doi: 10.1029/91JE02412.

Edgett, K.S., Williams, R.M.E., Malin, M.C. et al. (2003). Mars landscape evolution: influence of stratigraphy on geomorphology in the North polar region. *Geomorphology* 52 (3–4): 289–297. doi: 10.1016/S0169-555X(02)00262-3.

Feldman, W.C., Prettyman, T.H., Maurice, S. et al. (2004). Global distribution of near-surface hydrogen on Mars. *Journal of Geophysical Research* 109 (E9): doi: 10.1029/2003JE002160.

Fenton, L.K., Ewing, R.C., Bridges, N.T., and Lorenz, R. (2013). Extraterrestrial aeolian landscapes. In: *Treatise on Geomorphology* (ed. J.F. Shroder), 287–312. San Diego: Academic Press.

Fenton, L.K. and Hayward, R.K. (2010). Southern high-latitude dune fields on Mars: morphology, aeolian activity and climate change. *Geomorphology* 121 (1–2): 98–121. doi: 110.1016/j.geomorph.2009.1011.1006.

Flasar, F.M., Baines, K.H., Bird, M.K. et al. (2010). Atmospheric dynamics and meteorology. In: *Titan from Cassini-Huygens* (ed. R.H. Brown, J.P. Lebreton and J. Waite), 323–353. Berlin: Springer Science.

Fourrière, A., Claudin, P., and Andreotti, B. (2010). Bedforms in a turbulent stream: formation of ripples by primary linear instability and of dunes by nonlinear pattern coarsening. *Journal of Fluid Mechanics* 649: 287–328. doi: doi.org/10.1017/S0022112009993466.

Franklin, E.M. and Charru, F. (2011). Subaqueous barchan dunes in turbulent shear flow. Part 2: Fluid flow. *Journal of Fluid Mechanics* 675: 199–222.

Geissler, P.E. (2014). The birth and death of transverse aeolian ridges on Mars. *Journal of Geophysical Research, E: Planets* 119: 2583–2599. doi: 10.1002/2014JE004633.

Gillette, D.A. and Sinclair, P.C. (1990). Estimation of suspension of alkaline material by dust devils in the United-States. *Atmospheric Environment. Part A. General Topics* 24 (General Topics 24): 1135–1142.

Greeley, R., Arvidson, R.E., Elachi, C. et al. (1992). Aeolian features on Venus: preliminary Magellan results. *Journal of Geophysical Research, Planets* 97 (E8): 13319–13345. doi: 10.1029/92JE00980.

Greeley, R., Bender, K., Saunders, R.S. et al. (1997). Aeolian processes and features on Venus. In: *Venus II* (ed. S.W. Gougher, D.M. Hunten and R.J. Phillips), 547–589. Tucson: University of Arizona Press.

Greeley, R., Bender, K., Thomas, P.E. et al. (1995). Wind-related features and processes on Venus: summary of Magellan result. *Icarus* 11 (2): 399–420. doi: 10.1006/icar.1995.1107.

Greeley, R. and Iversen, J.D. (1985). *Wind as a Geological Process on Earth, Mars, Venus and Titan*. Cambridge: Cambridge University Press.

Greeley, R., Squyres, S.W., Arvidson, R.E. et al. (2004). Wind-related processes detected by the Spirit Rover at Gusev Crater, Mars. *Science* 305 (5685): 810–821. doi: 10.1029/92JE02580.

Grotzinger, J.P., Arvidson, R.E., Bell, J.F. et al. (2005). Stratigraphy and sedimentology of a dry to wet eolian depositional system, burns formation, Meridiani Planum, Mars. *Earth and Planetary Science Letters* 240 (1): 11–72. doi: 10.1016/j.epsl.2005.09.039.

Grotzinger, J.P., Hayes, A.G., Lamb, M.P., and McLennan, S.M. (2013). Sedimentary processes on earth, Mars, titan, and Venus. In: *Comparative Climatology of Terrestrial Planets* (ed. S.J. Mackwell, A.A. Simon-Miller, J.W. Harder and M.A. Bullock), 439–472. Tucson: University of Arizona Press.

Hayward, R., Mullins, K., Fenton, L.K. et al. (2007). *Mars Digital Dune Database: MC2-MC29*. USGS Open-File Report.

Head, J.W., Mustard, J.F., Kreslavsky, M.A. et al. (2003). Recent ice ages on Mars. *Nature* 426: 797–802. doi: 10.1038/nature0211.

Hersen, P., Douady, S., and Andreotti, B. (2002). Relevant length scale of barchan dunes. *Physical Review Letters* 89: 264301. doi: 10.1103/PhysRevLett.89.264301.

Holstein-Rathlou, C., Gunnlauggson, H.P., Merrison, J.P. et al. (2010). Winds at the Phoenix landing site. *Journal of Geophysical Research, Planets* 115 (E00E18): doi: 10.1029/2009JE003411.

Iversen, J.D. and White, B.R. (1982). Saltation threshold on Earth, Mars and Venus. *Sedimentology* 29 (1): 111–119. doi: 10.1111/j.1365-3091.1982.tb01713.x.

Jackson, D.W.T., Bourke, M.C., and Smyth, T. (2015). The dune effect on sand transporting winds on Mars. *Nature Communications* 9796 (2015): doi: 10.1038/ncomms9796.

Jia, P., Andreotti, B., and Claudin, P. (2017). "Giant ripples on comet 67P/Churyumov–Gerasimenko sculpted by sunset thermal wind." *Proceedings of the National Academy of Sciences* 114 (10): 2509–2514.

Kahn, R.A., Martin, T.Z., Zurek, R.W., and Lee, S.W. (1992). The Martian dust cycle. In: *Mars* (ed. H. Kieffer, B.M. Jakosky, C.W. Snyder and M.S. Matthews), 1017–1053. Tucson: University of Arizona Press.

Kinch, K.M., Sohl-Dickstein, J., Bell, J.F. III et al. (2007). Dust deposition on the Mars Exploration Rover Panoramic Camera (Pancam) calibration targets. *Journal of Geophysical Research, Planets* 112: E06S03. doi: 10.1029/2006JE002807.

Kok, J.F. (2010a). Difference in the wind speeds required for initiation versus continuation of sand transport on Mars: implications for dunes and dust storms. *Physical Review Letters* 104: 074502. doi: 10.1103/PhysRevLett.104.074502.

Kok, J.F. (2010b). An improved parameterization of wind-blown sand flux on Mars that includes the effect of hysteresis. *Geophysical Research Letters* 37: L12202. doi: 10.1029/2010GL043646.

Kok, J.F., Parteli, E.J.R., Michaels, T.I., and Karam, D.B. (2012). The physics of wind-blown sand and dust. *Reports on Progress in Physics* 75: 106901. doi: 10.1088/0034-4885/75/10/106901.

Kreslavsky, M.A. and Head, J.W. (2002). Mars: nature and evolution of young latitude-dependent water-ice-rich mantle. *Geophysical Research Letters* 29 (15): 14–21. doi: 10.1029/2002GL015392.

Kroy, K., Sauermann, G., and Herrmann, H.J. (2002). Minimal model for aeolian sand dunes. *Physical Review. E, Statistical, Nonlinear, and Soft Matter Physics* 66: 031302. doi: 10.1103/PhysRevE.66.031302.

Laskar, J., Correia, A.C.M., Gastineau, M. et al. (2004). Long term evolution and chaotic diffusion of the insolation quantities of Mars. *Icarus* 170 (2): 343–364. doi: 10.1016/j.icarus.2004.04.005.

Le Gall, A., Janssen, M.A., Wye, L.C. et al. (2011). Cassini SAR, radiometry, scatterometry and altimetry observations of Titan's dune fields. *Icarus* 213 (2): 608–624. doi: 10.1016/j.icarus.2011.03.026.

Livingstone, I. and Warren, A. (1996). *Aeolian Geomorphology: An Introduction*. London: Longman.

Lorenz, R.D., Lunine, J.I., Grier, J.A., and Fisher, M.A. (1995). Prediction of aeolian features on planets: application to Titan paleoclimatology. *Journal of Geophysical Research* 100: 26377–26386.

Lorenz, R.D. and Myers, M.J. (2005). Dust devil hazard to aviation. A review of United States air accident reports. *The Journal of Meteorology (Trowbridge)* 30 (299): 178–183.

Lorenz, R.D. and Zimbelman, J.R. (2014). *Dune Worlds: How Windblown Sand Shapes Planetary Landscapes*. London: Springer.

Lorenz, R.D., Wall, S., Radebaugh, J. et al. (2006). The sand seas of Titan: Cassini RADAR observations of longitudinal dunes. *Science* 312 (5774): 724–727. doi: 10.1126/science.1123257.

Lorenz, R., Claudin, P., Andreotti, B., Radebaugh, J., and Tokano, T. (2010). "A 3 km atmospheric boundary layer on Titan indicated by dune spacing and Huygens data." *Icarus* 205: 719–721.

Lorenz, R.D., Bridges, N.T., Rosenthal, A.A., and Donkor, E. (2014). Elevation dependence of bedform wavelength on Tharsis Montes, Mars: atmospheric density as a controlling parameter. *Icarus* 230: 77–80. doi: 10.1016/j.icarus.2013.10.026.

Lucas, A., Rodriguez, S., Narteau, C. et al. (2014). Growth mechanisms and dune orientation on Titan. *Geophysical Research Letters* 41 (17): 6093–6100. doi: 10.1002/2014GL060971.

Mandt, K.E., de Silva, S.L., Zimbelman, J.R., and Crown, D.A. (2008). Origin of the Medusae Fossae Formation, Mars: insights from a synoptic approach. *Journal of Geophysical Research, Planets* 113 (E12): doi: 10.1029/2008JE003076.

Mandt, K., de Silva, S., Zimbelman, J., and Wyrick, D. (2009). Distinct erosional progressions in the Medusae Fossae Formation, Mars, indicate contrasting environmental conditions. *Icarus* 204 (2): 471–477. doi: 10.1016/j.icarus.2009.06.031.

Marshall, J.R. and Stratton, D. (1999). Computer modeling of sand transport on Mars using a compartmentalized fluids algorithm (CFA). *Proceedings of the Lunar and Planetary Science Conference* 30: 1229.

Massé, M., Bourgeois, O., Le Mouélic, S. et al. (2010). Martian polar and circum-polar sulfate-bearing deposits: sublimation tills derived from the North Polar Cap. *Icarus* 209 (2): 434–451. doi: 10.1016/j.icarus.2010.04.017.

McCauley, J.F. (1973). Mariner 9 evidence for wind erosion in the equatorial and mid-latitude regions of Mars. *Journal of Geophysical Research* 78 (20): 4123–4137. doi: 10.1029/JB078i020p04123.

Mitchell, J.L. (2008). The drying of Titan's dunes: Titan's methane hydrology and its impact on atmospheric circulation. *Journal of Geophysical Research, Planets* 113: E08015. doi: 10.1029/2007JE003017.

Moore, H.J. (1985). The Martian dust storm of sol 1742. In: Proceedings of the 16th Lunar and Planetary Science Conference, I. *Journal of Geophysical Research: Atmospheres* 90(D, Supplement): 163–174, doi:10.1029/JB090iS01p00163.

Mulholland, D.P., Read, P.L., and Lewis, S.R. (2013). Simulating the interannual variability of major dust storms on Mars using variable lifting thresholds. *Icarus* 223 (1): 344–358. doi: 10.1016/j.icarus.2012.12.003.

Mustard, J.F., Cooper, C.D., and Rifkin, M.K. (2001). Evidence for recent climate change on Mars from the identification of youthful near-surface ground ice. *Nature* 412: 411414. doi: 10.1038/35086515.

Newman, C.E., Lewis, S.R., Read, P.L., and Forget, F. (2002). Modeling the Martian dust cycle, 1. Representations of dust transport processes. *Journal of Geophysical Research, Planets* 107 (E12): 5123. doi: 10.1029/2002JE001910.

Ojha, L., Wilhelm, M.B., Murchie, S.L. et al. (2015). Spectral evidence for hydrated salts in recurring slope lineae on Mars. *Nature Geoscience, advance online publication* doi: 10.1038/ngeo2546.

Pähtz, T., Kok, J.F., Parteli, E.J.R., and Herrmann, H.J. (2013). Flux saturation length of sediment transport. *Physical Review Letters* 111: 218002. doi: 10.1103/PhysRevLett.111.218002.

Paillou, P., Seignovert, B., Radebaugh, J., and Wall, S. (2016). Radar scattering of linear dunes and mega-yardangs: application to Titan. *Icarus* 270: 211–221. doi: 10.1016/j.icarus.2015.07.038.

Parteli, E.J.R., Durán, O., and Herrmann, H.J. (2007). Minimal size of a barchan dune. *Physical Review. E, Statistical, Nonlinear, and Soft Matter Physics* 75: 011301. doi: 10.1103/PhysRevE.75.011301. (Comment: Andreotti, B. and Claudin, P. 2008. PRE- SNSMP 76, 063302, doi:10.1103/PhysRevE.76.063302. Reply P.D.H. 2008. PRE- SNSMP76, 063302, doi:10.1103/PhysRevE.76.063302).

Parteli, E.J.R. and Herrmann, H.J. (2007). Dune formation on the present Mars. *Physical Review. E, Statistical, Nonlinear, and Soft Matter Physics* 76: 041307. doi: 10.1103/PhysRevE.76.041307.

Radebaugh, J., Lorenz, R.D., Farr, T. et al. (2010). Linear dunes on Titan and Earth: initial remote sensing comparisons. *Geomorphology* 121 (1–2): 122–132. doi: 10.1016/j.geomorph.2009.02.022.

Read, P.L. and Lewis, S.R. (2004). *The Martian Climate Revisited: Atmosphere and Environment of a Desert Planet*. Berlin: Springer.

Reffet, E., Courrech du Pont, S., Hersen, P., and Douady, S. (2010). Formation and stability of transverse and longitudinal sand dunes. *Geology* 38 (6): 491–494. doi: 10.1130/G30894.1.

Silvestro, S., Vaz, D.A., Ewing, R.C. et al. (2013). Pervasive aeolian activity along rover Curiosity's traverse in Gale Crater, Mars. *Geology* 41 (4): 483–486. doi: 10.1130/G34162.1.

Skinner, J.A. Jr., Tanaka, K.L., and Platz, T. (2012). Widespread loess-like deposit in the Martian northern lowlands identifies middle Amazonian climate change. *Geology* 40 (12): 1127–1130. doi: 10.1130/G33513.1.

Sullivan, R., Arvidson, R., Bell, J.F. et al. (2008). Wind-driven particle mobility on Mars: insights from Mars Exploration Rover observations at "El Dorado" and surroundings at Gusev Crater. *Journal of Geophysical Research, Planets* 113: E06S07. doi: 10.1029/2008JE003101.

Sullivan, R., Banfield, D., Bell, J.F. et al. (2005). Aeolian processes at the Mars Exploration Rover Meridiani Planum landing site. *Nature (London)* 436 (7047): 58–61. doi: 10.1038/nature03641.

Sullivan, R., Greeley, R., Kraft, M. et al. (2000). Results of the imager for Mars Pathfinder windsock experiment. *Journal of Geophysical Research* 105 (E10): 24547–24562. doi: 10.1029/1999JE001234.

Sutton, J.L., Leovy, C.B., and Tillman, J.E. (1978). Diurnal variations of the Martian surface layer meteorological parameters during the first 45 sols at two Viking Lander sites. *Journal of the Atmospheric Sciences* 35 (12): 2346–2355. doi: 10.1175/1520-0469 (1978)035<2346:DVOTMS>2.0.CO;2.

Tanaka, K.L. (2000). Dust and ice deposition in the Martian geologic record. *Icarus* 144 (2): 254–266. doi: 10.1006/icar.1999.6297.

Taylor, K.C., Lamorey, G.W., Doyle, G.A. et al. (1993). The "flickering switch" of late Pleistocene climate change. *Nature* 361: 432–436. doi: 10.1038/361432a0.

Telfer, M.W., Parteli, E.J.R. Radebaugh, J., Beyer, R.A., Bertrand, T., Forget, F., Nimmo, F., Grundy, W.M., Moore, J.M., Stern, S.A., Spencer, J., Lauer, T.R., Earle, A.M., Binzel, R.P., Weaver, H.A., Olkin, C.B., Young, L.A., Ennico, K., and Runyon, K. (2018). "Dunes on Pluto." *Science* 360 (6392): 992–997.

Thomson, B.J., Bridges, N.T., and Greeley, R. (2008). Rock abrasion features in the Columbia Hills, Mars. *Journal of Geophysical Research, Planets* 113: E08010. doi: 10.1029/2007JE003018.

Tirsch, D., Jaumann, R., Pacifici, A., and Poulet, F. (2011). Dark aeolian sediments in Martian craters: composition and sources. *Journal of Geophysical Research, Planets* 116 (E03002): doi: 10.1029/2009JE003562.

Tokano, T. (2008). Dune-forming winds on Titan and the influence of topography. *Icarus* 194 (1): 243–262. doi: 10.1016/j.icarus.2007.10.007.

Tokano, T. (2010). Relevance of fast westerlies at equinox for the eastward elongation of Titan's dunes. *Aeolian Research* 2 (2–3): 113–127. doi: 10.1016/j.aeolia.2010.04.003.

Ward, A.W. (1979). Yardangs on Mars: evidence of recent wind erosion. *Journal of Geophysical Research: Solid Earth* 84 (B14): 8147–8166. doi: 10.1029/JB084iB14p08147.

Wolff, M.J. (2003). Constraints on the size of Martian aerosols from Thermal Emission Spectrometer observations. *Journal of Geophysical Research, Planets* 108 (E9): 5097. doi: 10.1029/2003JE002057.

Wolff, M.J., Smith, M.D., Clancy, R.T. et al. (2006). Constraints on dust aerosols from the Mars Exploration Rovers using MGS overflights and mini-TES. *Journal of Geophysical Research, Planets* 111: E12S17. doi: 10.1029/2006JE002786.

Zimbelman, J.R., Bourke, M.C., and Lorenz, R.D. (2013). Recent developments in planetary Aeolian studies and their terrestrial analogs. *Aeolian Research* 11: 109–126. doi: doi.org/10.1016/j.aeolia.2013.04.004.

Zurek, R.W. and Martin, L.J. (1993). Interannual variability of planet-encircling dust storms on Mars. *Journal of Geophysical Research, Planets* 98 (E2): doi: 10.1029/1092JE02936.

12

Application

Andrew Warren

University College London, London, UK

12.1   Introduction

Aeolian geomorphology has been applied to: contemporary arable fields, rangeland, mine-dumps, nature conservation, coastal dunes, public health, recreation, road and rail engineering, climate change, warfare, and more. The spatial scale of enquiry has ranged from the trajectories of a particle of soil as it is blown over 'roughness elements' in an agricultural field, to bodies in our solar system, beyond earth (e.g. helping in the selection of landing sites on Mars). Application has successfully deployed methods as various as trial-and-error, field experiment, wind-tunnel simulation, the dating of sediments, remote-sensing, numerical modelling, among others. The first application was probably *c.*7000 years ago (Box 12.1); Mars is the new frontier (Chapter 11). Applied aeolian geomorphology has never been as active.

This chapter can do little justice to that list. Four examples have been chosen to cover a range of environments, scales, and users: (i) the Dust Bowl on the Great Plains of the USA in the 1930s (the classic example and the most thoroughly researched application), with a short survey of recent developments; (ii) 'Desertification' in the Sahel; (iii) the control of sand encroachment in extreme deserts; to speculate about the earliest days of application (Box 12.1) and to introduce

the idea of synergy between physical and financial objectives; and (iv) recreation and nature conservation on sandy coasts, as an example of applied biogeomorphology and as a softer target for application. As mistakes give as good lessons as successes (Syed, 2015), three 'mistakes' are searched for their lessons (large shelterbelts, desertification, and the introduction of marram beyond Europe).

12.2   Wind Erosion and Dust Production from Agricultural and Grazing Land

This section begins with a description of the role of field-scale aeolian science in the amelioration of the Dust Bowl in the Western Plains of the USA in the 1930s (Figure 12.1). It then looks at other field-scale research into wind erosion in the Sahel of Africa, before discussing research into wind erosion at greater scales.

12.2.1   The Dust Bowl

There is no doubt that the Dust Bowl was associated with a severe drought. There had been droughts of at least that severity in the pre-settlement period, to judge by the dunes

Box 12.1 The First Application of Aeolian Geomorphology

Fences to protect arable crops from moving sand were almost certainly the first application of aeolian science, even if there are large uncertainties about when and where the first of such fences appeared. The reservations include the location and date of the domestication of the date palm, which was and is the principal crop to be protected, and the rate of the diffusion of the palm from the site of its first domestication. The consensus, notwithstanding, is that the first domestication of the date palm was in Iraq, about 8000 years ago and that it diffused from there to the Arabian peninsula, to other drylands in south-western Asia and to the eastern Sahara shortly afterwards.

There are fewer uncertainties about when sand fences were first deployed. Focus falls on the Western Desert of Egypt. In the 1930s most Egyptologists believed that an exodus from the Western Desert had stimulated the start of civilization in the Nile Valley, and speculated that the cause of the exodus was the desiccation of what is now Western Desert. Later Egyptologists have discovered many more of these exoduses, but new carbon dating has given a date of 7000 BP for the start of a major spurt of desiccation (Kuper and Kröpalin 2006). Desiccation, of course, also released large amounts of sand into the wind.

Bagnold (1941, p. 219; Figure 1 of Box 1.1) injected some aeolian geomorphology into this debate. He noted that many 'chains' or 'streamers' of dunes had their origin in the Northern Desert, and had been blown in broad arcs that reached to some 700 km to the south (under the influence of the Trade Winds (see Chapter 1)). The chains are the most obvious dune pattern on the eastern side of the Western Desert; search around and zoom into 26°27″N; 30°26″E on Google Earth. These chains are made up, almost wholly, of mobile barchans (Chapter 6). Using his model of the relation between the size and rate of movement (or 'celerity') of dunes like these (Chapter 6), Bagnold used some contemporary estimates of wind speed to calculate how long it had taken for an average-length streamer to extend from the upwind source of its sand, to its advancing outliers in the 1930s. The answer was 7000 years. He could find no direct field evidence for when the sand for the barchans had been released, but noted that this date corresponded to the estimate of the date of the desiccation by the Egyptologists of his time (as discussed above). Thus, the first invasions of wind-blown sand, and the domestic date palm could have reached the oases in Egypt at about the same time, which loosely defines the date at which palm gardens needed to be protected from encroaching sand.

El Kharga (located on Figure 1.14) is one of these oases. Most of its palm gardens, given the unidirectional nature of the Trade Winds, need protection from sand coming only from the north (as at 25°37′31″N; 30°39′22″E; and 25°11′40″N; 30°31′12″E, eye altitude 1 km and many other examples). Most of the fences are short and little more than hedges of palm fronds. Many have been repeatedly swamped by sand and abandoned (as have the fields they once protected), to be replaced by hedges and fields further downwind, suggesting that they are positioned by trial and error. They are thus not totally effective, but, the technique having survived for millennia, they could be seen as adequate. In the zone of the palm gardens, longer, angled fences, designed to divert the sand, are few, and seem to be built of more durable materials, and are therefore likely to be a recent innovation.

they mobilised (Muhs and Holliday 1995), and earlier post-settlement droughts that had forced farmers off the land, but none that displaced as many people as the drought of the 1930s.

There is also little doubt that farming practice was a major, or as will be seen, even the primary contributor, to the Dust Bowl. Beginning in the late nineteenth century, farmers had replaced the indigenous

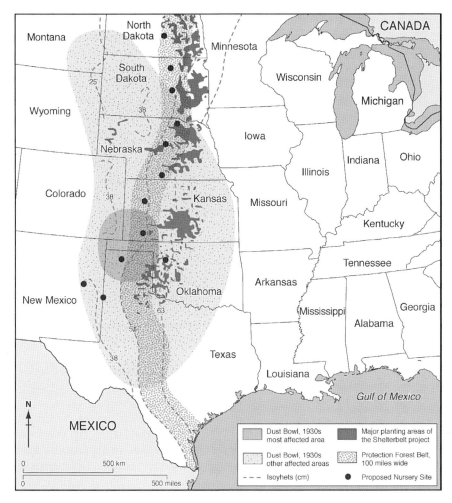

Figure 12.1 Composite figure of: (i) the location of the Dust Bowl (never precisely defined) and other areas affected by agricultural wind erosion in the 1930s; (ii) map drawn by Zon and Bates in 1933 showing an early draft of a map of shelterbelt planting. *Source:* from Droze (1977), p. 51; (iii) the distribution of areas later planted with field-boundary shelterbelts (Zon 1935, Figure 1).

grassland of the western Plains with thirstier, later-maturing crops. Many farmers disturbed the soil with ploughs and harrows, just as temperatures (and sometimes the windspeed) rose in the spring; some fields were left bare throughout the growing season, in an attempt to store moisture for the next year. It was the role of such practices in the Dust Bowl that prompted the Roosevelt federal administration to fund the United States Department of Agriculture (USDA) to commission research on farming techniques. After 1947, the research was centred at the

Wind Erosion Research Unit (WERU) in Manhattan, Kansas (Lyles 1985).

But, were agricultural practices and the drought independent factors? A new study claims that they were not. Most climatologists accept that drought on the Plains follows anomalous sea-surface temperatures in the Pacific (as part of the El Niño Southern Oscillation (ENSO) cycle, see Chapter 1), but the ENSO anomaly in the Dust Bowl years was not severe (Cook et al. 2009). When Cook and his colleagues fed the expansion of bare ground, as the Prairies were ploughed up, and the dust that this released, into a Global Circulation Model, it produced temperature and rainfall data that were similar to those in the instrumental record for the mid-1930s. Thus, the desperation to produce crops may have exacerbated and prolonged what would otherwise have been a quotidian drought.

Another set of explanations of the Dust Bowl is socio-economic. In the best known of these explanations, the Dust Bowl was triggered by the growing world market in cereals, which had begun with the First World War, the introduction of mechanisation to arable farming, and the introduction of instalment plans ('hire purchase' in British usage), which drove farmers to produce crops no matter the weather (Worster 1979). Hurt (1981, Chapter 2) discussed other socio-economic and political interpretations of the Dust Bowl. A socio-economic explanation was also applied to the Sahel, as will be seen.

12.2.2 Field-Scale Application of Aeolian Research After the Dust Bowl

The foundation of WERU was the rebirth of applied aeolian research (the first application occurred thousands of years before, see Box 12.1). WERU's first focus was wind erosion in a single field, for which it needed to invent research tools and pioneer experiments in a series of laboratory and field studies (Zingg 1951; Chepil et al. 1952). It began its laboratory studies on a reproduction of Bagnold's (Box 1.1) wind tunnel, adapting it later for its own purposes.

On the Plains, the findings from the small-scale application of WERU's findings were used to recommend practices that should reduce erosion, such as: (i) roughening the soil surface, which both stills the wind (Chapter 2) and catches more flying particles of soil than does a level surface (Chepil 1950; Figure 12.2); (ii) tillage to reduce erosion, such as ploughing at right angles to the expected wind (which increases the roughness of the surface and so

decreases the near-surface wind); (iii) alternating strips of bare ground where the crop has yet to mature with strips of grass, both to still the wind near the ground and to trap any saltating particles released from the bare strips (McConkey et al. 1990) (49°31′46″N; 111°21′38″W, eye altitude 4 km[1]); as well as other practices which were recommended by much later studies in the Sahel. WERU also invigorated research on shelter (Woodruff and Zingg 1953), with implications that will be dissected shortly. The Wind Erosion Prediction System (WEPS), a twenty-first-century product of WERU, is also described below.

12.2.3 Field-Scale Research into Wind Erosion in the Sahel of West Africa

The socio-economic argument that underlay concern about 'a new Dust Bowl' or 'desertification' in the Sahel, like the original Dust Bowl, was triggered by a severe drought in the West African Sahel in the early 1970s. As in the early years of research after the Dust Bowl story, it was also asserted that Sahelian small farmers used poor agricultural techniques, aggravated by a growing population. These claims were effective in marshalling an international conference (Warren and Maizels 1977) and became a theme in other international environmental conferences for some years after. Whether this was one of this chapter's mistakes is discussed later.

Field research in the Sahel has tested these claims and has discovered a more complex and, in general, a much less gloomy picture. As to claims of population pressure, far from there being a stimulus to over-use soils, some research showed that there was an effective shortage of labour at critical times in the farming year (Warren et al. 2003). As to farming practice, the local bush-fallow system, practised on much smaller fields, leaves land to revert to bush for a few years after a few years of cultivation. The research found that the system allowed soils to recover most

Figure 12.2 One way in which roughness can control wind erosion. *Source:* After Chepil (1950), Figure 4.

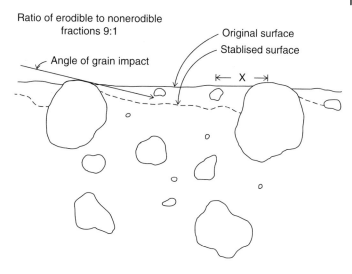

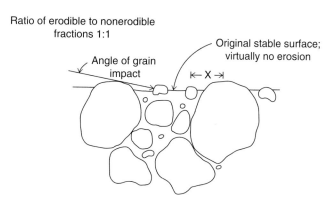

of their lost fertility, both from inputs of dust in the Harmattan (see Chapter 1) and from soil blown off fields in active cultivation (Rajot et al. 2008). In plots close to a village, domestic waste and the debris from cooking fires were also used to boost fertility (Bielders et al. 2004). Most of the Sahelian farmers were well aware of the need and the means to conserve fertility (Sterk and Haigis 1998). 'Desertification' is added to a list of mistakes, at the end of this chapter.

12.2.4   Application at Expanding Scales

Enlarging the spatial horizon (from a single or group of fields) implies enlarged time-horizons, new types of land use, new sources of information, new techniques of acquiring and analysing information, and serves more end-users.

12.2.4.1   More Types of Land Use

The most extensive of the new land-uses that is ushered in by a change in scale is pastoral land, which occupies a much bigger area than arable farming, both in total and as individual holdings. Rangeland is more susceptible to wind than to water erosion, because although it is more intense, rainfall heavy enough to create runoff is less common than the occurrence of winds strong enough to raise dust (Field et al. 2011). But the evidence for wind erosion on and dust production from rough grazing land is mixed. In Australia, the most dust-producing areas

Erodible Fraction (%)

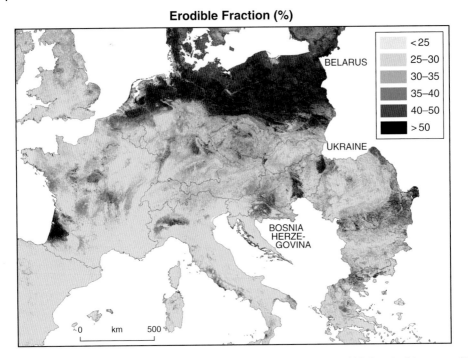

Figure 12.3 The 'Erodible Fraction' (the soil size-fraction that is most erodible by wind) in a part of Europe (using the relationship developed by Fryrear et al. 2000; Borrelli et al. 2014a, 2014b, 2015). Comparison of this figure with Figure 1.11 shows the coincidence of most of the main area of high wind-erosion in central Europe susceptibility coincides with the 'European Sand Belt'.

are rough grazing land on the margins of the 'Dry Heart' (see Figure 4.3, Chapter 4). But dry rough grazing country in the west of the United States, in general, yields much less dust than agricultural land.[2]

12.2.4.2 New Sources of Information

An essential tool for the assessment of wind erosion as the scale expands is the soil map (the coverage of the soil map has been diligently spreading across the US since 1896). In particular, soil maps reveal the location of vulnerable soils, particularly sandy and other 'light' soils, which have been shown, by finer-scale research, to be more vulnerable to wind erosion than coarser-textured soils, as in Europe (Figure 12.3). Sandy soils dry and drain quickly, and allow less capillary rise of water from below the surface than do fine-textured soils (Cornelis et al. 2004); sandy aggregates (clods) are more loosely held together, and thus have less resistance to

wind-erosion, than clods of finer-textured soils. There are sandy soils in many parts of the world, but they are especially extensive in semi-arid climates, where most are a legacy of dry periods in the Pleistocene and Holocene, as on the High Plains in the USA, and the Sahel, and even in south-eastern England.

The most widely used of these other tools is remote sensing (RS). One study using RS showed that in Texas and neighbouring areas, dust was raised mostly in the spring when fields were bare, before most centre-pivot irrigation systems had been switched on, but also when there was little rain and winds were often fierce (Lee et al. 1994). Another RS study, using an image taken by the Moderate Resolution Imaging Spectroradiometer (MODIS) image of an extensive dust storm in the southern High Plains of the USA and Mexico, found of 146 distinct sources of the dust, 58 from cropland, 49 from rangeland, and 30 from

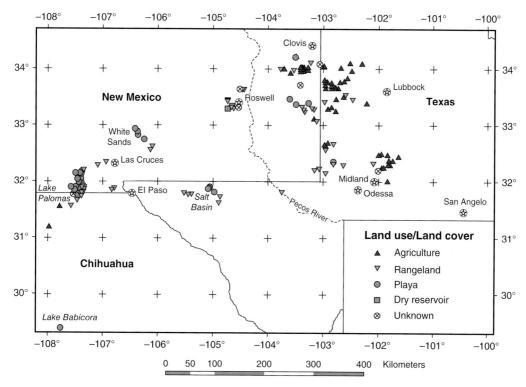

Figure 12.4 Land use classification of sites of the origins of dust plumes during a storm in the Texas area, detected on MODIS satellite imagery. *Source:* Lee et al. (2009), Figure 2.

playa basins (further discussed in Chapters 1 and 4) (Lee et al. 2009; Figure 12.4). A third example of applied RS concerns the management of pastoral land by burning, which showed that early-season fires released the most dust, and that rain took less sediment than the wind, because of its greater frequency (Sankey et al. 2013).

12.2.4.3 Larger Scales Bring New End-Users
Examples of these include those who need to ensure that subsidies achieve the maximum return, say, in food production, and public health authorities, who are responsible for the control of the health hazard in dust (Pérez and Künzli 2011).

12.2.4.4 Modelling
Modelling is a tool that has application at scales from the farm to beyond, for example, in measurement, diagnosis, and mitigation. At least three models exist. First, here, is the 'WEELS' Model, one of the outputs from the research project 'Wind Erosion of European Light Soils', funded by the European Union. One of the outputs from that model is shown in Figure 12.5. The model was applied to a site between Culfordheath and Barnham, in Suffolk, in south-eastern England (centred at 52°20′38″N; 00°51′21″E, eye altitude 3 km) (Böhner et al. 2003). The model has five main modules: (i) the erosivity of the wind (defined in Chapters 1 and 2); (ii) the erodibility of the soil; (iii) soil moisture; (iv) surface roughness; and (v) land use. Data for the model included: hourly wind speed, precipitation and temperature data from a local weather station; maps of soil characteristics (particularly texture); topography; the position and size of all shelterbelts (Vigiak et al. 2003), and land use data (the last in a few simplified categories).

The data allowed the model to run from 1970 to 1998. It calculated wind-speed, wind

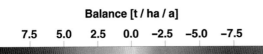

Figure 12.5 Wind erosion in part of Suffolk calculated for the period 1970–1998. *Source:* Böhner et al. (2003), Figure 4. See insert for colour representation of this figure.

direction, and surface roughness, all of which allowed it to compute the distribution of shear velocity (u_*, explained in Chapter 2) and erosion, field-by-field and hour-by-hour for the full period. The calculation of the amount of soil moved by the wind used Bagnold's model of sand movement by wind (Chapter 2). An extraordinary validation of the model was the discovery that a farmer had filmed a 'soil blow', on the some of the same fields, on the same day, hour, and direction (unusually north-easterly) that the WEELS model

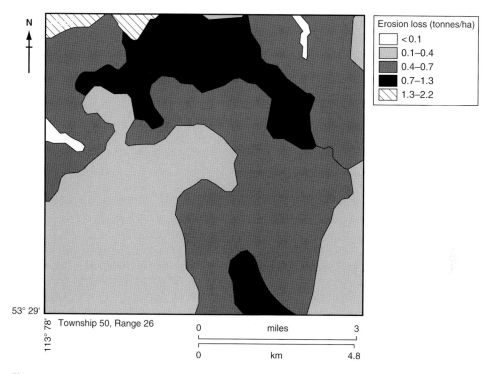

Figure 12.6 A wind-erosion susceptibility map created by using Wind Erosion Prediction System (WEPS) of a township in Alberta. Note: decimalised coordinates. *Source:* After Coen et al. (2004), Figure 3.

postdicted. The daily, even hourly, output from the model, suggested that it was indeed the wind that had taken the soil, rather than another farmer's suggestion that it had been taken on the beets to the local sugar factory in Bury St. Edmunds (sometimes called 'cultivation erosion').

WEPS[3] is a much more fully researched and more widely applied model than WEELS. It was developed by WERU (introduced above). WEPS can work on a field-by-field and hour-by-hour basis (or other choice of area or time period) and runs on a similar range of data as does WEELS, for US sites, all of which are now codified and web-available. WEPS's output is also more farmer-directed than WEELS, because it is designed specifically to be used to suggest ways in which wind erosion could be reduced. In the first few decades after the Dust Bowl, the USDA advised and subsidised dry-land farmers to adopt the kinds of prescriptions developed in the early days of WERU (details above), but now the emphasis is on advice about the 'set aside' of vulnerable fields (Uri 2001). Set-aside advice is now dependent on the WEPs model.

It is true that a field test found that WEPS generally under-predicted (Hagen 2004), but this does not invalidate its use, even in its first form. It has now been thoroughly tested in many other ways (van Pelt et al. 2007) and applied in Argentina, Canada, and Germany (Funk et al. 2004; Buschiazzo and Zobeck 2008; Li et al. 2013). Figure 12.6 is a map of wind erosion susceptibility for a township in Alberta which was calculated using WEPS. The Revised Wind Erosion Equation (REQ) is a simpler-to-use alternative, also field-tested, but is not as well supported as WEPS (Fryrear et al. 2000). AUSLEM is yet another model with the same purpose as WEELS and WEPs, designed for use in Australia (Webb et al. 2006).

12.3 Shelter, Wind Erosion, and Dust Production: A Possible Confusion of Scale?

This section is an overview of research and practice in shelterbelts. It first examines recent research into the effects of shelter on wind erosion at the field scale. These results are then used, with other data, to examine the history of a massive increase in scale: the 'Plains Shelterbelt'. This chapter closes with a brief review of the subsequent history of proposals for massive shelterbelts. Are they another of this chapter's 'mistakes'?

Many of the elements in the design of shelterbelts, such as tree species, length, orientation, etc. are not examined here. The focus here is on the effect of the height of a shelterbelt on wind erosion (Figure 12.7), which is probably the most important of their characteristics. The height of a shelter has two kinds of effect on erosion. The first is dominant: the reduction in the velocity of the wind in the lee. Figure 12.7 shows that, in one widely used model (supported, in broad outline, by many hundreds of modelling, field, and wind-tunnel studies), the downwind distance over which there is any reduction in wind-speed is 20–25h ('h' being the height of the shelterbelt). The second, the 'fetch effect' is explained in Chapter 2, and has little relevance to the argument here.

The results of experimental studies and experience in the field have produced many rule-of-thumb recommendations for the spacing of shelterbelts. One is that shelterbelts on dry, sandy soils should be planted 100 m apart, assuming a shelterbelt that is 10 m high (Holm 1975, quoted by de Jong and Kowalchuk 1995). Many dry-land soils are sandy, as explained above, although 100 m is somewhat greater than the width of most fields on the Plains (~75 m), leaving the downwind end of a sandy field somewhat vulnerable to erosion in high winds. Shelterbelts at closer spacing are only cost-effective if the crop has high value, or if it is very susceptible to wind damage, or in a very windy environment.

Shelterbelts have other advantages. They control injury to crops by sand-blast and vibration (Skidmore 1966); they reduce the deposition of dust on leaves (although see Prajapati 2012); and they shield livestock and homesteads from hot winds or blizzards. Some other functions of shelterbelts are

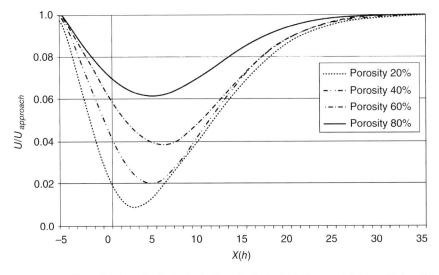

Figure 12.7 The reduction of velocity in the lee of a shelterbelt. The x-axis is in multiples of the height of the shelterbelt; the y axis is the ratio between the velocity of the approaching wind and the velocity in the lee. *Source:* a simplification of Wang Hao et al. (2001), Figure 7.

independent of that reduction: capturing carbon dioxide, improving landscape, and harbouring wildlife, whether game or pest.

Shelterbelts, however, have many other drawbacks. They take time to mature, so that other measures may be necessary for that period; they are costly to install and maintain; they remove land from crop production; they compete with crops for water and sunlight; they allow snow to lie late; and they have roots that snag ploughs and harrows. But these effects are confined to a much smaller zone downwind than the reduction of wind-speed itself. On some farms the economic balance may favour shelterbelts (Brandle et al. 1992). But arguments in their favour have not, manifestly, convinced all Plains farmers, many of whose shelterbelts, 40–60 years after they were planted, were found to be dead, dying, felled, or grubbed up (Sorenson and Marotz 1977; Marotz and Sorenson 1979; David and Rhyner 1999).

12.3.1 The Plains Shelterbelt

On the stump in the dry western Great Plains in 1932, Franklin D. Roosevelt, the President-to be, is said to have conceived a plan for a massive belt of trees, claiming it later as 'my baby' (Droze 1977). One of the arguments for the scheme in his eyes was the prospect that the Federal Government might buy abandoned farms from the numerous casualties of the Dust Bowl and plant them with trees.

Droze's account of the debate on the Shelterbelt project, as it then developed in Washington, DC, has remarkable detail: panics over lost documents and all. It showed that, when the proposal for a 'Plains Shelterbelt' first became public, it was a 100-mile-wide curving zone from the Canadian border, perhaps to the shore of the Gulf of Mexico (see Figure 12.1).

Figure 12.7 demonstrates, fairly conclusively, the futility of massive shelter belts (other characteristics of a shelterbelt, such as its width are very secondary considerations (Wang and Takle 1996)). Even in the 1930s, there was enough evidence that the Shelterbelt, as then proposed, would have

little effect on dust production, and this kind of evidence was used to the full in an extensive and acrimonious debate within the federal government and beyond. There were many other elements in the debate. Would the Shelterbelt do any of the following?

- survive in the dry climate of the selected zone?;
- ameliorate the climate of the Plains (a highly contested claim)?;
- halt the march of the desert, aka desertification? (a role for shelterbelts that was claimed in later proposals, as will be seen);
- be aligned to the most damaging winds? (Anonymous 1934; Chapman 1934: a summary of the opinions of 44 professional foresters).

Washington later bowed to most of the criticism, and, by the time planting began, had transformed the project into a scattering of short, narrow shelterbelts, in areas dotted across the Prairies, where soils were more favourable to trees, and up to 200 km east of the first proposal, thus in a wetter climate, (Figure 12.1; Zon 1935). By 1942, 29 000 km of shelterbelts had been planted, but few in south-western Kansas, and the Texas and Oklahoma panhandles, where aridity made their establishment difficult (Lee et al. 1999). Yet Figure 12.1 shows that this area was very much part of the 'Dust Bowl'. Indeed, it has been called the 'epicentre' (Worster 1979).

12.4 Blown Sand in the Desert

This section starts with an exploration of the early history of applied aeolian science. It then describes how environmental science and economics can be examined in parallel.

12.4.1 Oasis Agriculture

The most common item in all installations that aim to control encroaching sand is a fence or hedge or wall that deflects or collects sand that might reach a vulnerable target. Box 12.1 is a short early history of sand fences. It includes a description of their use in an Egyptian oasis.

12.4.1.1 Contemporary Sand Fences

The highway from Riyadh to the Dammam on the Gulf Coast of the Arabia (Route 40) passes, unavoidably, over more than 300 km of the Al Dahna sand sea (25°14′35″N; 47°44′52″E, eye altitude 2 km), where the predominantly unidirectional northerly wind (the *Shamal*, see Chapter 1) is confirmed by the dominantly south-facing slip faces of the dunes. Route 40 is protected by up to four lines of fences parallel to the road. Their purpose is to divert or to trap the oncoming sand. While they share basic elements with the sand-deflecting fences in El Kharga in Egypt (Box 12.1), some of these fences are very much more securely grounded in scientific research, and careful accountancy.

There is a large body of wind-tunnel research into fences. It aims to produce designs that have the optimal layout, duplication, angle to the oncoming wind, and porosity, in different wind and sand environments. In this cash-driven system, an important element in design is the optimal mix of characteristics to suit both their environment and the purse of the client. There is a large literature on these issues. One of these studies has gone further than most, by showing sand-fence research can move from simple recommendations about design to the optimal deployment of investment (Dong et al. 2011). Basing their work on wind-tunnel studies, this research group dubbed the design that optimised the reduction of wind-speed and turbulence alone the 'absolute optimal porosity', porosity being the proportion of holes to the area of the fence. But, when cost is considered, they showed that the task was to discover the porosity at which the cost of shelter would rise sharply, this being the 'practical optimal porosity'. If there was no constraint on the area protected, they found that the absolute porosity for shelter alone was 0.2, whereas the practical porosity was 0.4. If the requirement was to protect a wide area, the absolute porosity was in the range 0.5–0.1, while the practical porosity was 0.1.

The material in a fence must be fitted to the environment. Considerations include: UV levels (causing different rates of degradation), windiness (controlling sand blast), and salinity (near a coast or a salt lake). Jute, wood, and plastics are widely used in fences, depending on their resistance to different environments. In choosing the material, balance must be struck between cost and the urgency with which sand must be trapped. The priorities that govern the designs of fences on coastal dunes, are different again. Other ways to guard against blown sand are listed by Goudie (2010).

Systems that integrate these various methods are the next frontier, as on trans-Taklimakan highway (28°24′N; 83°12′E various eye altitudes) (Dong et al. 2004), and on the peripheries of some oases (Figure 12.8). The extent of the problem of sand control in that particular area is very visible at 39°19′31″N; 100°10′33′E, various eye altitudes.

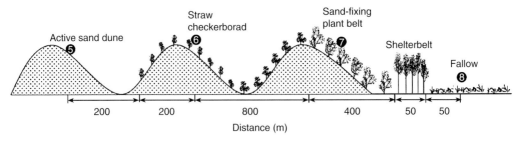

Credit: Zhibin He, Chinese Academy of Sciences, Beijing

Figure 12.8 Integrated management of encroaching sand in the Hexi Corridor, north-west China. *Source:* After Wenzhi et al. (2008).

12.5 Blown Sand on the Coast

This section examines three very different and potentially conflicting demands on coastal dunes: recreation, nature conservation, and coastal protection (not discussed in this chapter). In many places, the three requirements can be met with little conflict. In others, conflicts arise, as will be seen. The section also includes a description of the third 'mistake' promised in the introduction to this chapter. The morphology and dynamics of coastal dunes are much more fully explained in Chapter 7.

12.5.1 Blown Sand and Recreation

On coastal dunes near densely populated parts of the world, the environmental, and the social are more intimately mixed than in any of the other examples in this chapter, and, in general, with happier outcomes (seaside holidays, nature conservation).

The most coordinated national management of coastal dunes has been in The Netherlands, where coastal dunes are a crucial component in the maintenance of a continuously sinking coast (de Ruig 1998), and where recreation and conservation are taken seriously. Elsewhere, research and its application are usually undertaken by local-government bodies.

Soft engineering (sand fences and planting) is much kinder to recreation and ecology and much cheaper than hard engineering (such as massive sea walls or closable barriers), which is only appropriate where high-value installations are threatened by rapid coastal retreat, but has been employed sparingly, even in The Netherlands. The objectives of fences (for any of purpose) on coastal dunes are generally quite the reverse of those in the desert, whether in the oases or in commercially-driven installations. In the desert, the emphasis is on deflecting sand; on coasts it is usually on hoarding it

(an exception follows in Section 12.5.2), or, by discouraging the footfalls of visitors, to allow plants to grow and to capture yet more sand. There are many spatial patterns of fences: the Lancashire coast shows a pattern that is the result of an elaborate plan to defend and encourage the growth of vegetation on some very eroded dunes (53°33′14″N; 03°06′04″W, eye altitude 400 m) (Skelcher 2008); whereas the Long Island shore shows a single fence, with a single purpose: to keep people off the vegetation (40°35′28″N; 73°31′04″W, eye altitude 630 m).

There are many choices of fencing material. A dense network of small branches (brushwood) is both effective at trapping sand and cheaper than most alternatives, but decays quickly (Anthony et al. 2007). Wooden slat fences are more expensive, but as effective, and last longer. Synthetic fabrics may be as effective, but many decay more quickly than wood and may degrade in sunlight, as in the desert (Miller et al. 2001).

Planting or encouraging vegetation is less intrusive than erecting fences, more aesthetic, more natural, better for encouraging wildlife, and cheaper. On most beaches, the three objectives must be served by the same fence. As early as the sixteenth century, coastal dunes in Denmark and The Netherlands were being controlled by planting *marram grass* and/or by discouraging its destruction (Skarregaard 1989). Marram is the common English name of the grass *Ammophila arenaria*, which is a native of north-western European shores, where it is adapted to the salinity and recurrent drought of sandy coastal environments, and, more to the point, where its growth can keep abreast of burial by sand at a rate of $0.5\,\mathrm{m\,yr^{-1}}$, although the growth rate probably varies with climate (Carter 1990). In effect, marram can build a dune, either if merely protected (as by a fence), or, if necessary, planted or both. Marram is now very widely used as a means of coastal dune defence (Arens et al. 2001). This is marram at its best (Figure 7.5 shows *Ammophila brevigulata*, the North American species).

12.5.2 Coastal Protection

This section includes a description of the last exemplary mistake demonstrating marram at its worst. Where nature conservation is the priority, and on shores far from north-western Europe, marram can be a pest.

Among those coastal dune managers from around the world who saw first-hand what the planting of marram could achieve on the coast of the Landes in France in the mid-nineteenth century (Marsh 1885), marram quickly acquired a reputation as the saviour of coastal dunes, and, by extension, the protection of land behind the dunes. Enthusiasts from beyond Europe soon discovered another reason for endorsing this view: for them it was an introduced species, and as such less likely to be attacked by parasites and thus more able to fix sand. The use of marram spread quickly across the globe; in many places dunes were built where there had been none before. When marram was introduced to the coasts of Oregon and Washington State, its ecological advantages allowed it to out-compete native species of dune grass, especially one of same genus (*Ammophila breviligulata*, Figure 7.5) and marram did accumulate new dunes (Buell et al. 2002), but, with determined help from conservationists, *A. breviligulata* has fought back and now dominates quite large areas (Seabloom and Wiedemann 1994). On the other hand marram is now being seen as a threat to other dunes species even on dunes where it is indigenous, as in Wales.

The story of the introduction of marram has not been as happy in the Australia and New Zealand, where it was introduced in the late nineteenth century, with the express intention of encouraging the growth of coastal dunes as defence from the waves (see Maiden 1896). But, once introduced, it drastically changed the geomorphology of the dunes (Figure 12.9; Hilton et al. 2006). Low frontal dunes under the native vegetation were replaced by much larger ones, which, for the first time, developed blowouts and parabolic dunes, on the coast at Mason Bay (Figure 12.9).

Where coastal protection is or was the priority in managing coastal dunes, marram was introduced to shores beyond Europe, and for some years the introduction was seen as a success (bigger dunes, better able to withstand rising sea-levels). But even on many European shores, coastal protection is not the only priority with a significant exception, described shortly).

Where nature conservation is the priority, as in some coastal nature reserves in Australia and New Zealand today, including Stewart Island off South Island (Figure 12.9), marram and some other introduced plant species are a major threat. In addition to their threat to the physical environment. These plants have severely damaged the habitat for many native animal species (especially three species of penguin and the circum-Pacific migrant, the sooty shearwater). New Zealand is leading the response. There are already programmes to eliminate marram from some shores, including those of Stewart Island.

12.6 Conclusion: Learning from Past Mistakes

The essential lesson from the first proposal for the 'Plains Shelterbelt' is that proposals should be very widely assessed before implementation. The lesson applies mainly to the first draft of the Shelterbelt, not to its later application, but the temptation of massive shelterbelts lives on. Two even bigger shelterbelts were proposed later in the twentieth century, both aimed, not at an essentially measurable target like dust, but at a much less definable objective: 'desertification'. The first of these was to be a 'Green Dam', designed to hold back the desert in North Africa and the Sahel (African Forestry Commission 1971). A small part of the northern Green Dam was planted on the southern slopes of the Atlas Mountains in Algeria (Ballais 1994), but little else. The second proposal was floated by the Chinese government: a 'Great Green Wall'

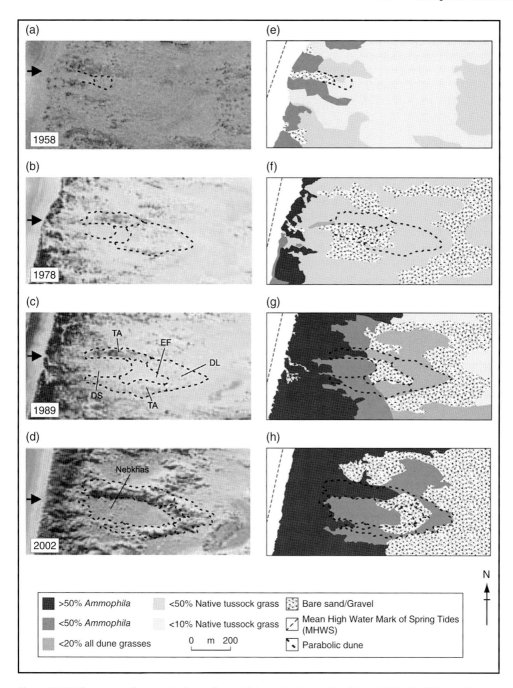

Figure 12.9 Changes in the morphology of coastal dunes in Mason Bay, Stewart Island, off the southern tip of South Island, New Zealand, after the introduction of marram grass. The labels on Figure 'c':1989 refer to characteristics of the new parabolic dune: 'TA': 'trailing arm; EF: erosional face; DL: deflation lobe; DS: deflation surface. (General view: 46°55′S; 167°46′E, eye altitude 2 km). Most of these features are visible on the Google-Earth scene. *Source:* Hilton et al. (2006).

(Academia Sinica 1977). The Chinese project was quickly downsized to the 'Three Norths' programme of more modest but still extensive plantations (41°47′43″N; 113°17′32″E, eye altitude 330 m); although that programme, even in its altered form, has now been criticised (Wang Xunming et al. 2010). A yet more recent project to build a green wall, running across northern Africa has begun, although to trenchant criticism (Pearce 2016).

'Desertification' has, at last, been seen for the gross oversimplification it always was. Studies using RS found that there were small areas where there had been loss of plant cover, but as a whole the idea of land degradation in semi-arid Africa had little foundation (Prince et al. 1998). The same message comes from a wide-ranging survey of studies of agricultural communities in semi-arid Africa (Behnke and Mortimore 2016). (Unfortunately, the Wikipedia article on desertification has not caught up with this research.) The management of blowing sand in the desert, given its costs and the recent volatility in the price of oil, may weaken the funding for the protection of roads and other installations from the encroachment of sand, but might also encourage yet more innovative research into the cost-effectiveness of and design of protection.

The introduction of marram to beaches beyond north-western Europe is in a common class of environmental mistake, in which the negative effects of an intervention are not apparent until many decades after its introduction. In these situations, hindsight is futile. The only feasible response to invasive marram, now, is the kind of small-scale rearguard action in Oregon a decade ago, and the more recent projects in New Zealand.

The management of blowing sand in the desert, given its costs and the recent volatility in the price of oil, may weaken the funding for the protection of roads and other installations from the encroachment of sand, but might also encourage yet more innovative research into the cost-effectiveness of and design of protection.

The management of dune coasts faces major challenges: rapid environmental change (especially rising sea-levels, and probably also windiness and storminess; Chapter 1); changes in tastes for, and the economics of, recreation. These changes will occur in different combinations on different beaches, each of which will demand its own approach to management, and with it, new challenges to the application of aeolian geomorphology.

Notes

1 References like these are to coordinates of an example which can be input into Google Earth.
2 See http://www.nrcs.usda.gov/wps/portal/ nrcs/detail/sc/technical/dma/nri/?cid= stelprdb1041887.

3 See https://infosys.ars.usda.gov/ WindErosion/weps/wepshome.html. The WEPS user guide, which can be downloaded from that site, is a very full introduction to the problems of wind erosion on agricultural land.

Further Reading

The application of aeolian geomorophology in several contexts is covered by Warren (2002; agricultural land), Warren (2013; dune systems, particularly in deserts) and Nordstrom (2000; beaches and coastal dunes).

References

Academia Sinica, Lanzhou Institute of Desert Research (1977). The Great Green Wall of China. UNESCO Courier 30 (July): 32–33.

African Forestry Commission. (1971). Windbreaks for controlling further advancement of the Sahara. Food and Agriculture Organization of the United Nations (FAO), Rome, Forestry Department.

Anonymous (1934). Pros and cons of the shelterbelt. A symposium of pointed comments on the proposal to tree-plant the prairies. *American Forests* 40 (11): 528–529. 545–546.

Anthony, E.J., Vanhée, S., and Ruz, M.-H. (2007). An assessment of the impact of experimental brushwood fences on foredune sand accumulation based on digital elevation models. *Ecological Engineering* 31 (1): 41–46. doi: 10.1016/j.ecoleng. 2007.05.005.

Arens, S.M., Jungerius, P.D., and van der Meulen, F. (2001). Coastal dunes. In: *Habitat Conservation: Managing the Physical Environment* (ed. A. Warren and J.R. French), 229–272. Chichester: Wiley.

Bagnold, R.A. (1941). *The Physics of Blown Sand and Desert Dunes*. London: Methuen (reprinted 1954, 1060, by Methuen; and 2005, by Dover, Mineola, NY).

Ballais, J.-L. (1994). Aeolian activity, desertification and the 'Green Dam' in the Ziban Range, Algeria. In: *Environmental Change in Drylands; Biogeographical and Geomorphological Perspectives* (ed. A.C. Millington), 177–198. Chichester: Wiley.

Behnke, R. and Mortimore, M. (2016). *The End of Desertification?: Disputing Environmental Change in the Drylands*. London: Springer Earth System Sciences.

Bielders, C.L., Rajot, J.-L., and Michels, K. (2004). L'érosion éolienne dans le Sahel nigérien: influence des pratiques culturales actuelles et méthodes de lutte. *Sécheresse (Montrouge)* 15 (1): 19–32.

Böhner, J., Schäfer, W., Conrad, O. et al. (2003). The WEELS model: methods, results and limitations. *Catena* 52 (3–4): 289–308. doi: 10.1016/S0341-8162(03):00019-5.

Borrelli, P., Ballabio, C., Panagos, P., and Montanarella, L. (2014a). Wind erosion susceptibility of European soils. *Geoderma* 232: 471–478. doi: 10.1016/j.geoderma. 2014.06.008.

Borrelli, P., Panagos, P., Ballabio, C. et al. (2014b). Towards a pan-European assessment of land susceptibility to wind erosion. *Land Degradation & Development* 27: 1093–1105. doi: 10.1002/ldr.2318.

Borrelli, P., Panagos, P., and Montanarella, L. (2015). New insights into the geography and modelling of wind erosion in the European agricultural land: application of a spatially explicit indicator of land susceptibility to wind erosion. *Sustainability* 7: 8823–8836. doi: 10.3390/ su7078823.

Brandle, J.R., Johnson, B.R., and Akeson, T. (1992). Windbreaks – are they economical? *Journal of Production Agriculture* 5 (3): 393–398. doi: 10.2134/jpa1992.0393.

Buell, A.C., Pickart, A.J., and Stuart, J.D. (2002). History and invasion patterns of *Ammophila arenaria* on the north coast of California. *Conservation Biology* 9 (6): 1587–1593. doi: 10.1046/j.1523-1739.1995.09061587.x.

Buschiazzo, D.E. and Zobeck, T.M. (2008). Validation of WEQ, RWEQ and WEPS wind erosion for different arable land management systems in the Argentinean pampas. *Earth Surface Processes and Landforms* 33 (12): 1839–1850. doi: 10.1002/esp.1738.

Carter, R.W.G. (1990). The geomorphology of coastal dunes in Ireland. In: *Dunes of the European Coasts* (ed. T.W.M. Bakker, P.D. Jungerius and J.A. Klijn), 31–39. Catena Supplement 18.

Chapman, H.H. (1934). Digest of opinions received on the shelterbelt project. *Journal of Forestry* 32 (9): 952–972.

Chepil, W.S. (1950). Properties of the soil which influence wind erosion. I. The governing principle of surface roughness.

Soil Science 69 (2): 149–162. doi: 10.1097/00010694-195102000-00008.

Chepil, W.S., Englehorn, C.L., and Zingg, A.W. (1952). The effect of cultivation on erodibility of soils by wind. *Soil Science Society of America Proceedings* 16 (1): 19–21. doi: 10.2136/sssaj1952.03615995001600010007x.

Coen, G.M., Tatarko, J., Martin, T.C. et al. (2004). A method for using WEPS to map wind erosion risk of Alberta soils. *Environmental Modelling and Software* 19 (2): 185–189. doi: 10.1016/S1364-8152(03)00121-X.

Cook, B.I., Miller, R.L., and Seager, R. (2009). Amplification of the North American dust bowl drought through human-induced land degradation. *The Proceedings of the National Academy of Sciences of the United States of America* 106 (13): 4997–5001. doi: 10.1073/pnas.0810200106.

Cornelis, W.M., Gabriels, D., and Hartmann, R. (2004). A conceptual model to predict the deflation threshold shear velocity as affected by near-surface soil water: II. Calibration and verification. *Soil Science Society of America Journal* 68 (4): 1162–1168. doi: 10.2136/sssaj2004.1154.

David, C.A. and Rhyner, V. (1999). An assessment of windbreaks in Central Wisconsin. *Agroforestry Systems* 44 (2–3): 313–331. doi: 10.1023/A:1006271215871.

De Jong, E. and Kowalchuk, T.E. (1995). The effect of shelterbelts on erosion and soil properties. *Soil Science* 159 (5): 337–345. doi: 10.1097/00010694-199505000-00007.

De Ruig, H.J.M. (1998). Coastline management in the Netherlands: human use versus natural dynamics. *Journal of Coastal Conservation* 4 (1): 127–134. doi: 10.1007/BF02806504.

Dong, Z., Chen, G., He, X. et al. (2004). Controlling blown sand along the highway crossing the Taklimakan Desert. *Journal of Arid Environments* 57 (3): 329–344. doi: 10.1016/j.jaridenv.2002.02.001.

Dong, Z., Luo, Y.W., Qian, G., and Wang, H. (2011). Evaluating the optimal porosity of fences for reducing wind erosion. *Sciences in Cold and Arid Regions* 3 (1): 0001–0012. doi: 10.3724/SP.J.1226.2011.00001.

Droze, W.H. (1977). *Trees, Prairies and People: A History of Tree Planting*. Denton: Texas Women's Institute.

Field, J.P., Breshears, D.D., Whicker, J.J., and Zou, C.B. (2011). Interactive effects of grazing and burning on wind-and water-driven sediment fluxes: rangeland management implications. *Ecological Applications* 21 (1): 22–32. doi: 10.1890/09-2369.1.

Fryrear, D.W., Bilbro, J.D., Saleh, A. et al. (2000). RWEQ: improved wind erosion technology. *Journal of Soil and Water Conservation* 55 (2): 183–189.

Funk, R., Skidmore, E.L., and Hagen, L.J. (2004). Comparison of wind erosion measurements in Germany with simulated soil losses by WEPS. *Environmental Modelling and Software* 19 (2): 177–183. doi: 10.1016/S1364-8152(03)00120-8.

Goudie, A.S. (2010). Dune migration and encroachment. In: *Geomorphological Hazards and Disaster Prevention* (ed. I. Alcantara-Ayala and A.S. Goudie), 199–202. New York: Cambridge University Press.

Hagen, L.J. (2004). Evaluation of the Wind Erosion Prediction System (WEPS) erosion submodel on cropland fields. *Environmental Modelling and Software* 19 (2): 171–176. doi: 10.1016/S1364-8152(03)00119-1.

Hilton, M., Harvey, N., Hart, A. et al. (2006). The impact of exotic dune grass species on foredune development in Australia and New Zealand: a case study of *Ammophila arenaria* and *Thinopyrum junceiforme*. *Australian Geographer* 37 (3): 313–334. doi: 10.1080/00049180600954765.

Holm, H.M. (1975). *Save the Soil. A Study in Soil Conservation and Erosion Control*. Regina SK: Saskatchewan Department of Agriculture.

Hurt, R.D. (1981). *The Dust Bowl: An Agricultural and Social History*. Chicago: Nelson-Hall.

Kuper, R. and Kröpelin, S. (2006). Climate-controlled Holocene occupation in the Sahara: motor of Africa's evolution. *Science* 313 (5788): 803–807. doi: 10.1126/science.1130989.

Lee, J.A., Allen, B.L., Peterson, R.E. et al. (1994). Environmental controls on blowing dust direction at Lubbock, Texas, U.S.A. *Earth Surface Processes and Landforms* 19 (5): 437–449. doi: 10.1002/esp.3290190505.

Lee, J.A., Attebury, J.K. and Skylstad, P.L. (1999). Soil erosion control methods on the Southern High Plains, 1930s–1980s. In: Traylor, I.R. Jr., Dregne, H. and Mathis, K. (eds), Desert development – the endless frontier, Proceedings of the 5th International Conference on Desert Development, International Center for Arid and Semiarid Land Studies (ICASALS), Texas Tech University, Lubbock, ICASALS Publication 99-2, pp. 710–714.

Lee, J.A., Gill, T.E., Mulligan, K.R. et al. (2009). Land use/land cover and point sources of the 15 December 2003 dust storm in southwestern North America. *Geomorphology* 105 (1–2): 18–27. doi: 10.1016/j.geomorph.2007.12.016.

Li, C., Zhao, H., Han, B., and Zhipeng, B. (2013). Combined use of WEPS and Models-3/CMAQ for simulating wind erosion source emission and its environmental impact. *Science of the Total Environment* 466–467: 762–769. doi: 10.1016/j.scitotenv.2013.07.090.

Lyles, L. (1985). Predicting and controlling wind erosion. *Agricultural History* 59 (2): 205–214.

Maiden, J.H. (1896). Marram grass (*Psamma arenaria*, R. et S.): a valuable sand-stay. *The Agricultural Gazette of New South Wales* 6 (1): 7–12.

Marotz, G.A. and Sorenson, C.J. (1979). Depletion of a Great Plains resource: the case of shelterbelts. *Environmental Conservation* 6 (3): 215–224. doi: 10.1017/S0376892900003088.

Marsh, G.P. (1885). *The Earth as Modified by Human Action: A Last Revision of "Man and Nature"*. New York: Charles Scribner's Sons.

McConkey, B.G., Zentner, R.P., and Nicholaichuk, W. (1990). Perennial grass windbreaks for continuous wheat production on the Canadian Prairies. *Journal of Soil and Water Conservation* 45 (4): 282–295.

Miller, D.L., Thetford, M., and Yager, L. (2001). Evaluation of sand fence and vegetation for dune building following overwash by Hurricane Opal on Santa Rosa Island, Florida. *Journal of Coastal Research* 17 (4): 936–948. doi: 10.2112/JCOASTRES-D-13-00031.1.

Muhs, D.R. and Holliday, V.T. (1995). Evidence of active dune sand on the Great Plains in the 19th century from accounts of early explorers. *Quaternary Research* 43 (2): 198–208. doi: 10.1006/qres.1995.1020.

Nordstrom, K.F. (2000). *Beaches and Dunes of Developed Coasts*. Cambridge University Press: Cambridge.

Pearce, F. (2016). A wall of trees across the Sahara is cool – but we don't need it? *New Scientist* 7 September.

Pérez, L. and Künzli, N. (2011). Saharan dust: no reason to exempt from science or policy (editorial). *Occupational and Environmental Medicine* 68 (6): 389–390. doi: 10.1136/oem.2010.063990.

Prajapati, S.K. (2012). Ecological effect of airborne particulate matter on plants. *Environmental Skeptics and Critics* 1 (1): 12–22. On line.

Prince, S., Brown de Colstoun, E., and Kravitz, L. (1998). Evidence from rain use efficiencies does not support extensive Sahelian desertification. *Global Change Biology* 4 (4): 359–374. doi: 10.1046/j.1365-2486.1998.00158.x.

Rajot, J.L., Formenti, P., Alfaro, S. et al. (2008). AMMA dust experiment: an overview of measurements performed during the dry season special observation period (SOP0) at the Banizoumbou (Niger) supersite. *Journal of Geophysical Research: Atmospheres* 113: D00C14. doi: 10.1029/2008JD009906.

Sankey, J.B., Wallace, C.S.A., and Ravi, S. (2013). Phenology-based, remote sensing of post-burn disturbance windows in rangelands. *Ecological Indicators* 30 (7): 35–44. doi: 10.1016/j.ecolind.2013.02.004.

Seabloom, E.W. and Wiedemann, A.M. (1994). Distribution and effects of *Ammophila breveligulata* Fern (American beachgrass) on the foredunes of the Washington coast. *Journal of Coastal Research* 10 (1): 178–188.

Skarregaard, P. (1989). Stabilization of coastal dunes in Denmark. In: *Perspectives in Coastal Dune Management, European Symposium, Leiden, Netherlands, September 7–11, 1987* (ed. F. van der Meulen, P.D. Jungerius and J. Visser), 151–162. The Hague: SPB Academic BV.

Skelcher, G. (2008). Fylde sand dunes management action plan. A report produced on behalf of the Fylde Sand Dune Project Steering Group. On line

Skidmore, E.L. (1966). Wind and sandblast injury to seedling green beans. *Agronomy Journal* 58 (3): 311–315. doi: 10.2134/agronj1966.00021962005800030020x.

Sorenson, C. and Marotz, G. (1977). Changes in shelterbelt mileage statistics over four decades in Kansas. *Journal of Soil and Water Conservation* 32 (6): 276–281.

Sterk, G. and Haigis, J. (1998). Farmers' knowledge of wind erosion processes and control methods in Niger. *Land Degradation and Development* 9 (2): 107–114. doi: 10.1002/(SICI)1099-145X(199803/04)9:2<107:AID-LDR285>3.0.CO;2-5.

Syed, M. (2015). *Black Box Thinking: Marginal Gains and the Secrets of High Performance.* London: Hachette.

Uri, N.D. (2001). A note on soil erosion and its environmental consequences in the United States. *Water, Air, and Soil Pollution* 129 (1–4): 181–197. doi: 10.1023/A:101033503152.

Van Pelt, R.S., Zobeck, T.M., Ritchie, J.C., and Gill, T.E. (2007). Validating the use of 137Cs measurements to estimate rates of soil redistribution by wind. *Catena* 70 (3): 455–464. doi: 10.1016/j.catena.2006.11.014.

Vigiak, O., Sterk, G., Warren, A., and Hagen, L.J. (2003). Spatial modeling of wind speed around windbreaks. *Catena* 52 (3–4): 273–288. doi: 10.1016/S0341-8162(03)00018-3.

Wang, H. and Takle, E.S. (1996). On three-dimensionality of shelterbelt structure and its influences on shelter effects. *Boundary-Layer Meteorology* 79 (1): 83–105. doi: 10.1007/BF00120076.

Wang, H., Takle, E.S., and Shen, J.M. (2001). Shelterbelts and windbreaks: mathematical modeling and computer simulations of turbulent flows. *Annual Review of Fluid Mechanics* 33: 549–586. doi: 10.1146/annurev.fluid.33.1.549.

Wang, X., Caixia, Z., Eerdun, H., and Dong, Z. (2010). Has the Three Norths Forest Shelterbelt Program solved the desertification and dust storm problems in arid and semiarid China? *Journal of Arid Environments* 74 (1): 13–22. doi: 10.1016/j.jaridenv.2009.08.001.

Warren, A. (ed.) (2002). Wind erosion on agricultural land in Europe. Office for Official Publications of the European Communities, European Commission, EUR 20370.

Warren, A. (2013). *Dunes: Dynamics, Morphology, History.* Oxford: Wiley-Blackwell.

Warren, A. and Maizels, J.K. (1977). Ecological change and desertification, Background Document A.Conf 74/7, U.N.E.P., Nairobi.

Warren, A., Osbahr, H., Batterbury, S., and Chappell, A. (2003). Indigenous views of soil erosion at Fandou Béri, southwestern Niger. *Geoderma* 111 (3–4): 439–456. doi: 10.1016/S0016-7061(02)00276-8.

Webb, N.P., McGowan, H.A., Phinn, S.R., and McTainsh, G.H. (2006). AUSLEM (Australian Land Erodibility Model) a tool for identifying wind erosion hazard in Australia. *Geomorphology* 78 (3–4): 179–200. doi: 10.1016/j.geomorph.2006.01.012.

Woodruff, N.P. and Zingg, A.W. (1953). Wind tunnel studies of shelterbelt models. *Journal of Forestry* 51 (3): 173–178.

Worster, D.E. (1979). *Dust Bowl: The Southern High Plains in the 1930s.* London: Oxford University Press.

Wenzhi, Z., Hu, G., Zhihui, Z., and He, Z. (2008). Shielding effect of oasis-protection systems composed of various forms of wind break on sand fixation in an arid region: a case study in the Hexi Corridor, Northwest China. *Ecological Engineering* 33 (2): 119–125. doi: 10.1016/j.ecoleng.2008.02.010.

Zingg, A.W. (1951). A portable wind tunnel and dust collector developed to evaluate the erodibility of field surfaces. *Agronomy Journal* 43 (4): 189–191.

Zon, R. (1935). Shelterbelts – futile dream or workable plan? *Science* 81 (2104): 391–394. doi: 10.1126/science.81.2104.391.

Index

Note: Items in **BOLD** are the principle entries in a list